# CELL BIOLOGY

## A Comprehensive Treatise

**Volume 1**

**Genetic Mechanisms of Cells**

# CELL BIOLOGY
## A Comprehensive Treatise

## Volume 1
## Genetic Mechanisms of Cells

*Edited by*

LESTER GOLDSTEIN

DAVID M. PRESCOTT

Department of Molecular, Cellular and Developmental Biology
University of Colorado
Boulder, Colorado

**Academic Press** *New York San Francisco London 1977*
*A Subsidiary of Harcourt Brace Jovanovich, Publishers*

ACADEMIC PRESS, INC.
111 Fifth Avenue, New York, New York 10003

*United Kingdom Edition published by*
ACADEMIC PRESS, INC. (LONDON) LTD.
24/28 Oval Road, London NW1

Library of Congress Cataloging in Publication Data

Main entry under title:

Cell biology.

    Includes bibliographies and index.
    1.   Cytology—Collected works.   I.   Goldstein,
Lester.   II.   Prescott, David M.
QH574.C43         574.8'7       77-74026
ISBN 0-12-289501-0

*34,618*

PRINTED IN THE UNITED STATES OF AMERICA

80 81 82   9 8 7 6 5 4 3 2

*It is fitting that we initiate this multivolume treatise on Cell Biology by dedicating this first volume to Dan Mazia, who initiated both of the editors into the excitement and rewards of research into the cell*

# Contents

## 1  Defining the Gene by Mutation, Recombination, and Function

*E. D. Garber and M. S. Esposito*

## 2  Gene Conversion, Paramutation, and Controlling Elements: A Treasure of Exceptions

*Michael S. Esposito and Rochelle E. Esposito*

## 3  The Onset of Meiosis

*Peter B. Moens*

## 4 Chromosome Imprinting and the Differential Regulation of Homologous Chromosomes

*Spencer W. Brown and H. Sharat Chandra*

## 5 Use of Mutant, Hybrid, and Reconstructed Cells in Somatic Cell Genetics

*Nils R. Ringertz and Thorfinn Ege*

## 6 Cytogenetics

*E. D. Garber*

# 7 Cytoplasmic Inheritance

*Ruth Sager*

# 8 Inheritance of Infectious Elements

*Louise B. Preer and John R. Preer, Jr.*

# 9 Non-nucleic Acid Inheritance and Epigenetic Phenomena

*Janine Beisson*

# List of Contributors

Numbers in parentheses indicate the pages on which the authors' contributions begin.

**Janine Beisson** (375), Centre de Génétique Moléculaire, C. N. R. S., Gif-sur-Yvette, France

**Spencer W. Brown**\* (109), Department of Genetics, University of California, Berkeley, California

**H. Sharat Chandra** (109), Microbiology and Cell Biology Laboratory, and Centre for Theoretical Studies, Indian Institute of Science, Bangalore, India

**Thorfinn Ege** (191), Institute for Medical Cell Research and Genetics, Medical Nobel Institute, Karolinska Institutet, Stockholm, Sweden

**Michael S. Esposito** (1, 59), Department of Biology, Erman Biology Center, The University of Chicago, Chicago, Illinois

**Rochelle E. Esposito** (59), Department of Biology, Erman Biology Center, The University of Chicago, Chicago, Illinois

**E. D. Garber** (1, 235), Barnes Laboratory, The University of Chicago, Chicago, Illinois

**Peter B. Moens** (93), Department of Biology, York University, Downsview, Ontario, Canada

**John R. Preer, Jr.**† (319), Department of Zoology, Indiana University, Bloomington, Indiana

**Louise B. Preer**‡ (319), Department of Zoology, Indiana University, Bloomington, Indiana

**Nils R. Ringertz** (191), Institute for Medical Cell Research and Genetics, Medical Nobel Institute, Karolinska Institutet, Stockholm, Sweden

**Ruth Sager** (279), Department of Microbiology and Molecular Genetics, Harvard Medical School, and Sidney Farber Cancer Institute, Boston, Massachusetts

\* Deceased.
† Present address: Department of Biology, Yale University, New Haven, Connecticut.
‡ Present address: Department of Biology, Yale University, New Haven, Connecticut.

# Preface

Almost two decades have passed since the appearance of the first volume of "The Cell" edited by Jean Brachet and Alfred E. Mirsky (Vols. I–VI, 1959–1964, Academic, New York), the only comprehensive treatment of cell biology to date. In the intervening years this field has advanced enormously. Cell biology is now a truly mature science; research that a decade or two ago appeared to be aimless gropings into the unknown has now to a great extent been replaced by theoretically well-grounded investigations that provide useful answers to questions on cell function. Moreover, contemporary investigations can draw on a large resource of well-documented facts about cells. Knowing all this, we agreed that this was a propitious time to compile an up-to-date comprehensive treatise on cell biology that would serve for the next decade or two as a single source of information on many areas of this discipline.

We planned this multivolume treatise as a primary source of fundamental knowledge for graduate students, investigators working in peripheral areas, and for anyone else in need of information on some particular phase of cell biology. Thus, we asked authors to write chapters emphasizing reasonably well-established facts and concepts, but not to attempt the more traditional up-to-the-minute reviews that investigators working in specialized fields count on. A measure of the maturity of cell biology also became evident from the fact that it has been a relatively simple matter to construct each volume around a single, unified theme.

As a reflection of the fundamental role of genetics in cell biology, as in most areas of biology, the introductory volume of this treatise is devoted to genetic mechanisms. Since it was impossible to cover the full range of these mechanisms in one volume, we have focused on subjects that have, or will have, particular relevance for cell biology and that are not likely to be dealt with adequately in traditional textbooks. The emphasis in these contributions has been on the nonmolecular aspects of genetics, but obviously no one can do justice to any aspect of genetics without some discussion of some of the molecular features of the mechanisms under consideration. Fuller discussions of the molecular biology of gene function will appear in subsequent volumes. Because of what we consider to be a certain degree of neglect of nonnuclear aspects of cell heredity in most other general treatments of genetic mechanisms, we have placed what some might consider to be a disproportionate emphasis on these subjects. However, we expect that the average reader will have had a solid background in the more traditional areas of genetics.

We are saddened to note that prior to the appearance of this volume Gordon Tomkins, of our Advisory Board, and Spencer W. Brown, coauthor of Chapter 4 of this volume, died prematurely. Both were making valuable scientific contributions at the time of their deaths; they will be sorely missed by the scientific community.

Lester Goldstein
David M. Prescott

# Titles of Other Volumes

# 1

# Defining the Gene by Mutation, Recombination, and Function

*E. D. Garber and M. S. Esposito*

## I. INTRODUCTION

Within four decades after the rediscovery of the Mendelian principles of heredity, geneticists had erected an elaborate theoretical structure of the gene based on mutation, recombination, and function. Although these approaches to understanding the nature of the gene often yielded confused or contradictory pictures, it was possible to attribute certain characteristics to the gene. By the fourth decade, the assault on the gene was stalled, indicating that new experimental organisms and methods would be required (Muller, 1945). Transmission (Mendelian) genetics was confronted by three major problems whose common denominator was the mysterious gene: (1) the physicochemical organization of the genetic material, (2) the origin and nature of genic mutation, and (3) the cellular machinery involved in transforming genotype into phenotype. The convergence of several developments was responsible for the explosive breakthrough and activity that characterized the next two decades of genetic research: (1) the introduction of microbial species, particularly prokaryotes, into the study of transmission genetics, (2) the availability of isotopically labeled precursors of two types of macromolecules, proteins and nucleic acids, (3) new physical and chemical techniques for extracting, separating, and characterizing these molecules, and (4) an active collaboration among geneticists, biochemists, and molecular biologists. Cell biologists were the beneficiaries of the new genetics, and their contributions resulted in a better understanding of the basic unit of life, the cell.

The double helix model for the structure of deoxyribonucleic acid (DNA), previously identified as the hereditary material, represents the talisman of genetics (Watson and Crick, 1953a,b). The model provided a basis for understanding and integrating the empirically determined characteristics of the gene and a springboard for the rational investigation of protein synthesis. The recently trained generation of cellular biologists did not experience the impact of the double helix model on the course of genetics. The torrent of publications and symposia devoted to the new genetics that appeared during the decade following the Watson–Crick publications indicates frenetic activity and headlong progress.

It has become fashionable to equate transmission genetics with classical genetics and molecular genetics with modern genetics. The terminology and concepts of transmission genetics, however, are imbedded in molecular genetics and used by cell and molecular geneticists. Transmission genetics produced models which did not demand molecular evidence to merit serious consideration. Mutation, recombination, and function represent the three operational methods to define the gene. Cell biologists and molecular geneticists provided the structural and molecular foundations for the concepts and terminology of transmission genetics.

## II. MUTATION

Mendel (1866) explained the inheritance of contrasting characters in the garden pea by proposing two forms, later termed *alleles*, of a hereditary factor, later termed *gene*, which segregate during the formation of pollen and eggs. He did not seem to be concerned with the origin of alleles or the physicochemical nature of the gene. The Mutation Theory proposed by de Vries (1901) accounted for the abrupt appearance of a relatively few individuals with an altered, heritable character in large populations of the evening primrose *(Oenothera lamarckiana)*. While the mutants appeared to explain the origin of alleles, the theory was based almost completely on the presence of an extra chromosome and not on mutated genes in the mutant plants. The sudden appearance of a transmissible altered phenotype constitutes a mutation. Mutations *in sensu lato* can result from an altered gene, structural or numerical chromosomal aberrations, or an extranuclear event involving such organelles as chloroplastids and mitochondria, or plasmids and episomes. Consequently, mutation is an operational term and must be defined in each case by Mendelian ratios, cytological observations, and the consequences of reciprocal crosses between mutant and nonmutant parents. Intrachromosomal aberrations, particularly minute missing chromosome segments, termed *deficiency* or deletion, can simulate point (genic) mutations, as determined by breeding tests.

The early transmission geneticists obtained mutant alleles by screening many progenies from parents collected from the wild or by noting occasional spontaneous mutants in progenies from homozygous wild-type parents. Early attempts to induce mutations by physical or chemical agents failed because the experimental design could not distinguish spontaneous from induced mutations. Muller (1927) used X-rays as the mutagenic agent to induce sex-linked lethal mutations in the X chromosomes of *Drosophila melanogaster* males which were then mated to females. This ingenious experimental design relied on the sex-ratio in progeny from females with one X-rayed X chromosome and a sex-linked lethal mutation in the other X chromosome. The control progeny had twice as many females as males; a female with an induced sex-linked lethal mutation, however, gave no male progeny. Many of these sex-linked lethals were probably chromosome deficiencies, but genic mutations have also been obtained by X-ray treatment. Auerbach and Robson (1947) induced mutations in *D. melanogaster* by chemical mutagens, indicating that point mutation could involve a chemical alteration in the genetic material. Microbial species eventually provided the appropriate experimental material for mutagenic studies. Large populations of unicellular individuals are treated with an agent that readily reaches the genetic material, and the occasional mutant can be easily detected. When the microbial species has a sexual stage or its equivalent, mutants can be tested to determine whether the mutation is genic.

Transmission geneticists developed empirical criteria to characterize genic mutation: (1) relative stability of the mutant phenotype, (2) finite occurrence, measured as mutation rate or mutant frequency, and (3) mutability to other alleles, spontaneously or after treatment with mutagens. These criteria were supported by breeding data and cytological observations whenever the organism possessed chromosomes suitable for study by light microscopy. By a remarkably fortuitous coincidence, the exquisite banding of the polytene chromosomes in the salivary glands of larvae of *D. melanogaster,* the genetically best studied organism until the 1950's, provided suitable material to detect minute deficiencies that could not be detected in other species.

Chemical mutagenesis was expected to furnish insight into the nature of the gene. Although these studies failed in their purpose, they provided techniques to obtain mutants and, later, to test models explaining genic mutation at the molecular level. Transformation in pneumococcus by extracted DNA was viewed as an interesting mutational phenomenon but failed to excite contemporary geneticists (Beadle, 1948). Although mounting circumstantial evidence from diverse sources implicated DNA as the genetic material (Olby, 1974), Hershey and Chase (1952) presented the convincing evidence from experiments with phage T2. They used radioactive tracers to follow DNA and protein from cell infection to lysis and the release of phage particles, demonstrating that DNA and not protein was indeed the genetic material.

## A. Circumstantial Evidence for Triplet Codons

The now familiar double helix of polynucleotide chains furnished the key in understanding the gene as the unit of mutation, recombination, and function. The four bases along the polynucleotide chain provide the units that code the 20 amino acids in proteins. Combinations of two adjacent bases are insufficient to code for 20 amino acids; three adjacent bases yield more than enough, 64 combinations. Serious attempts to formulate a code of bases for the different amino acids centered in base triplets. The question of overlapping or nonoverlapping triplets was raised and received theoretical and experimental answers, favoring nonoverlapping triplets. Colinearity between nucleotide sequences in the polynucleotide chain and amino acid sequences in the corresponding polypeptide was a reasonable but unproved assumption when Brenner (1957) analyzed the known sequences of amino acids in a number of polypeptides. The diversity of amino acids preceding or following a particular amino acid demanded 70 different overlapping triplets to account for the 29 amino acids, that is, more than the available 64 triplet combinations, later termed codons. Furthermore, one nucleotide substitution in one triplet would be expected in certain cases to yield three amino acid replacements for overlapping triplets. Ingram (1956) had already demonstrated the genic mutation for sickle cell hemoglobin was responsible for a single amino acid replacement in the $\beta$-globin chain.

Experimental evidence for a nonoverlapping triplet code emerged from an investigation of proflavin-induced frame-shift mutations in the $B$ cistron of the $r$II locus in phage T4 (Crick *et al.*, 1961). Proflavin causes the addition or subtraction of a nucleotide during the replication of the polynucleotide chains of the double helix of the phage, thereby altering the sequence of nucleotides from the site of the added or deleted nucleotide. Proflavin-induced mutations in the $r$II locus are detected by the large, sharply defined plaques on strain B of *Escherichia coli* and by their ability to kill, but not lyse cells of strain K (Benzer, 1955). Furthermore, these frame-shift mutations can be assigned to a segment at one end of the $B$ cistron by appropriate complementation and recombination tests. One proflavin-induced $r$II$B$ mutant gave spontaneous revertants which lysed K cells but did not yield precisely the same type of plaques as the wild-type phase. The pseudowild-type revertants were crossed with wild-type phage and the progeny included, wild–type, pseudowild-type and mutant phage. The mutant phage, however, represented mutations at two different but nearby sites in the $B$ cistron: one site for the original proflavin-induced mutation and the second site for the spontaneous mutation, termed an *intragenic suppressor*. Because the second mutation behaved as an intragenic suppressor in combination with the proflavin-induced $r$II$B$ mutation, it was reasonable to argue that the intragenic suppressor was also a frame-shift mutation. According to this argument, the first mutation was responsible for the addition (+) of a base pair and the second mutation for the loss (−) of a base pair. If the reading of the cistron to produce the corresponding polypeptide begins from a fixed point and proceeds along the putative, nonoverlapping triplets to the end of the cistron, then the addition (+) of a base pair would alter the triplet sequence from the site of the addition. In the pseudowild revertant, the deleted (−) base pair restores the proper triplet sequence from the site of loss, thereby yielding a functional polypeptide. If the interval between the added and deleted base pairs had been too long, the polypeptide would not likely be functional and a pseudowild revertant would not have occurred. Finally, repeated treatment of mutant phage with proflavin eventually yielded pseudowild revertants. Crosses between such revertants and wild-type phage gave progeny with mutations at three close sites in the *r*II*B* cistron, indicating that the pseudowild revertants resulted from three added $(+++)$ or deleted $(---)$ base pairs from successive frame-shift mutations.

Terzaghi *et al.* (1966) determined the amino acid sequence for the lysozyme produced by wild-type phage T4 and in a plus-minus revertant and found a difference of five amino acids in sequence at positions 36 to 40, thereby confirming the view that the intragenic suppressor restores the frame reading of the cistron.

## B. *In Vitro* Evidence for Codons

Nirenberg and Matthei (1961) used a cell-free protein-synthesizing system from *E. coli,* labeled amino acids, and a synthetic polyribonucleotide. Polyribonucleotides containing only uracil produced labeled polypeptide from phenylalanine monomers. One codon or triplet of three nucleotides, therefore, could be identified as UUU = phenylalanine before the evidence for a nonoverlapping triplet code was available. Furthermore, codons were assigned to mRNA which is concerned with translation, the synthesis of the polypeptide coded by the corresponding segment of DNA (cistron). The codons in mRNA are translated for polypeptide synthesis by the appropriate anticodon in tRNA which recognizes the mRNA codon. Synthetic polyribonucleotides assembled from different proportions of two nucleotides yielded polypeptides with different ratios of specific amino acids that were statistically correlated with the ratios for the different possible triplets so that it was possible to assign triplets to specific amino acids (Lengyel *et al.,* 1961). The *in vitro* search for codon assignments to the 20 amino acids produced the genetic code (Nirenberg and Leder, 1964; Khorana *et al.,* 1967): 61 of the 64 triplets were associated with specific amino acids. Certain amino acids are coded by more than one triplet, a phenomenon unfortunately termed code degeneracy. The three unassigned triplets were later identified by *in vivo* studies as polypeptide chain terminators.

## C. *In Vivo* Confirmation of the Genetic Code

The demonstration of an amino acid substitution at a specific position in one polypeptide of sickle cell hemoglobin as a consequence of a genetic mutation in man (an unlikely experimental subject in an era of microbial experimental organisms) was a significant contribution to the *in vivo* confirmation of the genetic code.

Ingram (1956) was confronted with the formidable task of detecting and identifying the proposed replaced amino acid in sickle cell hemoglobin.

The technical problems were evaded by the development of the fingerprint technique. Normal and sickle cell globin was digested with trypsin to produce comparable peptide fragments which were subjected to paper chromatography and then high voltage electrophoresis at right angles to the plane of chromatography. A pattern or fingerprint of about 30 peptide fragment sites was obtained. Different sites for peptide fragment 4 distinguished the two globins. Sequencing amino acids in the different fragments indicated that the sixth amino acid, glutamic acid, in the fragment from normal globin had been replaced by valine in the fragment from sickle cell globin. The replacement site was eventually assigned to the sixth amino acid in the $\beta$-chain.

Mutant polypeptides from several microbial species furnished *in vivo* confirmation of the code: coat protein of tobacco mosaic virus (Tsugita and Fraenkel-Conrat, 1962; Wittman and Wittman-Liebold, 1967); the A polypeptide of tryptophan synthetase from *E. coli* (Guest and Yanofsky, 1965; Yanofsky *et al.*, 1967b); human hemoglobin (White, 1972); and iso-1-cytochrome *c* from yeast (Sherman *et al.*, 1975). In every case, a nucleotide substitution in the resident codon for the wild-type amino acid gave the appropriate codon for the substituted amino acid as exemplified in polypeptide A of *E. coli* tryptophan synthetase. Guest and Yanofsky (1965) showed that the amino acid substitutions at a specific position in the A polypeptide of tryptophan synthetase of *E. coli* in revertants or from recombination between mutants with substituted amino acids for the same resident amino acid were readily explained by a single nucleotide replacement within one codon (Fig. 1).

### D. Molecular Basis of Genic Mutation

The consequences of a nucleotide replacement in one codon can be predicted from the genetic code in terms of the codons assigned to different amino acids but not necessarily in terms of the enzymatic or other activity of the resulting polypeptide. Moreover, because the code is degenerate, a nucleotide replacement need not result in an amino acid substitution. Finally, 3 of the 64 possible codons do not correspond to an amino acid. These codons were termed *nonsense codons* to contrast with the other codons which make sense, that is, correspond to an amino acid. The substitution of one amino acid for another so that a mutant phenotype results has been termed a *missense mutation*.

Watson and Crick (1953b) suggested that spontaneous genic mutations might result when a base in the polynucleotide chain assumes one of its tautomeric forms during DNA replication. In such a situation (Fig. 2), a "wrong" base would be inserted into this position in the complementary

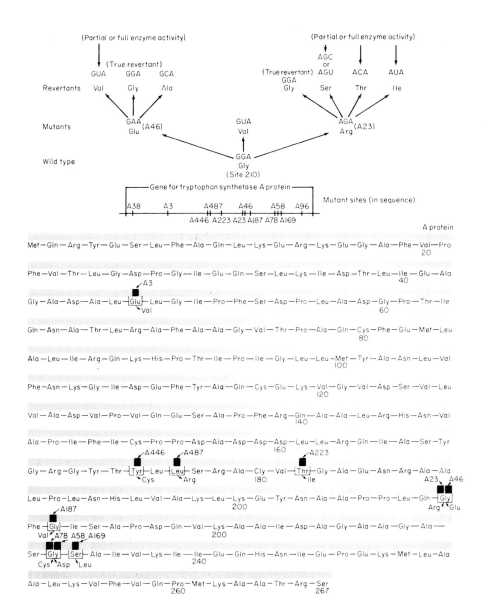

**Fig. 1.** Tryptophan synthetase polypeptide A from *Escherichia coli*: amino acid sequence, mutant amino acid substitutions for glycine at site 210 and amino acid substitutions restoring activity. The A polypeptide from two missense mutations representing an amino acid substitution for the resident glycine (site 210) had glutamic acid (A 46) or arginine (A23). Partially or fully active revertants from each mutant yielded A polypeptides with different, substituted amino acids. When the proper codon is assigned to glycine in the wild-type strain, a single nucleotide replacement by mutation can yield the appropriate codon for each substituted amino acid in the two mutants and in revertants from each mutant. The genetic fine structure map for missense mutations in the A polypeptide cistron can be correlated with the sites of amino acid substitutions in terms of sequences and of both map and polypeptide intervals (Yanofsky *et al.*, 1967a,b). (From John B. Jenkins, "Genetics." © 1975. Houghton Mifflin Company, Boston, Massachusetts. Reprinted by permission of the publisher.)

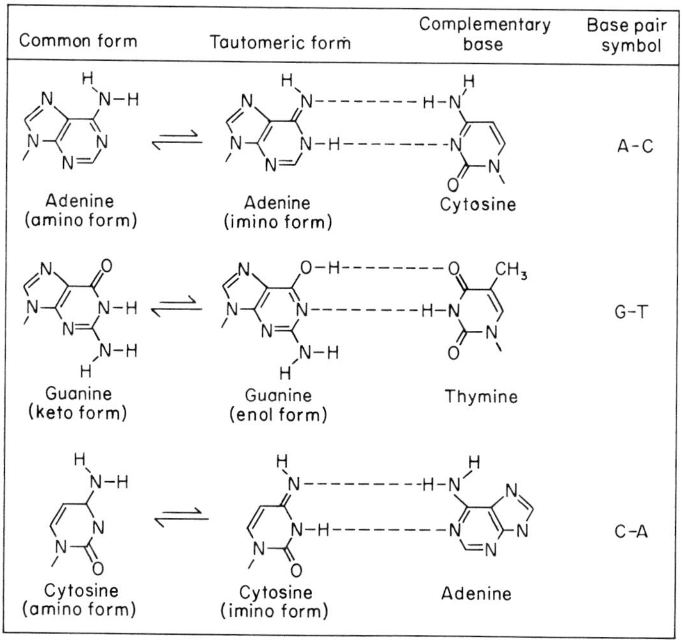

| Common form | Tautomeric form | Complementary base | Base pair symbol |
|---|---|---|---|
| Adenine (amino form) | Adenine (imino form) | Cytosine | A-C |
| Guanine (keto form) | Guanine (enol form) | Thymine | G-T |
| Cytosine (amino form) | Cytosine (imino form) | Adenine | C-A |

Fig. 2. Tautomeric forms of adenine, guanine, and cytosine and the bases with which these forms tend to establish stable hydrogen bonds. The unusual base pairs that result are presented in the right-hand column. (From Patt and Patt, "An Introduction to Modern Genetics," 1975. Addison-Wesley, Reading, Mass.)

chain, and a different sequence of bases would eventually be found in the polynucleotide chains of the double helix. After certain base analogs were shown to be incorporated into DNA as a replacement for the corresponding bases (Zamenhof and Griboff, 1954), Litman and Pardee (1956) demonstrated the mutagenicity of 5-bromouracil (5-BU) for phage T2. Using the rII locus in phage T4, Benzer and Freese (1958) showed that 5-BU was effective for certain sites within this locus, and these sites had a characteristic mutation frequency. Freese (1959) and Freese et al. (1961) introduced two terms to describe different types of base substitutions leading to genic mutation: (1) transition—a purine or pyrmidine substitution for the resident purine or pyrimidine, respectively, in one polynucleotide chain and a corresponding pyrimidine or purine substitution in the complementary polynucleotide chain; (2) transversion—a purine substitution for a resident pyrimidine or vice versa and a corresponding substitution for the resident base in the complementary chain. Base analogs increase the frequency of base substitutions during DNA replication compared with the spontaneous frequency due to rare tautomeric forms of the resident bases. Drake and Greening (1970), however, reported that a mutant DNA polymerase of phage T4 suppresses certain types of mutations produced by specific chemical mutagens, an observation implicating this enzyme in the selection of complementary nucleotides during DNA replication. Drake (1970) has presented a detailed account of the molecular basis of mutation.

## E. Suppressor Mutations

Revertants from a mutant phenotype to the wild type may result from a *back mutation* or a *suppressor mutation*. Suppressor mutations can be distinguished from back mutations by genetic analysis if the suppresor is separable by recombination from the mutation that it suppresses. The test for presence of a suppressor is made by examining the progeny of a cross of the revertant to the wild type. A suppressor mutation yields 75% wild-type progeny when the two mutations are not linked and significantly more wild-type than mutant progeny when the two mutations are linked. Furthermore, the mutant progeny recovered as segregants are indistinguishable from the original mutant. Although revertants resemble the wild type, a close examination usually reveals that the revertants are detectably different from the original wild type. For example, a mutant strain lacking activity for a specific enzyme can yield suppressor mutation-revertants but the specific activity of this enzyme in the revertants are markedly reduced compared to the specific activity for the wild-type enzyme.

Crick *et al.* (1961) obtained revertants for T4 *r*II mutants in the *B* cistron and demonstrated that many revertants simulated back mutations but generally resulted from a second mutation at a second site within the same cistron. This type of mutation represents an *intragenic suppression* for the frame-shift mutations produced by proflavin treatment. A bona fide back mutation results from the restoration of the resident nucleotide in the codon which was altered to produce the first mutation. A molecular explanation for *intergenic suppressors* required information on the amino acid sequence for a specific polypeptide and on the details of polypeptide synthesis, particularly the role of tRNA anticodons in recognizing mRNA codons.

Benzer and Champe (1962) exploited the deficiency mutant r589 in the *r*II locus of phage T4 to detect *r*II mutants caused by a premature termination of polypeptide synthesis (Fig. 3). A genetic study of these mutants furnished an explanation for intergenic suppressors. Mutant r589 has a deficiency straddling the junction of the *A* and *B* cistrons and complements any *r*II*B* cistron mutants with a wild-type *A* cistron when tested on the K

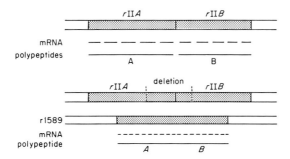

**Fig. 3.**   The rl589 deficiency mutation in phage T4 (Benzer and Champe, 1962). The rII+ locus probably produces one mRNA specifying the rIIA and rIIB polypeptides. The deficiency mutant lacks an essential part of the A cistron, the terminator codon at the end of the A cistron, and an unessential part of the B cistron. The one polypeptide from this mutant has no A activity and some B activity.

strain of *E. coli*. Because both wild-type A and B polypeptides coded by cistrons *A* and *B* are necessary for the lysis of strain K, the deficient segment of the *B* cistron in mutant rl589 is not essential for this function. Consequently, the missing segment of the *A* cistron appears to be essential for function. Genic mutations in the *B* cistron were detected by their ability to lyse the parental bacterial strain and inability to lyse a mutant of this strain. Double mutant phage strains including one of the genic mutations and the rl589 mutation were obtained and shown to lack the ability to lyse the parental bacterial strain, indicating that the genic mutations were responsible for the premature termination of B polypeptide synthesis. Because the mutant bacterial strain restored the activity of the viral genic mutants, it was reasonable to assume a common mechanism was responsible for the suppression of viral genic mutations by the host genotype.

The explanation of *intergenic suppression* considers an imaginative approach involving one mutant gene in the virus and another mutant gene in the bacterial cells and the role of tRNA in translation. The viral mutation represents a terminator codon which prematurely stops the synthesis of B polypeptide. When the viral mRNA arrives at the terminator codon, an amino acid is inserted and polypeptide synthesis continues, yielding in this case a functional B polypeptide. The mutation in the bacterial cell presumably modified a host tRNA or tRNA-activating enzyme so that the host tRNA anticodon fits the viral terminator codon. Consequently, the amino acid associated with this tRNA is inserted into the position of the terminator codon. Support for the tRNA anticodon hypothesis was provided by Garen and Siddiqi (1962), who investigated suppressor mutations for alkaline phosphatase-deficient mutants in *E. coli*, and by Sarabhai *et al.*

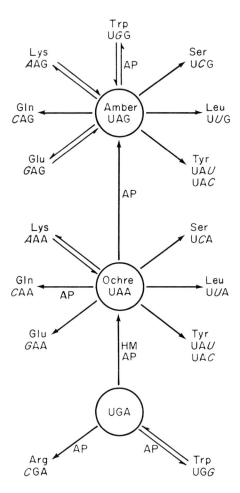

**Fig. 4.** The three nonsense codons inferred from reversion studies. The circled nonsense mutations occurred in wild-type individuals. When the resident amino acid in the wild-type polypeptide is known or inferred, an arrow points from that amino acid codon to the mutant nonsense codon. Revertants from the nonsense mutants were obtained and the specific amino acid is determined or inferred from genetic experiments. Revertants are indicated by arrows pointing from the nonsense codons to the respective amino acids. Chemical mutagens with a known mechanism of action were used to obtain revertants. Hydroxylamine (HM) is predominantly, if not exclusively, responsible for transitions (G–C→A–T) and 2-aminopurine (AP) only for transitions (G–C→A–T; A–T→G–C).

Codons assigned to the amino acids in polypeptides of revertants from the nonsense mutants are compared to determine the one codon yielding the pertinent codons by a single base substitution. For example, UAG (*amber*) is the only triplet producing the seven codons specifying each amino acid, and UAA (*ochre*), the only triplet producing the six codons for

(1964), who studied suppressor mutations of T4 mutants with defective viral protein.

Like nonsense mutants, missense mutants are also responsive to intragenic suppressors. Yanofsky *et al.* (1963) obtained two tryptophan-deficient mutants of *E. coli* with different missense mutations: *trpA* 446, glutamic acid substituted for glycine at position 210 in the tryptophan synthetase A polypeptide and *trpA* 446, cysteine substituted for tyrosine at position 174 in this polypeptide. When both mutations were included in the same cistron by recombination, a functional polypeptide was produced by intragenic suppression. Missense mutations also respond to intergenic suppressors. Missense mutations can be suppressed by tRNA mutations and the activity of missense and nonsense-suppressing tRNA's is responsive to the residual genotype of the cells with respect to alterations of ribosomal components (Gorini, 1970). In frame-shift intragenic suppression, an added nucleotide can produce a mutation which is suppressed by the loss of a nearby nucleotide.

## F. Nonsense Codons

*In vitro* methods failed to assign three codons to any amino acid. These codons were identified as premature terminators of polypeptide synthesis and mutations yielding any one of these codons were termed *nonsense mutations*. Each nonsense codon was identified by an *in vivo* method requiring revertants from a nonsense mutant. When the amino acid sequence of a polypeptide was established, it was possible to determine the site of the substituted amino acid in the revertants and to deduce the common denominator codon which would yield all of the codons for the substituted amino acid by a single nucleotide replacement.

Weigert and Garen (1965) analyzed twenty-one active revertants from an alkaline phosphatase-deficient *amber* (terminator) mutant of *E. coli* to infer from the possible amino acid codons that this nonsense mutation was coded by UAG (Fig. 4). An *amber* mutation in phage T4 was later shown also to result from a UAG terminator codon (Brenner *et al.*, 1965). Brenner

---

each amino acid. Nonsense mutations were detected and classified according to their correction by codon-specific suppressor mutations.

Genetic crosses between different nonsense mutants yield wild-type recombinants only for UAG × UGA. The recombinants presumably include tryptophan (UGG). Data for the amino acid substitutions were obtained by Sarabhai *et al.* (1964), Weigert *et al.* (1966), Sarabhai and Brenner (1967), and Brenner *et al.* (1967). (From Philip E. Hartman and Sigmund R. Suskind, "Gene Action," 2nd ed. © 1969. Reprinted by permission of Prentice-Hall, Inc., Englewood Cliffs, New Jersey.)

*et al.* (1965) identified a second type *(ochre)* of terminator mutation as UAA in T4, *r*II, using chemical mutagens to produce both *ochre* and *amber* mutants, revertants (suppressors) from these mutants, and *amber* mutants from *ochre* mutants. Brenner and Beckwith (1965) showed that mutants of *E. coli* suppressing *r*II *amber* mutants did not suppress *r*II *ochre* mutants but host mutations suppressing *ochre* mutants also suppressed *amber* mutants. The host *ochre*–suppressors recognize both *ochre* (UAA) and *amber* (UAG) codons while the host *amber*-suppressors recognize only the *amber* codon. Host-suppressors for *r*II terminator mutations could be assigned to *amber* or *ochre* by introducing the mutant virus strain into the hosts with the appropriate suppressor mutation. Sambrook *et al.* (1967) identified for *E. coli* and phage T4 the final unassigned triplet UGA as a third nonsense codon that was not suppressed by either the *amber* or the *ochre* suppressors. Finally, a mutant strain of *E. coli* with a specific and strong suppression for UGA in T4 has been isolated (Sambrook *et al.*, 1967).

Incomplete mRNA (transcription) or incomplete polypeptide (translation) synthesis from a complete mRNA could account for the action of terminator (nonsense) mutations. Brenner and Stretton (1965) used frame-shift mutations in the *r*II locus to demonstrate that *amber* and *ochre* mutations represented nonsense codons which prematurely terminated polypeptide synthesis (translation). An *amber* or *ochre* mutation was inserted by recombination into a site between the sites of a plus (+) and minus (−) frame-shift mutation of an active phage. The recombinant phage lysed K cells, indicating that the inserted nonsense codon had not prematurely terminated polypeptide synthesis. The first frame-shift mutation altered the sequence of nucleotides so that the inserted nonsense codon was misread and did not correspond to a terminator codon during the translation of a completed mRNA.

## III. RECOMBINATION

Mendel formulated the second law of heredity (independent assortment) to account for the breeding data from dihybrid parents (AABB × aabb). Bateson *et al.* (1905) discovered a departure from this law in the plant species *Lathyrus odoratus* by demonstrating a significant deviation from the expected Mendelian ratio for the four different phenotypes. The explanation for this unexpected observation came from later genetic studies with *Drosophila melanogaster* which correlated the linkage of genes with their location in the same chromosome.

The chromosome theory of sex-determination implicated the X chromosome at the determinant of sex in insects (Wilson, 1909). Morgan (1910) discovered the sex-linkage of the white-eye mutation in *D. melanogaster* by the routine practice of reciprocal crosses introduced by Mendel and correlated sex-linked inheritance with the transmission of the mutant gene in the X chromosome. The discovery of additional sex-linked mutations in this species which also exhibited linkage (nonindependent segregation) provided the experimental basis for constructing linkage maps.

Sturtevant (1913) converted the number of progeny with different phenotypes for a test cross of a female fly heterozygous for two sex-linked mutations ($X^{++} \times X^{ab} \times X^{abY}$) into recombination percentages: number of recombinant progeny/total number of progeny. This calculation not only allowed crosses with different number of progeny to be compared but also gave the significant clue for constructing a linkage map. The different recombination percentages of linkage values had an additive property. Mutant genes could be sequenced to yield a linear graph with numerical intervals between genes which corresponded to their linkage values. The site of a mutant gene in a linkage map was termed a *locus*. Extensive linkage maps have been constructed for a number of eukaryotic and prokaryotic species by this method. Somewhat unusual techniques, however, have been used for prokaryotes (Hayes, 1968).

The linkage map is based on breeding data and is essentially an abstraction, a linear probability graph. For the early transmission geneticists, the map resembled a string of beads in which each bead represented a gene. Crossing-over presumably occurred in the thread but not within the bead. Cytogeneticists were able to assign linkage groups to specific chromosomes in species with favorable pachytene chromosomes and to correlate each linkage map and its corresponding chromosome with respect to the chromosome arms and other prominent topological features. The rediscovery of the banded polytene chromosomes in the salivary glands of larvae of *D. melanogaster* provided the experimental material necessary to detect very small intrachromosomal structural aberrations and was in part responsible for the unique position of this species in genetic and cytogenetic research (Balbiani, 1881; Painter, 1934). Using a series of overlapping deficiencies including the locus for one mutant gene, Slizynska (1938) placed the locus of that gene within a segment containing a few bands in the X chromosome of *D. melanogaster* (see Chapter 2). Judd *et al.* (1972) placed one locus within one electron microscopically distinguishable chromomere. The experimental material was obtained by exploiting *pseudodominance,* the appearance of occasional recessive individuals in progenies which should include only dominant individuals. Females homozygous for sex-linked recessive alleles are mated with X-rayed dominant males. X-Rays produce small deficiencies, thereby eliminating one or

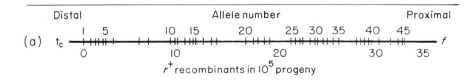

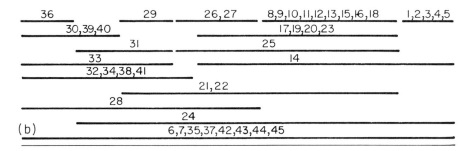

**Fig. 6.** The rudimentary locus in *Drosophila melanogaster*. (From Goodenough and Levine, "Genetics." Holt, Rinehart, and Winston, 1974. Reprinted with permission of the publisher. Carlson, 1971). (a) Fine structure map showing the loci for the flanking markers *tc* and *f*. Numbers 1 to 45 above the lines designate each allele and numbers below the line, map distances expressed as recombinants per $10^5$ progeny. (b) Complementation map in which the solid lines represent the 16 complementation groups; the numbers of the alleles in each group appear above each line.

The discovery of pseudoalleles by intragenic crossing-over in *D. melanogaster* required the screening of thousands of individuals to detect relatively few recombinant wild-type progeny. Although a locus might be complex, the processing of very large numbers of progeny presented a formidable technical problem. The discovery of nutritionally deficient mutants in *Neurospora crassa* (Beadle and Tatum, 1941) provided the experimental tool to demonstrate complex loci in fungal species with a sexual cycle. Roper (1953) demonstrated pseudoalleles for the *biotin* locus in *Aspergillus nidulans*. Recombinants from crosses between different mutants requiring this growth factor give prototrophic colonies on the minimal medium that does not support the growth of the nutritionally deficient parents and progeny. The frequency of prototrophic recombinants from *bi-1* × *bi-2* is approximately 1 per $2 \times 10^3$ ascospores and from *bi-1* × *bi-3*, approximately 1 per $5 \times 10^3$ ascospores. Pritchard (1955) used seven adenine-requiring mutants in this species to construct a linear fine structure map of the *ade-8* locus based on the frequency of prototrophic recombinants from different combinations of mutant alleles. Depending on the combination, the frequency of recombinant progeny ranged from 1 per

1250 ascospores to less than 1 per $15 \times 10^4$ ascospores. Furthermore, the sites of the heteroalleles were seriated by comparing recombination frequencies for the heteroalleles as well as for the heteroalleles and flanking genic markers.

## B. The Prokaryotic Gene

When bacteriologists encountered variants in ostensibly pure cultures, they attributed their origin to a number of different mechanisms including mutation, an inherent property of the Mendelian gene. The relatively few scanty reports prior to bacterial conjugation, existence of chromosomes, or mitosis detected by light microscopy were unreliable or contradictory. Consequently, geneticists did not view bacteria as appropriate experimental material. Transformation in pneumococcus was viewed as an unusual, inexplicable bacterial phenomenon presumably restricted to the one species.

The first acceptable experimental evidence for bacterial genes was presented by Luria and Delbrück (1943) for E. coli. A statistical analysis of the frequency of phage-resistant cells in independent cultures initiated by small inocula of a susceptible host strain indicated that mutation could be responsible for the fluctuation in the number of resistant cells in the cultures. Lederberg and Lederberg (1952) used replica-plating to detect mutants of E. coli resistant to different antibiotics and to phage prior to the exposure of a population sensitive to the selective agent. One argument to explain resistance to these agents had assumed a physiological rather than a genetic resistance which occurred in relatively few cells during exposure to the agent. Resistant mutants were identified in the replica-plate at sites corresponding to those for the survivors after exposure to the selective agent. Although such reports demonstrated that bacteria had genes, transmission geneticists accustomed to eukaryotic species could not devise appropriate experimental designs to detect recombination. An operational definition of sex as the source of recombination was needed: the pooling of genes from two parental cells for eventual distribution to progeny. This definition does not stipulate the mechanisms required for the pooling or the distribution of the genes. Consequently, an acceptable demonstration of recombination would constitute evidence for sex.

Lederberg and Tatum (1946) used two multiauxotrophic strains of E. coli with different nutritional requirements, which had been produced by repeated mutations from the prototrophic strain K12, to demonstrate genetic recombination. A culture containing cells from both multiauxotrophic strains was grown overnight in complete medium, centrifuged, thoroughly washed free of nutrients, and plated on a minimal medium

incapable of supporting the growth of either multiauxotrophic parental strain. Approximately one prototrophic cell (colony) per $10^7$ plated cells
was obtained. For example, prototrophic colonies were shown to originate
from one prototrophic cell and not from two adjacent mutant cells furnish-
ing the required nutrients for growth. Furthermore, appropriate sup-
plements in the minimum medium allowed auxotrophic mutants with a
nutritional requirement from each parental auxotrophic strain to produce
colonies which could then be shown to have the specific nutritional needs.
Lederberg (1947) constructed a linkage map by comparing the frequency
of recombinants for pairs of genes and determining significant deviations
from the expected equal frequencies for the four possible genotypes of
presumably independently segregating genes. This procedure represents
an application of the principles of transmission genetics without presuming
information on the mechanisms responsible for the pooling and distribution of the tested genes. Davis (1950) later demonstrated by the U-tube
experiment that physical contact (conjugation) between the cells of the
parental strains was necessary for genetic recombination. The U-tube contains a fritted glass filter at the bottom with pores too small for bacterial
transfer but sufficiently large for the rapid transfer of the nutrient broth.
Each arm was inoculated with a parental strain and, after incubation for
growth to turbidity, no prototrophic cells were recovered from either arm.
Forcing broth through the filter several times also gave negative results.

Hayes (1952) distinguished donor and recipient cells in crossing strains
of *E. coli*, indicating the unidirectional transfer of genes. When one multiauxotrophic streptomycin-resistant strain and a second multiauxotrophic
streptomycin-sensitive strain were mixed and plated on the antibiotic,
prototrophic colonies were obtained. When resistance and sensitivity were
reversed for these multiauxotrophic strains, no prototrophic colonies were
recovered. The streptomycin-sensitive recipient cells were inviable before
the transfer of the gene for streptomycin-resistance from the donor cells
was accomplished. In the reciprocal cross, the transfer of the gene for
streptomycin-resistance from the donor cells was accomplished even
though they were incapable of growth on the antibiotic.

To account for the conversion of recipient cells to donor cells following
contact of the two kinds of cells, Lederberg *et al.* (1952) and Hayes (1953)
postulated a replicating sex factor (F) in donor (F+) cells which was
absent in recipient (F−) cells. Recipient (F−) cells mixed with donor (F+)
cells were isolated, identified by their phenotype and then shown to behave as donor cells. A kinetic study of the infection of recipient cells
suggested that a self-replicating factor, termed F or fertility factor, was
transferred, after a physical contact, from a donor to a recipient cell.
Therefore, donor cells have the F factor and recipient cells lack this factor.

The chance isolation of substrains, termed HFr strains, characterized by their ability to yield a thousandfold greater number of recombinants than the parental F+ strain furnished the experimental material to elucidate the formal mechanism of genetic recombination in this species.

Wollman and Jacob (1955, 1958) mixed cells of a high frequency recombination (HFr) strain and a recipient strain, removed samples at intervals when the cells were conjugating, and subjected the samples to high-speed agitation prior to plating onto appropriately supplemented minimal medium to detect recombinant cells. The agitation separated conjugating cells without affecting the viability of the separated cells (interrupted mating). Samples removed prior to 8 minutes after mating gave no recombinants. Samples removed at intervals thereafter yielded recombinants which followed a genic sequence. For example, donor gene $A$ appeared at 8 minutes, donor gene $B$ at 9 minutes, and donor gene $C$ at 10 minutes (Fig. 7). Other genes appeared at later and irregular intervals. Finally, the frequency of each donor gene among the recombinants progressively in-

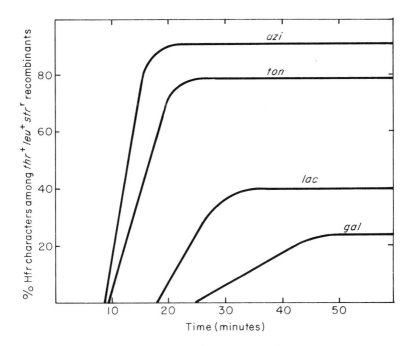

**Fig. 7.**   The time and level of recovery of unselected markers from the HFr donor strain among recombinants from the F− recipient strain after interrupted matings. Each marker is transferred to the F− recipient cells at a different interval and reaches a plateau value at a characteristic time (Wollman *et al.*, 1956).

creased from the time of its first appearance until reaching a value found by the usual crossing procedure.

The results of the interrupted mating experiment were explained by assuming that the HFr strain is a homogeneous population of donor cells which linearly transfer their specifically oriented chromosome so that only a designated end is the first to enter the recipient conjugant. Agitation breaks the chromosome as it transfers from donor to recipient and only the transferred segment can eventually replace the homologous chromosome segment in the recipient cell. The replacement presumably occurs by at least a double crossing-over between the donor segment and the resident segment. Interrupted mating yields a linear linkage map, linear sequence of genes, in which the unit of genetic distance between loci is designated one minute.

Different HFr strains were shown to produce different linkage maps with respect to the first entering gene (Fig. 8). This dilemma was resolved by comparing linkage maps and proposing a circular chromosome for *E. coli*. The F factor can be incorporated into one of a number of sites in the bacterial chromosome by crossing over. The integration of the F factor constitutes the HFr phenotype and the manner and site of integration

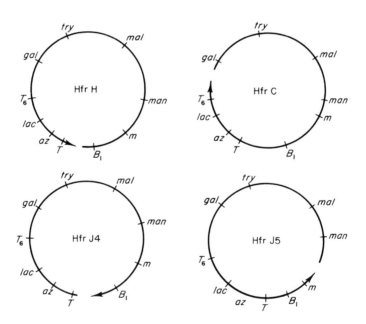

**Fig. 8.** Sequence of gene transmission for four independent HFr strains of *Escherichia coli* in terms of different sites of breakage by integration of the F factor into the circular chromosome so that the insertion of the donor linear chromosome progressively introduces those genes most distal (arrow) from the site of the integrated F factor.

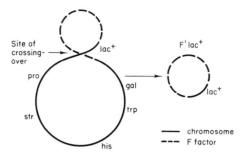

**Fig. 9.**   The origin of an F-prime (F') factor. An F factor is normally released from the bacterial process by the recombination process responsible for its integration into the chromosome. An occasional illegitimate pairing and crossing-over occur to generate an F' factor that includes an adjacent chromosomal segment.

specify the origin and orientation of the transferring linear donor chromosome. Because progeny from crosses involving HFr strains rarely displayed the HFr phenotype, it was assumed that the integrated F factor was transferred last. Cairns (1963) used electron microscopy and autoradiography to demonstrate the circular chromosome of *E. coli*, thereby correlating linkage data and bacterial cytology.

Adelberg and Burns (1960) found a new type of donor strain differing from the parental HFr strain in three characteristics. Although the direction of chromosome transfer did not differ from that for the parental strain, the efficiency of recombination was reduced by approximately 90%. While the transfer of the sex factor from donors to recipients was the same as for a typical F+ strain, the recipient cells were converted to the new donor type and not to a typical F+ strain. In one sense, the sex factor of the new donor type seemed to recall the parental HFr strain's integration site and orientation of chromosome transfer. Finally, treating cells of the new donor type with acridine orange to eliminate its sex factor and then introducing the sex factor from a typical F+ strain yielded a typical F+ strain. All of these unexpected observations could be explained by assuming the presence of a novel sex factor rather than the usual F factor in the new donor type.

The occasional reversion of HFr cells to F+ cells indicated that the F factor could spontaneously dissociate from its site in the bacterial chromosome. The novel sex factor presumably resulted from an unusual recombination involving the integrated sex factor and the bacterial chromosome so that part of the F factor with an adjacent segment of bacterial chromosome dissociated from the bacterial chromosome. The resident bacterial chromosome retained part of the F factor but was deficient for the adjacent segment (Fig. 9). The new type of F factor was designated F'

(F-prime). Selective techniques were developed to obtain F' factors from different HFr strains so that the captured chromosomal segment included different genes.

The interrupted mating technique provides a relatively efficient method to construct a linkage map for *E. coli*. The availability of different F' sex factors for this species gives the bacterial geneticist a method to construct cells that are diploid for selected genes. Consequently, heterozygotes with different combinations of linked genes, that is, one allele in the chromosome and another allele in the F' factor, or the reverse situation, can be constructed. The significance of these experimental manipulations familiar to transmission geneticists will be presented in the section on the development of the operon hypothesis.

## C. Transduction

Zinder and Lederberg (1952) tested combinations of monoauxotrophic mutants from a large collection of strains of *Salmonella typhimurium* and obtained prototrophic cells only from a specific combination of strains. The U-tube test to determine whether physical contact between cells was required for recombination revealed that prototrophs were present in one arm and lysed cells in the other arm of the tube. Appropriate tests indicated that donor alleles were transferred to recipient cells by a bacterial virus caused to be released by the donor cells which had been lysed following infection with a virulent bacterial virus released by the lysogenic recipient cells. A lysogenic bacterial strain harbors phage and occasional cells lyse spontaneously, releasing phage particles. These particles can be virulent for sensitive bacterial strains. Genetic recombination mediated by a virus was termed *transduction* and the vector virus, *transducing bacteriophage*. Only a relatively short chromosomal segment with the marker locus is involved in genetic recombination because the transferred segment is packaged within the transducing virus. Although transduction occurs with a low frequency (ca. $10^{-5}$–$10^{-7}$), substantial numbers of recombinants can be found because populations of $10^8$–$10^9$ cells can be screened for recombinants. Crossing-over is presumably involved in the replacement of the resident mutant allele in the bacterial chromosome by the donor segment with the dominant allele.

Stocker *et al.* (1953) used transduction to investigate the genetic basis of flagellar characteristics in *S. typhimurium*. Although contrasting phenotypes were transduced as expected, a pair of contrasting phenotypes, such as different nutritional requirements, can be occasionally transduced in a single event. Such cotransduction indicates that the transferred segment is sufficiently long to include at least two genetic loci. Further-

**TABLE I**

**Number of Prototrophic Recombinants from Transductional Crosses between Various Auxotrophic Mutants of *Salmonella typhimurium*** [a]

| Recipient | Donor | Number of wild-type recombinants |
|---|---|---|
| 1.[b] *trp*–D10 | *trp*–D10 | 0 |
| *trp*–D10 | wild–type | 1822 |
| 2.[c] *trp*–D10 | *met*–15 | 1617 |
| *trp*–D10 | *trp*–A8 | 208 |
| *trp*–D10 | *trp*–B4 | 602 |
| *trp*–D10 | *trp*–C3 | 207 |
| *trp*–C3 | *trp*–D10 | 88 |
| 3.[d] *trp*–D10 | *trp*–D1 | 4 |
| *trp*–D10 | *trp*–D6 | 2 |
| *trp*–D10 | *trp*–D7 | 7 |
| *trp*–D10 | *trp*–D9 | 12 |

[a] From Demerec and Hartman (1956).

[b] 1. Wild-type × mutant cross.

[c] 2. Crosses between phenotypically different or nonallelic mutants.

[d] 3. Crosses between phenotypically identical and allelic mutants. Capital letters indicate loci determining different biochemical functions and numerals signify independent mutations. Parameters: $8 \times 10^7$ recipient cells were infected with $4 \times 10^8$ phage particles from the donor and plated on minimal medium; prototrophic colonies were counted after 48 hours at 37°C.

more, the cotransduction of two phenotypes constitutes evidence of linkage and provides a method to construct a linkage map for this species.

Transduction was used by Demerec and Hartman (1956) to determine whether the four groups (A–D) of tryptophan-requiring mutants of *S. typhimurium* established by biochemical criteria could also be distinguished by a genetic criterion (Table I). Each group controls a different step in tryptophan biosynthesis (Brenner, 1955). Transduction studies yielded three significant observations (Fig. 10): (1) the frequency of pro-

| Mutant group | Conversion blocked | Substances accumulated | Growth stimulants | Cross-feeds mutant of groups: |
|---|---|---|---|---|
| A | → anthranilic acid | ⋯ | anthranilic acid indoleglycerol phosphate indole tryptophan | None |
| B | anthranilic acid → indoleglycerol phosphate | anthranilic acid | indoleglycerol phosphate indole tryptophan | A |
| C | indoleglycerol phosphate → indole | indoleglycerol phosphate | indole tryptophan | A, B |
| D | indole → tryptophan | indole and indoleglycerol phosphate | tryptophan | A, B, C |

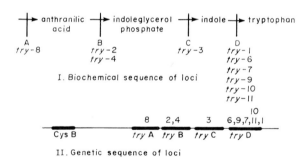

**Fig. 10.** Biochemical characteristics of different groups of tryptophan-deficient mutants of *Salmonella typhimurium* and the sequence of loci determining four sequential steps in the synthesis of tryptophan. Capital letters identify groups of mutants blocked in the same synthetic step and numerals indicate independent mutants (from Hayes, "The Genetics of Bacteria and Their Viruses" (second edition 1968) with permission of the publishers, Blackwell Scientific Publications, Oxford).

totrophic transductants for the *trp* mutants provides a measure of intragenic and intergenic recombination; (2) cotransduction tests involving *cysB* and pairs of *trp* mutants from different groups indicate very close linkage for the four *trp* loci, yielding a linear sequence for these loci; and (3) the sequence of the four loci corresponds with the sequence of biosynthetic steps controlled by these loci.

In fine structure analysis by transduction, crossing-over is assumed to be responsible for the replacement of resident genetic sites by donor sites. Consequently, closely linked sites are more likely to be included in a transferred segment and single crossover products are more likely to be recovered than double crossover products. Intragenic crossing-over is less likely to yield prototrophic transductants than intergenic crossing-over. The frequencies of prototrophic transductants for ten *trp* mutants were significantly reduced for mutants assigned to the same group compared with mutants in different groups. The recovery of prototrophic transductants for different mutants assigned to one genetic locus indicated that the bacterial gene locus was complex. Hartman (1956) and Hartman *et al.* (1960) determined by biochemical and complementation tests the site of block in histidine biosynthesis for more than 200 mutants requiring this amino acid. Seven functional loci were detected, each determining a specific biosynthetic step. For two loci, 31 of 34 mutants and 33 of 35 mutants gave prototrophic transductants for mutant alleles, indicating that bacterial genes include a number of mutational sites separable by genetic recombination. Furthermore, the detection of different mutational sites by recombination was dependent on the resolution obtainable by the technique to obtain recombinants.

Whenever cotransduction involves two different phenotypes, the transferred segment is sufficiently long to include both loci (Fig. 11). A three-point transduction test can be performed in situations where one phenotype results from mutations in different but contiguous cistrons (Fig. 12). The cotransduction of the *cysB* and each of the four *trp* loci indicated that these loci were closely linked. The fortuitous linkage of *cysB* (a locus determining a specific step in cysteine biosynthesis) and the contiguous *trp* loci furnished the means to sequence the loci by reciprocal three-point tests involving *cysB* and a pair of *trp* mutants at different loci. The parallel sequence of *trp* loci and of steps in tryptophan biosynthesis controlled by each locus was unexpected and unexplainable at the time.

An intensive genetic-biochemical study of 540 histidine-requiring mutants in *S. typhimurium* yielded ten loci by biochemical criteria and

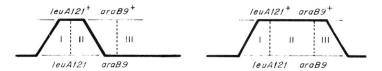

**Fig. 11.** Cotransduction for leucine-deficiency and arabinose-utilization in *Salmonella typhimurium*, using phage P22 from a *leu* + *ara* + donor strain and a *leu* A121 and *ara* B9 recipient strain. The *leu* + transductants are selected by plating on minimal medium and then tested for arabinose–utilization. A crossover must occur in region I (left) to obtain prototrophs and a second crossover in region III (right) to select arabinose-utilizing prototrophs.

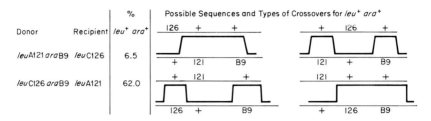

**Fig. 12.** Linkage sequence in *Salmonella typhimurium* by three-point transduction. The choice of possible sequence for the three mutant sites is determined by the relative probabilities of two events: a double crossover (more likely) or a quadruple crossover (less likely). The yield of *leu+ ara+* recombinants is greater when a double rather than a quadruple (two doubles) crossover is proposed to obtain these recombinants.

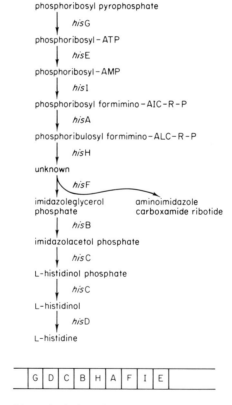

**Fig. 13.** Histidine biosynthesis in *Salmonella typhimurium*, showing the sequential steps, the genes specifying the enzymes catalyzing each reaction and the linkage map (bottom) of the histidine operon. The sequence of biochemical reactions does not agree with the sequence of loci in the operon.

transduction tests for the nine clustered genes in an operon (see below) controlling ten sequential steps in the biosynthesis of this amino acid (Loper *et al.*, 1964). In this case, the sequencing of the loci and of the pertinent biosynthetic steps are not parallel (Fig. 13). No explanation is available to account for the difference in sequencing of mutant sites for the tryptophan and histidine operons. Gene clustering appears to be relatively common in bacteria. Demerec (1964) noted that 70% of the 87 loci concerned with 18 biosynthetic pathways in *S. typhimurium* formed clusters of two or more genes. Although gene clusters are found in *E. coli* and *Bacillus subtilis*, they are rarely encountered in *Pseudomonas aeruginosa* (Fargie and Holloway, 1965). The significance of clusters of functionally related genes was not appreciated until Jacob and Monod (1961) investigated three closely linked genes in the *lac* (lactose) locus of *E. coli* and formulated the operon hypothesis to account for their genetic–biochemical observations.

## D. The *r*II Locus and the Cistron

The development of techniques to manipulate bacterial viruses of *E. coli* was ultimately responsible for the introduction of phage into the select group of genetically significant species (Ellis and Delbrück, 1939; Delbrück, 1940). The discovery of spontaneous mutants in the T-even phages provided the contrasting phenotypes required for transmission genetic studies. Two types of phenotypes were available: *r* (rapid lysis), mutant plaques obviously larger than *r*+ plaques, and *h* (host range), mutants virulent for bacterial strains resistant to lysis by the *h*+ (wild-type) virus. Bacterial cells infected with one *h* + *r* + and one *h r* viral particle lyse and yield viral progeny with the parental or recombinant *(h* + *r*, *h r*+)* genotypes. These results constitute evidence for sex in virus and allowed the methodology of transmission genetics to be applied to viruses (Delbrück and Bailey, 1946; Hershey, 1946; Hershey and Rotman, 1949).

A phage cross involves the addition of an appropriate number of viral particles with different genotypes to a suspension of bacterial cells with the proper genotype so that each cell is likely to be infected by two particles. Viral particles not bound to bacterial cells are removed; the cell suspension is rapidly diluted; and an appropriate number of cells is added to a lawn of bacterial cells with one or two appropriate genotypes before the infected cells lyse. Plaques are produced in the indicator lawn of cells, and the progeny particles can be recovered from each plaque and individually tested to determine the relative number of recombinant particles in each progeny. From an operational point of view, the cross represents the pooling of viral chromosomes within the bacterial cell, where they undergo replication and recombination. The viral chromosomes are eventually packaged for recovery within the viral particles released by the lysis of the bacteria and restricted to the plaque. In crosses between strains characterized by different types of plaques, the number of plaques with different morphology can be directly scored without recovering individual particles to determine their genotype. Hershey and Rotman (1949) used host-range and independent $r$ mutants in phage T2 to construct a linear linkage map with a sequence of $r$ mutants relative to the $h+$ locus and to each other. One cluster of very closely linked $r$ mutant sites was later identified as the $r$II locus. Streisinger et al. (1964) later demonstrated a circular linkage map for phage T4 by a three-point linkage test for a large collection of diverse types of mutants. Although the viral chromosome is physically linear, hydrogen bonding of terminal single complementary polynucleotide chains extending from each end of the linear viral chromosome produces a functionally circular chromosome after infection. Consequently, the genetic evidence for a circular linkage map formally agrees with the microscopic evidence for a transiently circular viral chromosome in the T-even phages.

Hershey and Chase (1952) discovered T2 phage particles heterozygous for certain $r+$ and $r$ alleles which were unstable in that cells infected with one of these particles yielded mainly $r+$ and $r$ segregants and a small number of heterozygous particles. The viral chromosome incorporated into the particle was shown to have the same nucleotide sequence at each end, a phenomenon termed terminal redundancy, which comprised approximately 1–3% of the genome (Streisinger et al., 1964). Furthermore, the phage particles include chromosomes with different terminally redundant segments that represent random circular permutations of one chromosome. While each chromosome within the particle is linear, any pair of markers can be linked in a majority of the particles but also will be found at the ends of the chromosome, yielding an apparently circular linkage map. These observations can be explained by assuming that the individual phage chromosomes are produced for packaging into each particle by cutting a long chain or concatenate of linked phage genomes in equal segments somewhat longer than complete genomes. Consequently, heterozygosity for the $r$ and $r+$ alleles results from the inclusion of an $r$ allele at one end and an $r+$ allele at the other end of certain terminally redundant phage particles.

Advances in the transmission genetics of the T-even phages were responsible for an experimental design that eventually extended fine structure analysis by recombination to its limits and led to an unexpected redefining of the gene. Benzer (1955, 1957, 1961) discovered one class of $r$ mutants in phage T4 which could be experimentally manipulated to detect

**TABLE II**

**Plague Morphology of *r* Mutants of Phage T4 Isolated on Strain B of *Escherichia coli* when Plated on Strains S and K–12**[*a*]

|  | Strain of *E. coli* | | |
|---|---|---|---|
| Phage | B | S | K |
| wild | wild | wild | wild |
| *r*I | *r* | *r* | *r* |
| *r*II | *r* | wild | — |
| *r*III | *r* | wild | wild |

[*a*] Strain S is a sensitive, nonlysogenic strain from strain K which is lysogenized by λ phage (Benzer, 1957).

recombination frequencies as low as $10^{-6}$. The *r* mutants in this strain were assigned to one of three groups determined by plaque morphology observed on each of three strains of *E. coli* (Table II). The *r*II mutants were distinguished from the *r*I and *r*II mutants because K12 (λ) cells infected with *r*II mutants are killed but not lysed, with the result that plaques are not observed. Crosses between *r* mutants from different groups indicated that each group occupied a different locus in the viral chromosome (Fig. 14). Infecting K cells with *r*+ and *r*II particles results in lysis and the release of both *r*+ and *r*II particles, revealing that *r*II phage multiplies within the host cell. While the biochemical details involved in killing but nonlysis are not yet clear, the results show that the specific function controlled by the *r*II locus does not operate in *r*II mutants.

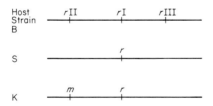

**Fig. 14.** The relationship between host strains B, S and K of *Escherichia coli* and the mapping of *r* mutations. When pairs of different *r* mutations are crossed by infecting B cells, recombinant wild-type (*r*+) phage are produced and recombinations frequencies are used to assign the mutants to one of three loci (Benzer, 1957).

The fine structure analysis of the $r$II locus requires infecting B strain cells with two $r$II mutants of independent origin, adding the infected cells to a lawn of K cells, and scoring plaques. While the $r$II mutants multiply within B cells which lyse, only the $r$II$B+$ virus multiplies within K cells to produce lysis. Each plaque that can form represents at least one recombinant $r+$ particle in the progeny from one cross. The number of plated B cells and of plaques provides a measure of the frequency of recombination for the two $r$II mutants. Very low plaque frequencies indicate very close linkage between the mutant sites.

The $r$II mutants selected for fine structure analysis were shown to be revertible, yielding spontaneous $r+$ phages. Unrevertible mutants were assumed to result from deficiencies within the $r$II locus. Crosses between specific combinations of the latter do not yield $r+$ recombinants, suggesting overlapping deficiencies that share a common missing segment. Furthermore, crosses between certain $r$II genic mutants and deficiency mutants also do not yield $r+$ recombinants. An ingenious system using a linear sequence of different deficiencies for crude and fine mapping genic mutations within the $r$II locus was developed to assign genic mutations unambiguously to a specific site. The large collection of independent genic $r$II mutants could not have been mapped without the technical advantage of the deficiency tests. More than 2,400 spontaneous genic mutants yielded a linear sequence of 428 sites. Adding linkage values gave an estimate of 8% recombination between the limits of the locus. Assuming a total linkage map length of approximately 700–800% for phage T4, the $r$II locus occupies approximately 1% of the map.

While the fine structure analysis of the $r$II locus was a *tour de force* of transmission genetics, the discovery of the cistron had a greater impact on genetic theory and terminology. When K cells are infected with certain combinations of $r$II mutants and lyse, the viral progeny are predominantly $r$II mutants. An extensive survey of $r$II mutant combinations revealed that the mutants could be assigned to one of two groups: $A$ and $B$. A combination of a group $A$ and group $B$ mutants yields mutant viral progeny; a combination of two group $A$ or two group $B$ mutants gave wild-type ($r$II$+$) recombinant progeny. Benzer (1957) proposed the following explanation for these observations. The $r$II locus determines a function requiring combined products, each controlled by one of two contiguous chromosome segments of the locus. The A segment includes the sites of the $r$IIA mutants and the B segment, the sites of the $r$IIB mutants. The geometry of mutant sites for combinations of $r$II mutants producing lysis of infected K cells is the same as that in the complementation or "cis–trans" tests previously developed for diploid eukaryotes. The A and B segment in the $r$II locus was termed a cistron, the unit of function specifying a gene product. The lysis of K cells by mutants in different cistrons ($r$IIA $+$

$r$II$B$ $\times$ $r$II$A$ $r$II$B$+) results from functional combined products determined by the $r$II$A$+ and $r$II$B$+ cistrons in each phage mutant. Consequently, lysis occurs but the released phage remain mutant. Two mutants in one cistron cannot produce the functional product determined by that cistron unless a crossing-over occurs between the mutant sites in the cistron.

The genetic map and molecular weight of the DNA double helix of the T4 chromosome furnished the basis for estimating the length of the $r$II locus and its cistrons in chemical terms. The linkage map extends for about 700 units and the minimal detectable map distance, i.e., crossing-over between mutant sites, is approximately 0.02. Two sites separated by 0.02 unit include $3 \times 10^{-5}$ (0.02/700) of the mappable genome. The DNA in phage T4 has a molecular weight equivalent to $2 \times 10^5$ nucleotide pairs. The minimal detectable recombination segment would be $3 \times 10^{-5} \times 2 \times 10^5$ or approximately six nucleotide pairs. While six nucleotide pairs would correspond to two codons, recombination can occur within one codon, indicating that six nucleotide pairs is not an indivisible unit of recombination. The genetic length of the $A$ and $B$ cistrons is approximately 6 and 4 units, respectively, using recombination frequencies for mutants at the outer limits for each cistron. Therefore, cistron $A$ includes approximately 1700 nucleotide pairs (6/700 $\times$ $2 \times 10^5$), a coding capacity of ca. 570 amino acids, and cistron $B$ approximately 1100 nucleotide pairs (4/700 $\times$ $2 \times 10^5$), a coding capacity of ca. 370 amino acids.

Three assumptions are needed in relating these genetic values to chemical values: (1) the linkage map represents the phage chromosome; (2) all of the phage genetic information is carried by the phage chromosome and none of the genetic information is duplicated; and (3) genetic recombination occurs with the same probability throughout the chromosome. The errors inherent in these assumptions should be considered in assessing the significance of the calculated values.

The cistron is defined as the unit of function in terms of complementation tests. The coincidence of the linkage and functional maps for the $r$II locus indicates that cistron might be a more precise term than gene. Complementation tests for mutations determining one enzyme, however, do not always define a cistron. Consequently, the term gene has not been replaced by cistron but both terms may be interchangeable when supported by experimental evidence. The one gene–one enzyme relationship, however, has been refined to the one cistron–one polypeptide relationship.

## IV. GENE FUNCTION

Mutant genes are detected by their impact on the phenotype of the organism at some stage of development. The syndrome of more than one altered character determined by one mutant gene was named *pleiotropy* and indicated that a basic error in the course of development could have many seemingly unrelated ramifications (Fig. 15). Transmission geneticists preferred to work with syndromes with at least one constant mutant character and identified the mutation with this character. The expressivity (degree of expression of a mutant phenotype) of many mutant phenotypes in higher plant and animal species was shown to be determined by environmental factors and, in some cases, could be sufficiently altered by manipulating the environment during development to simulate the wild-type phenotype. Transmission geneticists prefer mutants with complete penetrance, that is, all of the mutant progeny display the mutant phenotype in a standard environment. Attempts to understand the transformation of genotype into phenotype in cellular terms, however, were frustrated by the complexity of cellular metabolism and organismal development. Geneticists could not formulate appropriate experiments using the concepts and methodology available to contemporary biochemists concerned with cellular metabolism. Furthermore, geneticists were involved in transmission genetics and cytogenetics with easily manipulated species and biochemists with the unraveling of cellular metabolism with relatively primitive methods. Garrod's (1902, 1909) evi-

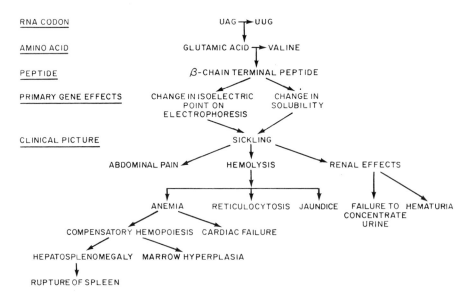

**Fig. 15.** The molecular origin of the sickle cell syndrome and its ramifications as an example of pleiotropy. (From "Heredity and Disease" by A. H. Porter. Copyright 1968, McGraw-Hill. Used with permission of McGraw-Hill Book Company.)

dence of a relationship between mutant genes and defective enzymes was prematurely presented to geneticists and biochemists with essentially parochial interests.

Sturtevant (1920) discovered a gynandromorph D. *melanogaster* which had a female half (XX) and a male half (X) on each side of a plane extending along the middle of the fly. The zygote was heterozygous for a number of sex-linked genes and presumably, during the first mitotic division, the X chromosome with the dominant alleles was lost from one daughter cell. While the female half displayed dominant phenotypes, the male half had the expected recessive phenotypes (pseudodominance) except for eye color. The eye on the male side was wild-type instead of the expected vermilion color. This observation was explained by proposing a compound produced in the wild-type side which diffused to the mutant male side during development and compensated for the metabolic lesion producing the vermilion eye. Beadle and Ephrussi (1937) transplanted embryonic eye discs from 26 different eye color mutants of D. *melanogaster* into wild-type larvae to determine the phenotype of the transplanted eye in the mature host. Only two transplants from vermilion and cinnabar acquired the wild-type phenotype and the other transplants retained their mutant phenotype. The vermilion *(v)* and cinnabar *(cn)* mutations were nonautonomous in that diffusible compounds produced in the wild-type flies repaired the metabolic lesion in the mutant flies. The other eye color mutations were autonomous, retaining their mutant phenotype.

Reciprocal transplants for the vermilion and cinnabar mutants revealed that *cn* transplant in *v* hosts developed autonomously but *v* transplants in *cn* hosts displayed the wild-type phenotype. These observations indicated a sequence of events involving two specific diffusible compounds in the development of eye color: $v+\rightarrow cn+$. The $v+$ allele is responsible for the production of a $v+$ compound and the $cn+$ allele for the production of a $cn+$ compound. Consequently, cinnabar flies furnish $v$ eye discs with the $v+$ compound but vermilion flies do not provide $cn+$ compound for $cn$ eye discs. This explanation assumes that vermilion produces a block at $v+$ but not at $cn+$ and supplying $v+$ compound, as cinnabar does, allows for the continued biosynthesis of wild-type eye pigments. The chemical nature of the immediate precursor of $v+$ compound and of both $v+$ and $cn+$ compounds was later identified (Ephrussi, 1942):

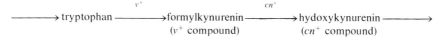

$$\longrightarrow \text{tryptophan} \xrightarrow{v^+} \underset{(v^+ \text{ compound})}{\text{formylkynurenin}} \xrightarrow{cn^+} \underset{(cn^+ \text{ compound})}{\text{hydoxykynurenin}} \longrightarrow$$

Although the eye disc transplant experiments in *Drosophila* did not furnish the general experimental tool to investigate the genetic control of biosynthetic pathways, they did stimulate the design of a direct approach to this problem by considering a negative phenotype which would reflect a genic mutation. Bcadle and Tatum (1941) used the fungal eukaryotic species *Neurospora crassa,* which had been used for transmission studies and would grow on a defined (minimal) medium containing an organic carbon source, biotin, and relatively few salts. The uninucleate conidia were exposed to X-rays and were first plated on a complex medium so that surviving conidia produced colonies. The individual colonies were then tested for their ability to grow on defined media. Three nutritionally deficient (auxotrophic) mutants, each determined by a single gene mutation, were obtained and found to require a specific compound (pyridoxine, thiamine, or *p*-aminobenzoic acid) for growth on the defined medium. The Beadle–Tatum technique has since been used to isolate auxotrophs requiring a specific amino acid, vitamin, growth factor, purine or pyrimidine, to mention the more common types in many fungal, algal, and bacterial species. Moreover, auxotrophic mutants have been the experimental tools in identifying the intermediate compounds for specific biosynthetic pathways (Srb and Horowitz, 1944), thereby relating genic mutations to specific enzymes, providing genic markers for the fine structure analysis of loci (Roper, 1953), and demonstrating genetic recombination in bacterial species (Lederberg and Tatum, 1946; Zinder and Lederberg, 1952).

Beadle (1945) speculated that each biochemical reaction was directed by a specific gene whose primary purpose was to control a reaction by affecting the configuration of the proteinaceous enzyme. This relationship was named the one gene–one enzyme hypothesis. Numerous apparent exceptions to this hypothesis were reported and later shown to reflect the complex interrelationships of biosynthetic pathways, inhibitory mechanisms, compartmentalization of enzymes in cellular organelles and membranes, and the quaternary structure of many enzymes. For example, one mutant in *N. crassa* requires both threonine and methionine for growth as the result of a point mutation. When the biosynthetic pathways leading to these different amino acids were established, the mutation was found to have a deficiency for homoserine, an amino acid required for the biosynthesis of both threonine and methionine. Adding only homoserine to the minimum medium restored growth for this mutant. In the monogenically determined, segregational petite mutant of *Saccharomyces cerevisiae,* several respiratory enzymes cannot be detected. The mutation in this case apparently impairs the mitochondria so that all of these enzymes which reside within this organelle do not function. The identification and characterization of proteins in intact cells, cellular extracts, and cell-free protein-synthesizing systems by sophisticated technology resulted in the definition of proteins and enzymes in physicochemical terms so that mutation could be related to molecular alterations (one gene–one enzyme, one cistron–one polypeptide).

Four hierarchial levels of enzyme and protein structure were established by biochemical and biophysical criteria. The primary structure refers to the sequence of amino acids in the linear polypeptide chain—the only level of protein structure directly determined by the information in genes. Although 20 amino acids (L-isomers) have been found in proteins, these amino acids can be modified to perform specific functions in the biological activity of the polypeptide by enzymatic reactions after incorporation into a polypeptide. In determining the amino acid sequence of beef insulin, Sanger (1956) demonstrated that a specific sequence, rather than a specific content of amino acids, characterized a particular polypeptide. The secondary structure refers to the alpha-helical coiling of certain segments and the uncoiled but folded segments of the polypeptide chain (Pauling and Corey, 1950). Although the kind, number, and sequence of amino acids in the primary structure offer an obvious source of polypeptide diversity, the intrahelical hydrogen bonding between amino acids in the helical segments introduces another magnitude of diversity. The tertiary structure refers to the three-dimensional configuration assumed by the folding of the polypeptide chain as a result of the amino acid interactions involving diverse types of covalent and noncovalent bonds. In folding, amino acids with hydrophobic side chains tend to occupy the interior and those with hydrophilic side chains the periphery of the globular configuration which offers maximal stability in aqueous solutions. The quaternary structure refers to the protein–protein interactions whereby the aggregation of two or more polypeptides forms a complex macromolecule, presumably by the same types of noncovalent bonds employed for the secondary and tertiary structures and largely a function of the properties of the polypeptides themselves. Most enzymes and many proteins have a quaternary structure, an observation with special significance in relating mutation to gene function. Consequently, the wellspring of the diversity and biological specificity of enzymes and proteins is found in the primary structure which ultimately determines the other hierarchies.

Transmission geneticists apply the terms *dominant, recessive,* and *codominant* to alleles on the basis of the phenotypes of homozygotes and heterozygotes. At the molecular level, the dominant allele specifies an active enzyme and the recessive allele, either an inactive or an absent enzyme. The dominant allele in the heterozygote usually yields sufficient enzyme to produce the dominant phenotype, with a perhaps reduced level of the enzyme if the enzyme's level is not regulated. For codominant alleles, each allele specifies an active enzyme yielding different products so that the heterozygote exhibits the phenotype of each homozygote.

*Modified dihybrid ratios* may be explained by referring to the relationship between mutant alleles and the pertinent biosynthetic pathway(s) involved in producing the wild-type phenotype (Fig. 16). While crosses between different mutant homozygotes can produce phenotypically identical hybrids in the $F_1$, the $F_2$ progeny may fit $9:3:3:1$, $9:7$ or $9:3:4$ ratios with four, two, or three phenotypic classes, respectively. In the progeny with four classes (cross 1), the recessive alleles block different biosynthetic pathways which lead to different phenotypes or different combinations of similar phenotypes, such as eye color in insects. The other ratios can be explained by assuming a single biosynthetic pathway with a different block by each gene. For the $9:7$ ratio, either recessive homozygote exhibits the mutant phenotype, and for the $9:3:4$ ratio, one recessive homozygote or the double homozygote displays one mutant phenotype and the other

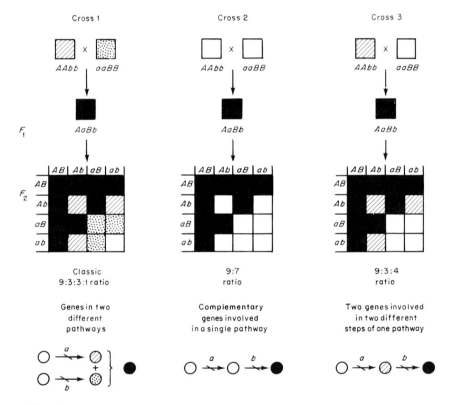

**Fig. 16.** Three different crosses between homozygous diploids producing hybrids heterozygous for two pairs of genes and yielding three different dihybrid ratios. The Punnett squares display the $F_2$ genotypes and phenotypes in each case. The biochemical interpretation for each ratio is presented beneath each cross. (From Philip E. Hartman and Sigmund R. Suskind, "Gene Action," 2nd ed. © 1969. Reprinted by permission of Prentice-Hall, Inc., Englewood Cliffs, New Jersey.)

homozygote mutant a second phenotype. Two dominant alleles in the modified ratios complement each other to produce the wild-type phenotype, indicating that each gene has a different function. The independent assortment of these genes also indicates different loci. If the genes were linked, crossing-over would also indicate different loci. Consequently, complementation and recombination would coincide in the conclusion that two different genes are responsible for the observations. If the two genes were so closely linked that recombination could not be detected, a complementation (cis–trans) test provides one means to test the different functions of two mutant genes yielding similar phenotypes.

In *epistasis,* one mutant gene inhibits the expression of another but nonallelic gene. For example, homozygous recessive alleles *(aa)* for albino prevent the expression of the dominant gene for color. Recessive albinism is a similar example of recessive epistasis. In maize, the dominant mutant gene I is epistatic, suppression dominant genes for seed color.

## A. Complementation Tests

Complementation tests assume that the wild-type phenotype results from a protein–protein and *not* from a gene–gene interaction (Benzer, 1957). The tests require two alternative arrangements of the wild-type and mutant alleles in the same chromosome when the genes are very closely linked: (1) *cis*—both mutant alleles (*ab*) in one chromosome and both wild-type alleles (+ +) in the other chromosome, and (2) *trans*—one mutant allele (*a* +) in one chromosome and the other mutant allele (+ *b*) in the homologous chromosome. If the *cis* arrangement is required, it implies a gene–gene interaction; if a *trans* arrangement suffices, a protein–protein interaction is indicated. Two technical problems may have to be solved to accomplish a complete cis–trans test: (1) obtaining the *cis* arrangement by recombination, and (2) introducing both chromosomes or pertinent chromosome segments into one cell. Complementation tests have been designed for haploid and diploid eukaryotes, bacteria and viruses (Fig. 17).

## B. Intergenic Complementation

Beadle and Coonradt (1944) demonstrated growth for a heterokaryon involving different tryptophan-requiring mutants of *N. crassa* on a minimal medium that would not support the growth of either mutant. Once the biosynthetic pathway for this amino acid was established, each mutant was shown to block a different step by determining the synthesis of an enzyme required for that step. Furthermore, genetic studies indicated that

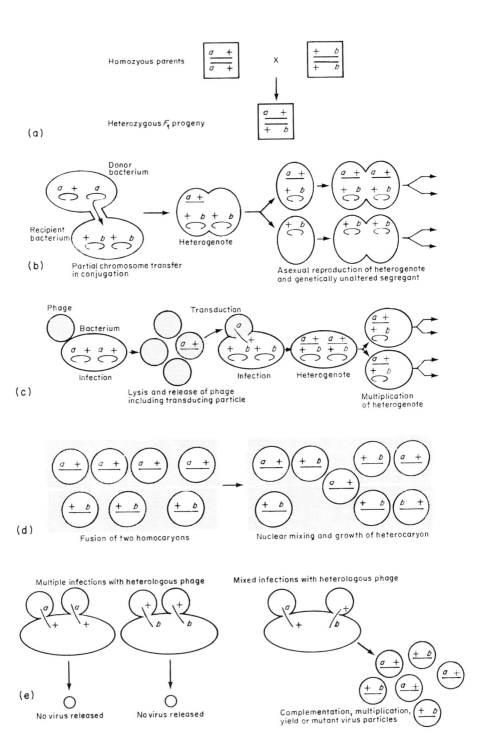

Homozyous parents

Heterozygous $F_1$ progeny

(a)

Donor bacterium

Recipient bacterium

Heterogenote

(b) Partial chromosome transfer in conjugation

Asexual reproduction of heterogenote and genetically unaltered segregant

Phage

Transduction

Bacterium

Infection

Lysis and release of phage including transducing particle

Infection

Heterogenote

Multiplication of heterogenote

(c)

(d)

Fusion of two homocaryons

Nuclear mixing and growth of heterocaryon

Multiple infections with heterologous phage

Mixed infections with heterologous phage

(e)

No virus released

No virus released

Complementation, multiplication, yield or mutant virus particles the mutant genes occupied different loci. Complementation tests for large numbers of independent mutations yielding the same phenotype provide significant information before resorting to other genetic studies, on the probable number of loci for the mutant genes.

## C. Intragenic (Interallelic, Intracistronic) Complementation

Fincham and Pateman (1957) and Giles *et al.* (1957) reported complementation for mutant alleles determining the activity of *N. crassa* enzyme which had been assigned to a single locus. Consequently, complementation for two mutant genes determining one phenotype does not necessarily discriminate between mutations in one cistron or between mutations in different cistrons. Catcheside and Overton (1959) speculated that the mutant nuclei in the heterokaryon specified the synthesis of altered polypeptides which aggregated so that the quaternary structure yielded an obviously functional enzyme unlike the aggregate of altered polypeptides produced by each homokaryon. In current terminology, the wild-type enzyme is a homomultimer including at least two identical polypeptides and the functional enzyme from intragenic complementation is a heteromultimer. This explanation for intragenic complementation has been termed the "hybrid enzyme hypothesis."

According to Fincham (1966), intragenic complementation can be distinguished by four characteristics: (1) most mutants in the locus are noncomplementing; (2) the complementing mutants produce enzymes with very low activity or an inactive protein immunologically identical to the active enzyme; (3) enzyme activity in the heterokaryon is low, rarely exceeding 25% of wild-type or intergenic activity; and (4) the hybrid enzyme often qualitatively differs from the wild-type enzyme and the poorly active mutant enzymes. The formation of functional hybrid enzymes in heterokaryons for mutants with no detectable enzyme activity presumes that each mutant produces a complete polypeptide with a substituted amino acid, that is, a missense mutation.

**Fig. 17.** General methods to include two mutations in the *trans* configuration for complementation tests in diploid or halloid eukaryotes, or in prokaryotes. (a) Heterozygous diploid after fertilization; (b, c) partial diploidy in a bacterial cell following conjugation and transfer of a chromosomal segment by F′ factor or abortive transduction; (d) heterokaryon produced after hyphal fusion to produce a population of genotypically different haploid nuclei in a common cytoplasmic domain; (e) mixed infection of a prokaryotic host cell by genotypically different viral particles under conditions where neither viral strain particles mature to produce lysis but after lysis, mutant viral particles are recovered. (From Philip E. Hartman and Sigmund R. Suskind, "Gene Action," 2nd ed. © 1969. Reprinted by permission of Prentice-Hall, Inc., Englewood Cliffs, New Jersey.)

Studies of alkaline phosphatase from *E. coli* furnished the first convincing *in vitro* evidence that hybrid enzymes indeed result from intragenic complementation (Schlesinger and Levinthal, 1963). The wild-type enzyme was dissociated, yielding two identical polypeptide monomers which had no enzymatic activity. Monomers from two intragenic complementing mutants were allowed to aggregate to produce active enzyme, and the active dimer was shown by several different methods to be hybrid, that is, a dimer with different polypeptide monomers.

Catcheside and Overton (1959) suggested that a failure to demonstrate intragenic complementation for a large number of independent mutants *may* indicate that the pertinent enzyme is monomeric, that is, a single polypeptide. Relatively few enzymes, however, are monomeric. The oligomeric (quaternary) structure for many enzymes suggests a selective value for this type of protein organization: (1) hybrid enzymes may regain a measure of activity not found in the mutants lacking activity; (2) conformational changes for allosteric enzymes serve a regulatory function; and (3) different isozymes can be produced in response to different physiological demands on the organism. The different isozymes of mammalian and avian lactic acid dehydrogenase in tissues and organs of an individual indicate considerable versatility inherent in this oligomeric enzyme.

## D. Complementation Maps

Intragenic complementation tests can be formally represented by a map, using a simple rule: noncomplementing mutants are represented by two overlapping lines and complementing mutants by nonoverlapping lines. While complementation tests for relatively few mutants can yield a linear map with several lines, extensive tests with many mutants can produce impressive geometric patterns (Fig. 18). The meaning of comple-

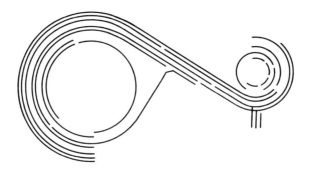

**Fig. 18.** Complex complementation map for the *ade-6* locus in *Schizosaccharomyces pombe*. (From Fincham and Day, ''Fungal Genetics,'' 3rd ed. 1971. With permission of the publishers, Blackwell Scientific Publications, Oxford; Leupold and Gutz, 1965.)

mentation maps has been the subject of much speculation in terms of possible tertiary and quaternary structures for the polypeptides and enzymes. The physical reality projected from such maps is highly problematic at this time but presumably reflects the structure of the gene product, not the gene (Fincham, 1966).

## E. Two Cistrons–One Enzyme

When two loci determine the function of one enzyme, it follows from the one cistron–one polypeptide relationship that each cistron is responsible for a different polypeptide in the quaternary structure. The genetic study of the *r*II locus in phage T4 showed that the locus was complex in that two adjacent cistrons were involved in determining the phenotype. For the alpha and beta polypeptides of hemoglobin, the two cistrons occupy clearly separated loci. Tryptophan synthetase is an interesting example of an enzyme with two A and two B polypeptides in *E. coli* but only one polypeptide in *N. crassa,* which nevertheless has the catalytic functions associated with each of the bacterial polypeptides. The fungal enzyme can be cleaved into two segments, each of which exhibits a different function. Although the separated A and B polypeptides of *E. coli* are clearly different and capable of catalyzing different enzymatic reactions *in vitro*, the interaction of the two polypeptides in the oligomeric enzyme is required for the maximum *in vivo* physiological activity in the wild-type cells.

## F. Cistron–Polypeptide Colinearity

The linear sequence (1) of mutant sites within a cistron, (2) of codons along the transcribing polynucleotide chain, and (3) of amino acids in the polypeptide suggests a *colinearity* for codons and the corresponding amino acids in the cistron and polypeptide, respectively. Defective mutants involving the A polypeptide of tryptophan synthetase in *E. coli* furnished first the appropriate experimental material to establish cistron-polypeptide colinearity (Yanofsky *et al.,* 1964, 1967a).

Wild-type tryptophan synthetase (T'ase) couples indoleglycerolphosphate and L-serine to produce L-tryptophan. The T'ase defective mutants were divided into two groups by cis–trans tests which identified two functional units (cistrons), as noted above. The dissociation of the monomers of the quaternary structure of wild-type T'ase yields two identical A and two identical B polypeptides. The B polypeptide catalyzes the *in vitro* indole to tryptophan reaction at a rate much lower than for the B polypeptide combined with wild-type or mutant A polypeptide. Furthermore, all *A* cistron mutants grow on indole because they still have the wild-type B polypeptide. These observations furnished the screen to obtain A polypeptide mutants that could be assigned to sites within the *A* cistron by genetic fine structure analysis. The A polypeptide mutants were then assigned to two groups by the CRM reaction (Fig. 19). Antibodies to wild-type A polypeptide react with mutant A polypeptide when the muta-

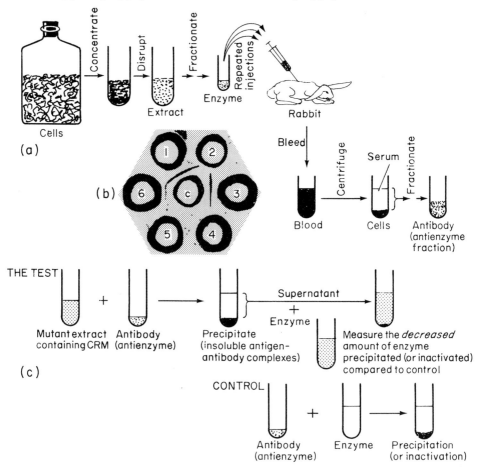

**Fig. 19.** Detection of cross-reacting material, CRM. (From Philip E. Hartman and Sigmund R. Suskind, "Gene Action," 2nd ed. © 1969. Reprinted by permission of Prentice-Hall, Inc., Englewood Cliffs, New Jersey.) (a) Preparation of immune serum to obtain antibody (antienzyme) for active enzyme; (b) agar diffusion (Ouchterlony) technique to detect CRM. Holes cut into an agar layer are filled with appropriate contents: antibody to center well, enzyme to well 1 and extracts of a *CRM+* mutant to wells 3 and 6. After diffusion of antibody, enzyme protein and CRM molecules, a band of precipitin forms between antibody and extract wells that coalesces with the band between antibody and enzyme wells. Extracts from a *CRM−* mutant added to wells 4 and 5 do not yield a band between these wells and the well with antibody; (c) antibody added to extract with enzyme produces an inactive antibody–enzyme precipitate.

tion has not significantly altered the tertiary conformation. Only *CRM*+ mutants lacking A activity were selected for fingerprint analysis of the A polypeptide, because these mutants were assumed to be the ones that result from an amino acid substitution (missense mutation).

The amino acid sequence for wild-type A polypeptide was determined and the amino acids in the *CRM*+ mutant A polypeptide identified the resident and substituted amino acid (Yanofsky *et al.*, 1967a). The A polypeptide has 267 amino acids and the corresponding cistron, 801 nucleotides. By comparing the fine structure linkage map and the sites in the various mutant polypeptides of substituted amino acids, colinearity of mutant sites (codons) and of substituted amino acids was established (Fig. 1). Moreover, closely linked mutant sites represented neighboring rather than distant amino acids in the polypeptide. For example, mutants *trpA* 446 and *trpA* 487 separated by a map distance of 0.04 (total cistron length = 4.32) involved amino acid substitutions at positions 174 and 176 in the polypeptide chain while mutants *trpA* 88 and *trpA* 446 separated by a map distance of 1.6 involved amino acids at positions 48 and 174, respectively.

Sarabhai *et al.* (1964) used phage T4 and different nonsense mutants for head-coat protein to demonstrate cistron–polypeptide colinearity by comparing the sequence of amino acids in the prematurely terminated mutant polypeptides. The first evidence for cistron–polypeptide colinearity in eukaryotic species was provided by Sherman *et al.* (1970) for iso-1-cytochrome *c* in yeast, utilizing essentially similar techniques.

## G. Regulation

Several different regulatory mechanisms determine whether a gene will or will not be permitted to function and produce the expected phenotype. Although genically controlled regulation (*a la* operon) has provided insight into the different mechanisms responsible for gene function, certain chromosomal aberrations alerted geneticists before the discovery of the operon to the possibility of regulation by altered gene expression.

## H. Position Effects

The Bar *position effect* first cast doubt on the structural integrity of the gene as the exclusive unit of function even though this "mutant gene" was later found to be a tandem duplication of the 16A segment in the X chromosome of *D. melanogaster* (Sturtevant, 1925). The number of ommatidia in the mutant's eyes is altered not by the total number of 16A segments but by the position of these segments in homologous chromosomes. Prior to this discovery, genes were thought to function regardless of their location in the chromosome. The $B/B$ eye has more ommatidia than a $BB/+$ eye. Although a formal model can be constructed by assuming that one gene in the 16A segment acquires control of eye development when the gene is in physical contact with a new, rather than the usual neighboring gene as in the wild-type 16A segment, no evidence is available for any reasonable explanation for this stable, or S-type, position effect.

A second type of chromosomal effect involving regulation of gene function also occurs in *D. melanogaster*. The variegated, or V-type, position effect refers to the functioning of a gene in direct proximity to constitutive heterochromatin. A female heterozygous for the white-eye mutation has uniformly wild-type red eyes. When the wild-type allele is relocated by inversion or reciprocal translocation to a heterochromatic region, the eyes display patches or red and white ommatidia. Although this type of position effect has been intensively studied, the mechanism whereby heterochromatin determines whether the wild-type allele is functional in specific cells has not yet been satisfactorily explained (Baker, 1968).

Facultative heterochromatin has also been shown to alter gene function in female mice in a manner similar to the variegated position effect in *Drosophila*. During embryogenesis one of the two X chromosomes undergoes heterochromatization, and genes in this chromosome do not appear to function (Lyon, 1962). Because either X chromosome can become heterochromatic and remain heterochromatic in clonal progeny cells, the female is a mosaic of cellular populations with respect to the X chromosome. A female heterozygous for a recessive autosomal allele for coat color has the wild-type phenotype. When the dominant allele is relocated by reciprocal translocation with an X chromosome that places the locus relatively close to the X chromosome segment, heterochromatization of the X chromosome segment has a spreading effect into the autosomal segment with the wild-type allele (Eicher, 1970). The wild-type allele in this clonal progeny does not function and the female has a variegated coat of wild-type mutant phenotypes. The mechanism responsible for gene control by facultative heterochromatin is probably similar to that for constitutive heterochromatin.

## I. The Operon Concept

The operon concept was formulated to account for constitutive and inducible enzymes, thereby introducing the notion of *structural* and *regulatory* enzymes and explaining how one mutation can be responsible for controlling the function of several genes. The Jacob-Monod (1961) model for genic regulatory mechanisms in the synthesis of proteins is responsible for viewing protein synthesis in terms of transcription and translation.

Enzymes had been identified as constitutive or adaptive (inducible), depending on their activity or inactivity in microbial species grown in the presence or absence, respectively, of the appropriate substrate in the medium. In *E. coli*, $\beta$-galactosidase had been characterized as an inducible enzyme, hydrolyzing lactose to glucose and galactose that appears in the cells only after lactose is added to the medium as the sole source of organic carbon. The discovery of a specific system for the uptake and accumulation of $\beta$-galactosides in *E. coli* and the application of the term permease to this system eventually led to the identification of the enzyme, galactoside permease, which can be responsible for phenotypically *lac* − mutants if the enzyme is defective. A third enzyme, galactoside transacetylase, catalyzing the transfer of an acetyl group from a donor molecule to the galactose moiety of $\beta$-galactosides was later discovered. These three enzymes concerned with lactose utilization were induced coordinately in the presence of lactose or other $\beta$-galactoside inducers.

Assays for the three enzymes in a number of *lac* − and *lac* + mutants of *E. coli* and heterozygotes for different combinations of mutants, complementation tests for different combinations of these mutants, and the mapping of mutants by conjugation and transduction yielded the observations which were explained by the operon concept (Jacob *et al.*, 1960). Furthermore, the operon concept accommodates mutants involving lactose utilization, which were not available during the initial investigation (Table III).

The significant *lac* mutants are (1) $Z- Y+ A+$, $\beta$-galactosidase $(Z)$ absent and galactosidase permease $(Y)$ and galactoside transacetylase $(A)$ present; (2) $Z+ Y+ A+$, only permease absent; (3) $Z+ Y+ A-$, only transacetylase absent; and (4) $I-$, constitutive rather than inducible for the three enzymes $(Z+ Y+ A+)$. Genetic mapping of the four genes reveals that $Z$, $Y$, and $A$ are very closely linked and in a contiguous sequence and that the $I$ locus is also linked but clearly separate from the other loci (Fig. 20).

Mating experiments with HFr $I+ Z+$ inducible donor cells grown in lactose-free medium and $I- Z-$ recipients in a lactose-free medium indicated that $\beta$-galactosidase synthesis commences after the alleles from the donor entered the cytoplasm of the recipients, continues for approximately one hour, and then ceases (Fig. 21). These observations were explained by assuming that the recipient cells could not synthesize $\beta$-galactosidase but would after the wild-type allele had entered their cytoplasm. Although enzyme synthesis continued in the absence of inducer (constitutive), the $I+$ allele eventually produced sufficient product, repressor (as we subsequently learned), to have a negative effect on enzyme synthesis. According to Pardee *et al.* (1959), the repressor "inhibits information transfer from structural gene (or genes) to protein," an idea

**TABLE III**

**Mutations in the Structural and Regulatory Genes of the Lactose Operon in *Escherichia coli* and their Effect on Gene Expression**[a]

| | Structural genes | | | |
| --- | --- | --- | --- | --- |
| | | Activity | | |
| Mutation | | $Z$ | $Y$ | $A$ | Comments |

| Mutation | $Z$ | $Y$ | $A$ | Comments |
| --- | --- | --- | --- | --- |
| Missense | | | | |
| $Z$ | 0 | $I^{++}$ | $I^{++}$ | No coordinate effect or effect on inducibility |
| $Y$ | $I^{++}$ | 0 | $I^{++}$ | |
| $A$ | $I^{++}$ | $I^{++}$ | 0 | |
| Nonsense | | | | |
| $Z$ | 0 | 0 | 0 | Suppressible by nonsense suppressors |
| $Z$ | 0 | $I^{+}$ | $I^{+}$ | |
| $Y$ | $I^{+}$ | 0 | $I^{+}$ | |

| | Regulatory genes | | | | |
| --- | --- | --- | --- | --- | --- |
| | | | Activity | | |
| Mutation | Effect | $Z$ | $Y$ | $A$ | Comments |
| $I^{-}$ | Repressor does not bind to operator | C | C | C | $I^{+}$ dominant to $I^{-}$ |
| $I^{S}$ | Repressor permanently binds to operator | 0 | 0 | 0 | $I^{S}$ dominant to $I^{+}$ |
| $O^{c}$ | Operator cannot bind aporepressor | C | C | C | $O^{c}$ *cis* dominant and *trans* recessive |
| $P^{-}$ | RNA polymerase binds inefficiently | 0 $I^{+}$ | 0 $I^{+}$ | 0 $I^{+}$ | $P^{+}$ *cis* dominant and *trans* recessive |

[a] $I^{++}$, strongly inducible, $I^{+}$, weakly inducible; C, constitutive; 0, no activity.

later developed into a scheme for the regulation of protein synthesis by genes. Finally, inducers function as inactivators of the repressor material, which is absent in the constitutive $I-$ mutant.

In formulating the operon concept, Jacob and Monod (1961) assumed that the trio of linked structural genes behaved coordinately in the $I+$ or $I-$ cells because the trio formed a functional unit, the operon, whose activity was determined by the status of a closely linked site, the operator (O). The discovery of a constitutive mutant isolated from an $I+$ strain and

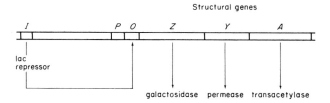

**Fig. 20.** The lactose operon in *Escherichia coli*. The three structural genes are transcribed as a single polycistronic messenger by RNA polymerase binding to the promotor (P) site and transcribing to the terminator site at the end of the last structural cistron. Binding of *lac* repressor-protein, specified by the *I* locus, to the operator (*O*) site prevents transcription. When a metabolic derivative of lactose binds and alters the conformation of the repressor-protein, the repressor-complex cannot bind to the operator site and the structural genes are transcribed.

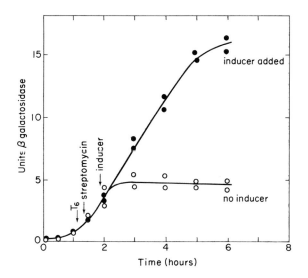

**Fig. 21.** Synthesis of β-galactosidase in a culture of *Escherichia coli* containing HFr *I*+ *Z*+ $T_6$ Sm donors grown in lactose-free medium and F− *I*− *Z*− $T_6{}^r$ $Sm^r$ recipients also grown in lactose-free medium. Donors are eliminated by adding phage $T_6$ and streptomycin so that only *Z*− recipients with donor *Z*+ can produce β-galactosidase. Enzyme is synthesized constitutively, becoming level at approximately 1–2 hours. Enzyme synthesis becomes inducible when *I*+ repressor is produced. (From M. W. Stickenberger, "Genetics." Macmillan, New York, 1968. Reprinted with permission. Pardee, 1959).

its linkage relationships with the *I* locus and those of the trio of structural genes was responsible for proposing the operator, a key element of the operon. This mutant is located not at the *I* locus but is closely linked with the trio of structural genes at a site immediately preceding the *Z* locus.

While $I+/I-$ cells were inducible with respect to expression of the $Z$ gene, the new mutant was found to be constitutive in the *cis* configuration ($C$ $Z+/C+Z-$) but inducible in the *trans* configuration ($C Z-/C+Z+$). The site was termed the operator locus and presumably the mutation ($O^c$) results from a deleted segment of the polynucleotide chain.

To account for the effect of the repressor, the repressor was assumed to combine with the intact operator site in the polynucleotide chain in such a way as to block operon function, that is, prevent enzyme synthesis. When the operator is not combined with the repressor, the operon functions and enzyme synthesis occurs. Inducibility represents the combining of the inducer, or a metabolic product from inducer, with the repressor so that the inducer-repressor complex cannot combine with the operator site. The $O^c$ mutation as a deficiency does not present a recognizable site for the repressor and therefore repressor does not combine at the operator site. A second mutation ($I^s$) at the $I$ locus was discovered when it was noted that the inducer did not function as expected and yet the mutant was not able to synthesize the enzymes. In this mutation, the configuration of the repressor was assumed to differ from the wild-type repressor and could no longer combine with the inducer. The repressor for the $\beta$-galactosidase system in *E. coli* has since been isolated using a mutant strain $I^Q$ which produced much more repressor than the wild-type $I+$ strains (Gilbert and Müller-Hill, 1966) and characterized as an oligomeric protein. The combining of inducer and repressor alters the conformation of the repressor as would be expected for an allosteric protein. The repressor specified by $I^s$ mutation apparently does not combine with inducer.

An additional genetic component of the *lac* operon in *E. coli* accounts for mutations which greatly diminish the maximal rate of enzyme synthesis in $I+$ cells growing in high inducer concentrations or in $I-$ mutant strains (Jacob *et al.*, 1964). These mutants were assigned by genetic tests to a site adjacent to the operator site. The added site was termed *promotor* and its function was explained in terms of transcription. In the $P+$ cells, RNA polymerase commences transcription of the operon at the promotor site. In cells with the mutant promotor site, the frequency of transcription is reduced and fewer than the normal number of messengers is produced.

The clustering of functionally related genes first detected by genetic methods in bacterial species had no acceptable explanation until the operon concept was proposed. Although operons with different numbers of structural genes are known in a number of bacterial and viral species, acceptable evidence for operons in eukaryotic species has not yet been presented although regulatory genes are known. Possible operons have been reported for closely linked, functionally related genes in *Neurospora* and yeast (Fincham and Day, 1971). An unequivocal decision is not yet available in these cases to support one of the following possible explanations: (1) one very long cistron coding for one product with multiple functions, for example, tryptophan synthetase in *Neurospora*, or (2) a group of closely linked cistrons with an operator as in a bacterial or viral operon.

## J. Semantic Problems in Defining Genes and Genic Mutations

The cistron includes a sequence of codons specifying particular amino acids. In considering transcription, a sequence of nucleotides preceding the first codon for an amino acid serves as a recognition site (promotor) for RNA polymerase and a sequence of nucleotides following the last amino acid codon as a terminator site. The mRNA for translation specifies the polypeptide determined by the cistron (gene). In defining a gene, recognition and termination sites could be viewed as a part of a gene because altered recognition and termination sites would be accepted as mutations for the function associated with the pertinent polypeptide. On the other hand, only DNA sequence corresponding to the translated polypeptide could be viewed as the gene, and only mutations altering the polypeptide should be recognized as genic mutations. The conflict in accepting either definition for a gene is related to attempts to isolate a cistron and then to synthesize the corresponding mRNA template for the corresponding polypeptide. Failure to produce the pertinent polypeptide from an isolated segment of DNA via the corresponding mRNA might be attributed to the absence of the associated initiating and terminating sequences of nucleotides in the isolated segment (Shapiro *et al.*, 1969).

With the codon as the unit specifying an amino acid and a nucleotide replacement yielding a different codon for a different amino acid or a nonsense codon, a genic mutation can be defined in terms of nucleotide substitutions. Frame-shift mutations represent an added or subtracted nucleotide in the cistron which in effect alters nucleotide sequences from the point of addition or subtraction. From an operational point of view, both frame-shift or nucleotide substitutions yield genic mutations. A nucleotide substitution within a codon, however, can have different consequences depending on the specific base and site of substitution and the effect of the new codon on the functioning of the polypeptide (Fig. 22). The degeneracy of the code allows for certain nucleotide substitutions in particular codons without any change in the specified amino acid. Missense mutations may or may not yield an overtly altered polypeptide, depending on the methods used to detect an altered polypeptide. A nonsense mutation near the end of a cistron may or may not result in an obviously altered polypeptide in terms of function. Missense and nonsense mutations producing obviously altered polypeptide function are the ones usually found.

The emphasis on transcription and translation to define genes in terms of

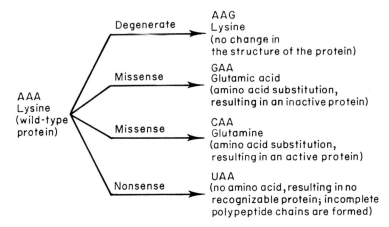

**Fig. 22.** Possible phenotypes resulting from one-letter nucleotide alterations of a hypothetical codon. (From Philip E. Hartman and Sigmund R. Suskind, "Gene Action," 2nd ed. © 1969. Reprinted by permission of Prentice-Hall, Inc., Englewood Cliffs, New Jersey.)

the one cistron–one polypeptide relationship should not obscure other relationships between gene and function. Genes are responsible for the production of specific tRNA's by transcription. Mutations resulting from a nucleotide substitution in the cistron would be directly reflected in the nucleotide sequence for the corresponding tRNA, which might affect many proteins. Furthermore, mutant sites in a nucleotide sequence can be detected by an altered phenotype which does not necessarily involve any altered gene product. For example, promotor and operator function as recognition sites for RNA polymerase and repressor/inducer, respectively. A mutation in these sites can prevent transcription by RNA polymerase. In the first and the final analysis, a genic mutation is operationally defined in terms of an altered phenotype.

## REFERENCES

Adelberg, E. A., and Burns, S. N. (1960). Genetic variation in the sex factor of *Escherichia coli*. *J. Bacteriol.* **79,** 321–330.

Auerbach, C., and Robson, J. M. (1947). The production of mutations by chemical substances. *Proc. R. Soc. Edinburgh* **62,** 271–283.

Baker, W. K. (1968). Position-effect variegation. *Adv. Genet.* **14,** 133–169.

Balbiani, E. G. (1881). Sur la structure du noyan des cellules salivaires chez les larves de *Chironomus. Zool. Anz.* **4,** 637–641 and 662–666.

Bateson, W., Saunders, E. R., and Punnett, R. C. (1905). Experimental studies in the physiology of heredity. *Rep. Evol. Comm. R. Soc.* **2,** 1–55 and 80–99.

Beadle, G. W. (1945). Genetics and metabolism in *Neurospora*. *Physiol. Rev.* **25**, 643–663.

Beadle, G. W. (1948). Genes and biological enigmas. *Am. Sci.* **36**, 71–74.

Beadle, G. W., and Coonradt, V. L. (1944). Heterocaryosis in *Neurospora crassa*. *Genetics* **29**, 291–308.

Beadle, G. W., and Ephrussi, B. (1937). Development of eye colors in *Drosophila:* Diffusible substances and their interrelations. *Genetics* **22**, 76–86.

Beadle, G. W., and Tatum, E. L. (1941). Genetic control of biochemical reactions in *Neurospora*. *Proc. Natl. Acad. Sci. U.S.A.* **27**, 499–506.

Benzer, S. (1955). Fine structure of a genetic region in bacteriophage. *Proc. Natl. Acad. Sci. U.S.A.* **41**, 344–354.

Benzer, S. (1957). The elementary units of heredity. *In* "The Chemical Basis of Heredity" (W. D. McElroy and B. Glass, eds.), pp. 70–93. Johns Hopkins Press, Baltimore, Maryland.

Benzer, S. (1961). On the topography of the genetic fine structure. *Proc. Natl. Acad. Sci. U.S.A.* **47**, 403–415.

Benzer, S., and Champe, S. P. (1962). A change from nonsense to sense in the genetic code. *Proc. Natl. Acad. Sci. U.S.A.* **48**, 1114–1121.

Benzer, S., and Freese, E. (1958). Induction of specific mutations with 5-bromouracil. *Proc. Natl. Acad. Sci. U.S.A.* **44**, 112–119.

Brenner, S. (1955). Tryptophan biosynthesis in *Salmonella typhimurium*. *Proc. Natl. Acad. Sci. U.S.A.* **41**, 862–863.

Brenner, S. (1957). On the impossibility of all overlapping triplet codes in information transfer from nucleic acids to proteins. *Proc. Natl. Acad. Sci. U.S.A.* **43**, 687–694.

Brenner, S., and Beckwith, J. R. (1965). Ochre mutants, a new class of suppressible nonsense mutant. *J. Mol. Biol.* **13**, 629–637.

Brenner, S., and Stretton, A. O. W. (1965). Phase shifting of *amber* and *ochre* mutants. *J. Mol. Biol.* **13**, 944–946.

Brenner, S., Stretton, A. O. W., and Kaplan, S. (1965). Genetic code: "nonsense" triplets for chain termination and their suppression. *Nature (London)* **206**, 994–998.

Brenner, S., Barnett, I., Katz, E. R., and Crick, F. H. C. (1967). UGA: A third nonsense triplet in the genetic code. *Nature (London)* **213**, 449–450.

Bridges, C. B. (1936). The Bar "gene" a duplication. *Science* **83**, 210–211.

Bridges, C. B. (1938). A revised map of the salivary gland X chromosome of *Drosophila melanogaster*. *J Hered.* **29**, 11–13.

Cairns, J. (1963). The bacterial chromosome and its manner of replication as seen by autoradiography. *J. Mol. Biol.* **6**, 208–213.

Carlson, P. S. (1971). A genetic analysis of the rudimentary locus of *Drosophila melanogaster*. *Genet. Res.* **17**, 53–81.

Catcheside, D. G., and Overton, A. (1959). Complementation between alleles in heterocaryons. *Cold Spring Harbor Symp. Quant. Biol.* **23**, 137–140.

Crick, F. H. C., Barnett, L., Brenner, S., and Tobin-Watts, R. J. (1961). General nature of the genetic code for proteins. *Nature (London)* **192**, 1227–1232.

Davis, B. D. (1950). Non-filterability of the agents of genetic recombination in *E. coli*. *J. Bacteriol.* **60**, 507–508.

Delbrück, M. (1940). The growth of bacteriophage and lysis of the host. *J. Gen. Physiol.* **23**, 643–660.

Delbrück, M., and Bailey, W. T. Jr. (1946). Induced mutations in bacterial viruses. *Cold Spring Harbor Symp. Quant. Biol.* **11**, 33–37.

Demerec, M. (1964). Clustering of functionally related genes in *Salmonella typhimurium*. *Proc. Natl. Acad. Sci. U.S.A.* **51**, 1057–1060.

Demerec, M., and Hartman, Z. (1956). Tryptophan mutants in *Salmonella typhimurium*. *Carnegie Inst. Washington Publ.* **612**, 3–17.

de Vries, H. (1901). "Die Mutationtheorie." Veit, Leipzig.

Drake, J. W. (1970). "The Molecular Basis of Mutation." Holden-Day, San Francisco, California.

Drake, J. W., and Greening, E. O. (1970). Suppression of chemical mutagenesis in bacteriophage $T_4$ by genetically modified DNA polymerases. *Proc. Natl. Acad. Sci. U.S.A.* **99**, 823–829.

Eicher, E. M. (1970). X autosome translocations in the mouse: Total inactivation versus partial inactivation of the X chromosome. *Adv. Genet.* **15**, 176–259.

Ellis, E. N., and Delbrück, M. (1939). The growth of bacteriophage. *J. Gen. Physiol.* **22**, 365–384.

Ephrussi, B. (1942). Chemistry of "eye color hormones" of *Drosophila*. *Q. Rev. Biol.* **17**, 327–338.

Fargie, B., and Holloway, B. W. (1965). Absence of clustering of functionally related genes in *Pseudomonas aeroginosa*. *Genet. Res.* **6**, 284–299.

Fincham, J. R. S. (1966). "Genetic Complementation." Benjamin, New York.

Fincham, J. R. S., and Day, P. R. (1971). Fungal Genetics," 3rd ed. Blackwell, Oxford.

Fincham, J. R. S., and Pateman, J. A. (1957). Formation of an enzyme through complementary action of "mutant" alleles in separate nuclei in a heterocaryon. *Nature (London)* **179**, 741–742.

Freese, E. (1959). The difference between spontaneous and base-analogue induced mutations on phage T4. *Proc. Natl. Acad. Sci. U.S.A.* **45**, 622–633.

Freese, E., Bautz, E., and Bautz-Freese, E. (1961). The chemical and mutagenic specificity of hydroxylamine. *Proc. Natl. Acad. Sci. U.S.A.* **47**, 845–855.

Garen, A., and Siddiqi, O. (1962). Suppression of mutations in the alkaline phosphatase structural cistron of *E. coli*. *Proc. Natl. Acad. Sci. U.S.A.* **48**, 1121–1127.

Garrod, A. E. (1902). The incidence of alkaptonuria: A study in chemical individuality. *Lancet* **2**, 1616–1620.

Garrod, A. E. (1909). "Inborn Errors of Metabolism," 2nd ed. Oxford Univ. Press, London and New York.

Gilbert, W., and Müller-Hill, B. (1966). Isolation of the *lac* repressor. *Proc. Natl. Acad. Sci. U.S.A.* **56**, 1891–1898.

Giles, N. H., Partridge, C. W. H., and Nelson, N. J. (1957). The genetic control of adenylosuccinase in *Neurospora crassa*. *Proc. Natl. Acad. Sci. U.S.A.* **43**, 305–317.

Goodenough, U, and Levine, P. R. (1974). "Genetics." Holt, New York.

Gorini, L. (1970). Informational suppression. *Annu. Rev. Genet.* **4**, 107–135.

Green, M. M. (1963). Pseudoalleles and recombination in *Drosophila*. In "Methodology in Basic Genetics" (W. J. Burdette, ed.), pp. 279–286. Holden-Day, San Francisco, California.

Green, M. M., and Green, K. C. (1949). Crossing-over between alleles at the lozenge locus in *Drosophila melanogaster*. *Proc. Natl. Acad. Sci. U.S.A.* **35**, 586–591.

Guest, J. R., and Yanofsky, C. (1965). Amino acid replacements associated with reversion and recombination within a coding unit. *J. Mol. Biol.* **12**, 793–804.

Hartman, P. E. (1956). Linked loci in the control of consecutive steps in the primary pathway of histidine synthesis in *Salmonella typhimurium*. *Carnegie Inst. Washington Publ.* **612**, 35–61.

Hartman, P. E., and Suskind, S. R. (1969). "Gene Action," 2nd ed. Prentice-Hall, Englewood Cliffs, New Jersey.

Hartman, P. E., Loper, J. C., and Serman, D. (1960). Fine structure mapping by complete transduction between histidine-requiring *Salmonella* mutants. *J. Gen. Microbiol.* **22**, 323–353.

Hayes, W. (1952). Recombination in *Bact. coli* K-12: Unidirectional transfer of genetic material. *Nature (London)* **169**, 118–119.

Hayes, W. (1953). Observations on a transmissible agent determining sexual differentiation in *Bact. coli. J. Gen. Microbiol.* **8**, 72–88.

Hayes, W. (1968). "The Genetics of Bacteria and their Viruses," 2nd ed. Blackwell, Oxford.

Hershey, A. D. (1946). Spontaneous mutations in bacterial viruses. *Cold Spring Harbor Symp. Quant. Biol.* **11**, 67–76.

Hershey, A. D., and Chase, M. (1952). Independent functions of viral proteins and nuclei and in growth of bacteriophage. *J. Gen. Physiol.* **36**, 39–56.

Hershey, A. D., and Rotman, R. (1949). Genetic recombination between host-range and plaque type mutants of bacteriophage in single bacterial cells. *Genetics* **34**, 44–71.

Ingram, V. M. (1956). A specific chemical difference between the globins of normal human and sickle cell anaemia haemoglobin. *Nature (London)* **78**, 792–794.

Jacob, F., and Monod, J. (1961). Genetic regulatory mechanisms in the synthesis of proteins. *J. Mol. Biol.* **3**, 318–356.

Jacob, F., Perrin, D., Sanchez, C., and Monod, J. (1960). L'opéron: Groupe de genes à l'expression coordinée par un opérateur. *C. R. Habd. Seances Acad. Sci.* **250**, 1727–1729.

Jacob, F., Ullman, A., and Monod, J. (1964). Le promoteur, élément génétique nécessaire à l'expression d'un opéron. *C.R. Habd. Seances Acad. Sci.* **254**, 3125–3128.

Jenkins, J. B. (1975). "Genetics." Houghton, Boston, Massachusetts.

Judd, B. H., Shen, M. W., and Kaufman, T. C. (1972). The anatomy and function of a segment of the X chromosome of *Drosophila melanogaster. Genetics* **71**, 139–156.

Khorana, H. G., Büchi, H., Ghosh, H., Gupta, N., Jacob, T. M., Kössel, H., Morgan, R., Nargong, S. A., Ohtsuka, E., and Wells, R. D. (1967). Polynucleotide synthesis and the genetic code. *Cold Spring Harbor Symp. Quant. Biol.* **31**, 39–49.

Lederberg, J. (1947). Gene recombination and linked segregations in *Escherichia coli. Genetics* **32**, 505–525.

Lederberg, J., and Lederberg, E. M. (1952). Replica plating and indirect selection of bacterial mutants. *J. Bacteriol.* **63**, 399–406.

Lederberg, J., and Tatum, E. L. (1946). Gene recombination in *Escherichia coli. Nature (London)* **158**, 558.

Lederberg, J., Cavalli, L. L., and Lederberg, E. M. (1952). Sex compatability in *E. coli. Genetics* **37**, 720–730.

Lengyel, P., Speyer, J. F., and Ochoa, S. (1961). Synthetic polynucleotides and the amino acid code. *Proc. Natl. Acad. Sci. U.S.A.* **47**, 1936–1942.

Leupold, U., and Gutz, H. (1965). Genetic fine structure in *Schizosaccharomyces. Genet. Today, Proc. Int. Congr., 11th, 1963* Vol. 2, pp. 31–35.

Litman, R. M., and Pardee, A. B. (1956). Production of bacteriophage mutants by a disturbance of deoxyribonucleic and metabolism. *Nature (London)* **178**, 529–531.

Loper, J. C., Grabner, M., Stahl, R. C., Hartman, Z., and Hartman, P. E. (1964). Genes and proteins involved in histidine biosynthesis in *Salmonella. Brookhaven Symp. Biol.* **17**, 15–50.

Luria, S. E., and Delbrück, M. (1943). Mutations of bacteria from virus sensitivity to virus resistance. *Genetics* **28**, 491–511.

Lyon, M. F. (1962). Sex chromatin and gene action in the mammalian X chromosome. *Am. J Hum. Genet.* **14**, 135–148.

Mendel, G. (1886). Versuche über Pflanzenhybriden. *Verh. Naturforsch. Ver. Brunn* **4**, 3–44.

Morgan, T. H. (1910). Sex limited inheritance in *Drosophila. Science* **32**, 120–122.

Muller, H. J. (1927). Artificial transmutation of the gene. *Science* **66**, 84–87.

Muller, H. J. (1945). The gene. *Proc. R. Soc. Biol.* **134**, 1–37.

Nirenberg, M. W., and Leder, P. (1964). RNA code words and protein synthesis. *Science* **145**, 1399–1407.

Nirenberg, M. W., and Matthei, J. H. (1961). The dependence of cell-free protein synthesis in *E. coli* upon naturally occurring or synthetic polyribonucleotides. *Proc. Natl. Acad. Sci. U.S.A.* **47**, 1588–1602.

Olby, R. (1974). "The Path to the Double Helix." Univ. of Washington Press, Seattle.

Oliver, C. P. (1940). A reversion to wild-type associated with crossing-over in *Drosophila melanogaster*. *Proc. Natl. Acad. Sci. U.S.A.* **26**, 452–454.

Painter, T. S. (1934). A new method for the study of chromosome aberrations and the plotting of chromosome maps in *Drosophila melanogaster*. *Genetics* **19**, 175–188.

Pardee, A. B., Jacob, F., and Monod, J. (1959). The genetic control and cytoplasmic expression of inducibility in the synthesis of $\beta$-galactosidase by *E. coli*. *J. Mol. Biol.* **1**, 165–178.

Patt, G. R. (1975). "An Introduction to Modern Genetics." Addison-Wesley, Reading, Massachusetts.

Pauling, L., and Corey, R. B. (1950). Two hydrogen-bonded spiral configurations of the polypeptide chain. *J. Am. Chem. Soc.* **71**, 5349.

Porter, I. H. (1968). "Heredity and Disease." McGraw-Hill, New York.

Pritchard, R. H. (1955). The linear arrangement of a series of alleles of *Aspergillus nidulans*. *Heredity* **9**, 343–371.

Roman, H. (1956). Studies of gene mutation in *Saccharomyces*. *Cold Spring Harbor Symp. Quant. Biol.* **21**, 175–185.

Roper, J. A. (1953). Pseudoallelism. *Adv. Genet.* **5**, 208–218.

Sambrook, J. F., Fan, D. P., and Brenner, S. (1967). A strong suppressor specific for UGA. *Nature (London)* **214**, 452–453.

Sanger, F. (1956). The structure of insulin. *In* "Currents in Biochemical Research" (D. E. Green, ed.). Wiley (Interscience), New York.

Sarabhai, A., and Brenner, S. (1967). A mutant which reinitiates the polypeptide chain after chain termination. *J. Mol. Biol.* **27**, 145–162.

Sarabhai, A. S., Stretton, A. O. W., Brenner, S., and Bolle, A. (1964). Colinearity of the gene with the polypeptide chain. *Nature (London)* **201**, 13–17.

Schlesinger, M. J., and Levinthal, C. (1963). Hybrid protein formation of *E. coli* alkaline phosphatase leading to *in vitro* complementation. *J. Mol. Biol.* **1**, 1–12.

Shapiro, J., MacHatlie, L., Eron, G., Ippen, K., and Beckwith, J. (1969). Isolation of pure *lac* operon DNA. *Nature (London)* **224**, 768–774.

Sherman, F., Stewart, J. W., Parker, J. H., Putterman, G. J., Agrawal, B. B. L., and Margoliash, E. (1970). The relationship of gene structure and protein structure of iso-1-cytochrome x from yeast. *Symp. Soc. Exp. Biol.* **24**, 85–107.

Sherman, F., Jackson, M., Liebman, S. W., Schweingruber, A. M., and Stewart, J. W. (1975). A deletion map of *cyc* 1 mutants and its correspondence to mutationally altered iso-1-cytochromes *c* of yeast. *Genetics* **81**, 51–73.

Slizynska, H. (1938). Salivary chromosome analysis of the white-facet region in *Drosophila*. *Genetics* **23**, 291–299.

Srb, A. M., and Horowitz, N. H. (1944). The ornithine cycle in *Neurospora* and its genetic control. *J. Biol. Chem.* **154**, 129–139.

Stocker, B. A. D., Zinder, N. D., and Lederberg, J. (1953). Transduction of flagellar characters in *Salmonella*. *J. Gen. Microbiol.* **9**, 410–433.

Streisinger, G., Edgar, R. S., and Denhardt, G. H. (1964). Chromosome structure in phage T4. I. Circularity of the linkage map. *Proc. Natl. Acad. Sci. U.S.A.* **51**, 775–779.

Strickberger, M. W. (1968). "Genetics." Macmillan, New York.

Sturtevant, A. H. (1913). The linear arrangement of six sex-linked factors in *Drosophila*, as shown by this mode of association. *J. Exp. Zool.* **14**, 43–59.

Sturtevant, A. H. (1920). The vermilion gene and gynandromorphism. *Proc. Soc. Exp. Biol. Med.* **17**, 70–71.

Sturtevant, A. H. (1925). The effects of unequal crossing-over at the Bar locus in *Drosophila*. *Genetics* **23**, 291–299.

Terzaghi, E., Okada, Y., Streisinger, G., Emrich, J., Inouye, M., and Tsugita, A. (1966). Change of a sequence of amino acids in phage T4 lysozyme by acridine-induced mutations. *Proc. Natl. Acad. Sci. U.S.A.* **56**, 500–507.

Tsugita, A., and Fraenkel-Conrat, H. (1962). The composition of proteins of chemically evoked mutants of TMV RNA. *J. Mol. Biol.* **4**, 73–82.

Watson, J. D., and Crick, F. H. C. (1953a). A structure of deoxyribonucleic acid. *Nature (London)* **171**, 737–738.

Watson, J. D., and Crick, F. H. C. (1953b). Genetical implications of the structure of deoxyribonucleic acid. *Nature (London)* **171**, 964–967.

Weigert, M. G., and Garen, A. (1965). Base composition of nonsense codons in *E. coli*. Evidence from amino-acid substitutions at a tryptophan site in alkaline phosphatase. *Nature (London)* **206**, 992–994.

Weigert, M. G., Gallucci, E., Lanka, E., and Garen, A. (1966). Characteristics of the genetic code *in vivo*. *Cold Spring Harbor Symp. Quant. Biol.* **31**, 145–150.

White, J. M. (1972). Haemoglobin variation. *In* "The Biochemical Genetics of Man" (D. J. H. Brock and O. Mayo, eds.). Academic Press, New York.

Wilson, E. B. (1909). Recent researches on the determination and heredity of sex. *Science* **29**, 53–70.

Wittman, H. G., and Wittman-Liebold, B. (1967). Protein chemical studies of two RNA viruses and their mutants. *Cold Spring Harbor Symp. Quant. Biol.* **31**, 163–172.

Wollman, E. L., and Jacob, F. (1955). Sur le mécanisme du transfer de matériel génétique au cours de la recombination chez. *E. coli* K-12 *C.R. Hebd. Seances Acad. Sci.* **240**, 2449–2451.

Wollman, E. L., and Jacob, F. (1958). Sur les processus de conjugaison et de recombinaison chez *E. coli*. V. Le méchanisme du transfert de matériel génétique. *Ann. Inst. Pasteur, Paris* **95**, 641–666.

Wollman, E. L., Jacob, F., and Hayes, W. (1956). Conjugation and genetic recombination in *Escherichia coli* K-12. *Cold Spring Harbor Symp. Quant Biol.* **21**, 141–162.

Yanofsky, C., Henning, U., Helinski, D. R., and Carlton, B. C. (1963). Mutational alteration of protein structure. *Fed. Proc., Fed Am. Soc. Exp. Biol.* **22**, 75–79.

Yanofsky, C., Carlton, B. C., Guest, J. R., Helinski, D. R., and Henning, U. (1964). On the colinearity of gene structure and protein structure. *Proc. Natl. Acad. Sci. U.S.A.* **51**, 266–272.

Yanofsky, C., Ito, J., and Horn, V. (1967a). Amino acid replacements and the genetic code. *Cold Spring Harbor Symp. Quant. Biol.* **31**, 151–162.

Yanofsky, C., Drapeau, G. R., Guest, J. R., and Carlton, B. C. (1967b). The complete amino acid sequence of the tryptophan synthetase A protein ($\alpha$ subunit) and its colinear relationship with the genetic map of the A gene. *Proc. Natl. Acad. Sci. U.S.A.* **57**, 296–298.

Zamenhof, S., and Griboff, G. (1954). Incorporation of halogenated pyrimidines into the deoxyribonucleic acids of *Bacterium coli* and its bacteriophages. *Nature (London)* **174**, 306–307.

Zinder, N. D., and Lederberg, J. (1952). Genetic exchange in *Salmonella*. *J. Bacteriol.* **64**, 679–699.

# 2

# Gene Conversion, Paramutation, and Controlling Elements: A Treasure of Exceptions

*Michael S. Esposito and Rochelle E. Esposito*

## I. INTRODUCTION

The topics discussed in this chapter represent exceptions to the commonly accepted rules of genetic transmission and expression. The history

of genetics attests to the value of Bateson's advice to treasure such exceptions (Bateson, 1928). In more than one instance, their study has led to
conceptual reorientation and further understanding of the organization and
functioning of the genome. Gene conversion, paramutation, and controlling elements do not exhaust the list of apparent violations of the ordinary
rules of genetics. One may add, for example, position-effect variegation
(cf. Baker, 1968), gene magnification (cf. Tartof, 1973), meiotic drive (cf.
Zimmering *et al.*, 1970), aspects of immunogenetics (cf. Hood, 1973),
directed mutation of mating-type alleles in yeast (Hawthorne, 1963), and
selfing in *Salmonella* (Demerec, 1963), to name a few. All of these suggest
that the current understanding of mutational and recombinational mechanisms is incomplete.

We have chosen to contrast gene conversion, paramutation, and controlling elements since these three phenomena seem to be manifestations of
different types of recombinational events. Of the three, gene conversion is certainly the best understood, and is now regarded as a central
feature of genetic recombination of eukaryotes, bacteria, and viruses. The
latter view of gene conversion, and progress in its understanding, reflects
in part the recognition of analogous events in systems amenable to study
at the molecular level. Paramutation and controlling elements, though less
well understood, are particularly intriguing since they involve both
mutation-like and recombination-like genomic changes, as well as regulation of gene expression.

## II. GENE CONVERSION

The term gene conversion was introduced into the literature of genetics
by Winkler (1930), who proposed that recombination of linked heterozygous markers (*AB/ab*), could occur by other than breakage and reunion if
one member of an allelic pair were replaced by its counterpart during
meiosis. Winkler found support for this view in reports of aberrant segregation following meiosis in tetrads of fungi and mosses, but the over-
whelming evidence in favor of reciprocal exchange between homologous
chromosomes caused this hypothesis to be laid aside. In 1953, Lindegren
reported instances of aberrant segregation ($3A : 1a$ and $1A : 3a$, rather
than $2A : 2a$), in tetrads of the yeast *Saccharomyces cerevisiae*, and proposed that gene conversion as envisaged by Winkler was the basis of these
rare exceptions. Since gene conversion constitutes an apparent violation
of Mendel's first law, it is not surprising that acceptance of its existence
required rigorous proof. Numerous standard genetic phenomena including
trisomy, polyploidy, genic modifiers, and mitotic recombination preceding

meiosis, were identified as contributors to departures from $2A : 2a$ meiotic segregation of monohybrids (reviewed by Roman, 1963).

In 1955, Mitchell documented the occurrence of gene conversion in *Neurospora crassa*, incorporating genotypes and genetic markers that excluded other explanations. Since that time gene conversion has been employed to describe rare instances of $3A : 1a$ and $1A : 3a$ and other exceptions to the expected $2A : 2a$ segregation of heterozygous markers in both lower and higher eukaryotes. It has also been observed in mitotic populations, and similar phenomena have been described in viral-prokaryote systems (reviewed by Radding, 1973; Stadler, 1973).

## A. Meiotic Gene Conversion

Gene conversion during meiosis has been studied in both heterozygous $(A/a)$ and heteroallelic\* $(a1\text{-}1/a1\text{-}2)$ diploids. The properties of meiotic gene conversion have emerged primarily from experiments employing ascomycetes, in which the products of an individual meiosis are contained within an ascus. These organisms include the yeasts *Saccharomyces cerevisiae* and *Schizosaccharomyces pombe,* which form asci containing four ascospores, and molds such as *Aspergillus nidulans, Ascobolus immersus, Neurospora crassa*, and *Sordaria fimicola*, which form eight-spored asci. In these latter organisms each of the four haploid products of meiosis undergoes an additional mitotic division before ascospores are delineated. Gene conversion has also been demonstrated in *Drosophila melanogaster* (Hexter, 1963; Chovnick *et al*, 1970; Carlson, 1971).

### 1. Aberrant Segregation and its Relation to Genetic Exchange

**a. Heterozygotes (A/a).** Gene conversion in heterozygotes has been documented for numerous types of mutations including those conferring auxotrophic requirements, fermentation markers, and ascospore morphology and color variants. The latter class is particularly useful since segregation patterns can be observed directly within the ascus. Gene conversion at the gray spore locus $(g)$, of *Sordaria fimicola,* reported by Kitani, Olive, and El-Ani (1962), is illustrated in Fig. 1. The wild type of *S. fimicola* produces black ascospores while spores bearing the $g$ mutation remain gray. The asci shown in Fig. 1 come from a cross incorporating a pair of heterozygous markers bracketing the $g$ locus.

Several features of gene conversion in fungi are exhibited by the data from the cross. The vast majority of the octads exhibit $4G : 4g$ segregation patterns expected from normal segregation at meiosis. The distribution

---

\* Heteroalleles are mutant alleles of the same gene in which different intragenic sites have been mutated; i.e., different nucleotides in the gene sequence have been altered.

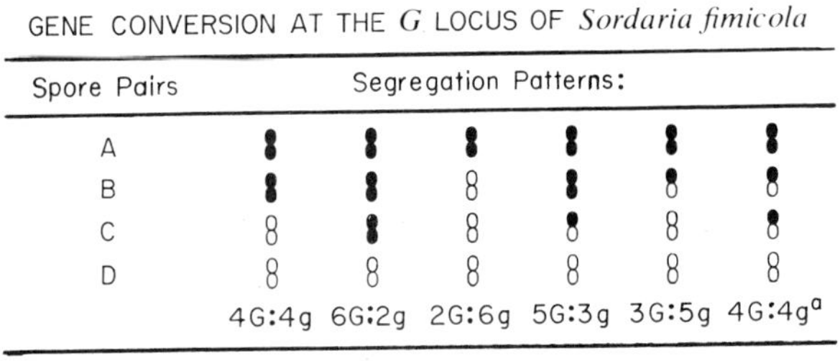

GENE CONVERSION AT THE *G* LOCUS OF *Sordaria fimicola*

| Spore Pairs | Segregation Patterns: | | | | | |
|---|---|---|---|---|---|---|
| A | ▉ | ▉ | ▉ | ▉ | ▉ | ▉ |
| B | ▉ | ▉ | ○ | ▉ | ○ | ○ |
| C | ○ | ▉ | ○ | ○ | ○ | ○ |
| D | ○ | ○ | ○ | ○ | ○ | ○ |
| | 4G:4g | 6G:2g | 2G:6g | 5G:3g | 3G:5g | 4G:4g[a] |

**Fig. 1.** Asci of *Sordaria fimicola*, from the cross of *G* × *g* (Kitani *et al.*, 1962). Unambiguous identification of spore pairs is made employing segregation of heterozygous markers bracketing the *G* locus. The number of asci in each class per $10^5$ asci was: $4G:4g$ $(9.99 \times 10^4)$; $6G:2g$ (47); $2G:6g$ (6); $5G:3g$ (52); $3G:5g$ (10); and aberrant $4G:4g$ (8).

(i.e., the linear order) of black and gray ascospores within these asci is influenced by the presence or absence of an exchange between the locus and its centromere, as well as the orientation of the spindles during the nuclear divisions. Five other types of asci were also encountered; these are due to gene conversion and exhibit $6G:2g$, $2G:6g$, $5G:3g$, $3G:5g$, and aberrant $4G:4g^a$ segregation. Tetrads of the $6G:2g$ and $2G:6g$ types represent conversion from mutant to wild type and wild type to mutant, respectively. Markers bracketing the *g* locus segregated in the normal $4A:4a$ pattern, indicating that aberrant segregation at *g* was not due to abnormal nuclear behavior. There was a significant excess of conversions of the mutant allele to wild type in comparison to conversions of wild type to mutant (47 per $10^5$ versus 6 per $10^5$ asci). Studies of conversion in several species have illustrated that conversion to mutant may sometimes predominate, whereas in other cases conversion occurs with equal frequency in both directions (Leblon, 1972a,b; Leblon and Rossignol, 1973; Hurst *et al.*, 1972). The most striking feature of the $6G:2g$ and $2G:6g$ asci, however, was the frequency with which conversion at the *g* locus was accompanied by reciprocal recombination of flanking markers only 4 map units apart. In these asci approximately 40% of the asci exhibited exchange in this interval.

The remaining three types of conversion asci exhibit postmeiotic segregation, sometimes referred to as half-chromatid conversion. In $5G:3g$ and $3G:5g$ octads, members of one spore pair resulting from the same postmeiotic mitosis differ in genotype at the involved locus. In $4G:4g^a$ postmeiotic segregation octads members of two spore pairs differ in genotype. This latter class can be distinguished when other segregating markers are present to identify the sister spore pairs (see Fig. 1). Postmeiotic segregation is quite frequent in *S. fimicola*, in comparison to other fungi, but it has been detected in *A. immersus* (Lissouba *et al.*, 1962), and even in the tetrad fungus *S. cerevisiae* (Esposito, 1971). Lack of parity of the $5G : 3g$ and $3G : 5g$ classes was in the same direction as observed in whole-chromatid conversion, though other studies have shown that this concordance is not a constant feature of conversion. The $5G : 3g$, $3G : 5g$, and $4G : 4g^a$ octads, like the $6G : 2g$ and $2G : 6g$ octads, exhibited enhanced recombination for outside markers.

The nonrandom coincidence of conversion and reciprocal exchange, as well as the demonstration of postmeiotic segregation have been the keys to construction of molecular models of recombination described in a later section. In *S. cerevisiae*, conversion-associated exchange has been examined in several regions of the genome; approximately half of conversion tetrads exhibit exchange (Hurst *et al.*, 1972). However, after subtraction of the exchange expected solely due to the length of the particular interval, the frequency of conversion-associated exchange is seen to vary considerably (Table I). The meaning of this variability remains to be explained. As in the data from *S. fimicola*, the vast majority of crossovers associated with conversion involve the strand which participates in the conversion event. For these and other reasons described below, gene conversion has been assumed to be an intimate part of the recombination process.

**b. Heteroallelic Diploids ($a1$-$1$/$a1$-$2$).** Meiotic recombination between mutant alleles of a gene was first reported in *Drosophila melanogaster* (Oliver, 1940), while mitotic heteroallelic recombination was first ob-

**TABLE I**

**Frequencies of Outside Marker Exchange Accompanying Conversion in *Saccharomyces cerevisiae*[a]**

| Site of conversion | Standard exchange frequency of outside markers (A) | Observed exchange frequency among convertant asci (B) | Increment due to conversion (B − A) |
|---|---|---|---|
| *his* 1-315 | 0.052 | 7/21 = 0.33 | 0.28 |
| *his* 1-7 | 0.248 | 232/452 = 0.51 | 0.26 |
| *his* 1-1 | 0.248 | 52/75 = 0.69 | 0.44 |
| *arg* 4-2,16,17[b] | 0.284 | 116/256 = 0.45 | 0.17 |
| *arg* 4-1,2,4,16,19[b] | 0.058 | 23/33 = 0.70 | 0.64 |
| *arg* 4-2,16[b] | 0.010 | 14/22 = 0.64 | 0.64 |
| *SUP* 6 | 0.232 | 155/301 = 0.51 | 0.28 |

[a] Calculated from data of Fogel and Hurst (1967) and Hurst, Fogel, and Mortimer (1972).

[b] Data from various two-point and multipoint heteroallelic crosses incorporating *arg* 4 alleles.

served in *Aspergillus nidulans* (Roper, 1950). Subsequent studies demonstrated that alleles of a locus recombine more often by gene conversion than by reciprocal exchange.

In 1955, Mitchell reported the results of a cross involving allelic pyridoxine (*pdx*) requiring mutants of *Neurospora crassa*. Ascospores from a cross in which the *pdx* alleles were flanked by markers approximately 5 map units apart, consisted of 0.2% $pdx^+$ prototrophic recombinants. The $pdx^+$ spores included both parental and recombinant types for the outside markers. Homoallelic crosses yielded substantially fewer $pdx^+$ spores indicating that simple reversion of one or the other allele was not the origin of the $pdx^+$ prototrophs resulting from the heteroallelic cross. Tetrad analysis of 585 asci, in which all four of the spore pairs germinated, yielded 4 asci containing a $pdx^+$ spore pair. The genotype at the *pdx* locus of the auxotrophic members of each of these asci was determined by mating to tester strains carrying the parental *pdx* alleles. After meiosis the ascospores from these matings were examined for the presence of prototrophs due to recombination (see Table II). If heteroallelic recombination resulted from reciprocal exchange the asci should have contained the reciprocal product, a spore pair carrying both parental *pdx* alleles. This genotype would have been recognized by its failure to generate 0.2% $pdx^+$ recombinants in crosses to both tester strains. The reciprocal recombinant was not found. The genotypes of the four asci reported by Mitchell are

**TABLE II**

**Identification of Heteroalleles by a Recombinational Test**[a]

| Genotype of mutant spores | $A^+$ spores after crosses to testers | |
|---|---|---|
| | × $a$1-1 | × $a$1-2 |
| $a$1-1 | − | + |
| $a$1-2 | + | − |
| $a$1-1, $a$1-2 | − | − |

[a] The genotype of mutant spores in tetrads from crosses of $a$1-1 × $a$1-2 may be determined by crossing each segregant to tester strains. Following meiosis, homoallelic crosses (i.e., $a$1-1 × $a$1-1 and $a$1-2 × $a$1-2), yield few or no $A^+$ ascospores, whereas heteroallelic crosses (i.e., $a$1-1 × $a$1-2), yields $A^+$ recombinants. Double mutant segregants (i.e., $a$1-1, $a$1-2), yield no $A^+$ recombinants with either tester.

**TABLE III**

**Gene Conversion at the *pdx* Locus of *Neurospora crassa*** [a]

Cross: $+ + pdxp\ co \times pyr\ pdx\ + +$

| Tetrad number | Genotype: spore pairs | | | | Tetrad number | Genotype: spore pairs | | | |
|---|---|---|---|---|---|---|---|---|---|
| 1 | + | + | pdxp | co | 3 | + | + | + | + |
| | + | + | + | co | | + | + | pdxp | co |
| | pyr | + | pdxp | + | | pyr | pdx | + | + |
| | pyr | pdx | + | + | | pyr | + | + | co |
| 2 | pyr | pdx | + | + | 4 | + | + | pdxp | co |
| | pyr | pdx | + | + | | + | + | + | co |
| | + | + | + | co | | pyr | pdx | + | + |
| | + | + | pdxp | co | | pyr | pdx | + | + |

[a] The asci shown were reported by Mitchell (1955). The flanking markers, *pyr* and *co*, are approximately 5 map units apart. These markers as well as the mating type alleles *A* and *a* (not shown), segregated in normal $2+:2-$ fashion. The alleles *pdx* and *pdxp* of the segregants are written as (*pdx* +) and (+ *pdxp*) respectively, to emphasize the contribution of $3+:1-$ segregation to the formation of prototrophic heteroallelic recombinants, i.e., (+ +). Note that in tetrad 3 both sites segregated $3+:1-$ resulting in two pyridoxine prototrophs. This is evidence for the formation of two heteroduplexes.

shown in Table III. In each instance, the $pdx^+$ reflects the result of a $6+:2-$ segregation at one or the other of the *pdx* sites. In more extensive studies in other systems, reciprocal recombination between heteroalleles has been observed, but it usually is a relatively minor source of intragenic recombinants (Lissouba *et al.*, 1962; Stadler and Towe, 1963; Fogel and Hurst, 1967; Kruszeswka and Gajewski, 1967).

## 2. Fidelity of Gene Conversion

Since gene conversion in hybrids occurs at frequencies orders of magnitude greater than spontaneous mutation of homozygotes, it is of considerable importance to determine whether gene conversion generates diversity, i.e., novel mutant alleles. The fidelity of gene conversion during meiosis was first extensively examined in *S. cerevisiae* by Roman (1956), who employed a test of recombination. A total of eight asci from *ADE/ade* monohybrids involving different *ade* loci were collected which exhibited 1*ADE*:3*ade* segregation. The three *ade* alleles within each ascus were backcrossed to the corresponding parental *ade* allele. These hybrids behaved as homoallelic diploids, i.e., they did not exhibit heteroallelic recombination. In each instance, conversion yielded an allele indistinguishable by the test of recombination from the input parental allele.

Further evidence that gene conversion does not generate novel alleles was provided by experiments of Fogel and Mortimer (1970), who devised a system capable of detecting single base pair differences among mutant alleles in conversion tetrads. They examined gene conversion in strains carrying the *ochre* nonsense suppressor *SUP*6, and *ochre* as well as *amber* mutant alleles. The *SUP*6 mutation of *S. cerevisiae* is an alteration in the anticodon recognition site in one of the structural genes specifying a tyrosyl transfer RNA (Capecchi *et al.*, 1975). The resulting altered tyrosyl tRNA (AUG→AUI) suppresses *ochre* mutations (UAA) by inserting tyrosine at the site of *ochre* nonsense codons during protein synthesis. Gene conversion at the *sup*6 locus was observed in 301/2,862 asci derived from an *SUP*6/*sup*6 hybrid. In approximately half of the conversion asci the supressor segregated 3*SUP*6 : 1*sup*6. These tetrads thus harbored three *ochre* suppressors. In no case did any of these suppressors display the ability to suppress *amber* (UAG) as well as *ochre* (UAA) mutations. This phenotype is theoretically possible if conversion at the *sup*6 locus results in a novel transfer RNA gene in which the anticodon is altered from AUG to AUU, instead of AUI specified by the *SUP*6 mutation. Tyrosyl transfer RNA's bearing the anticodon AUU would be expected to recognize both *ochre* and *amber* codons according to the codon–anticodon pairing rules proposed by Crick (1966). The tetrads that segregated 1*SUP*6 : 3*sup*6 likewise did not contain *sup*6 alleles capable of suppressing *amber* mutations. Such an *amber*-specific suppressor would require AUC at the anticodon recognition site, a configuration different from both parental alleles.

An additional test of fidelity involved aberrant segregation of the *ochre* mutation *arg*4-17. A diploid of genotype ARG4/arg4-17 yielded 22 conversion asci at the *ARG*4 locus among 317 tetrads analyzed. In 11 conversion asci 1*ARG*4 : 3*arg*4 segregation was observed. Each *arg*4 allele was shown to be identical to *arg*4-17 with respect to its response to a specific set of ochre nonsense suppressors (Fogel and Mortimer, 1970).

Prototrophic heteroallelic recombinants have also been studied to determine whether the functional enzyme resulting from the recombinational event differs from the wild type. No differences were found suggesting that conversion also does not generate novel wild-type alleles (Zimmermann, 1968).

### 3. Polarity

Polarity in gene conversion refers to the observation of a gradient of conversion frequencies along the length of a gene. The most extreme example of polarity was recorded by Lissouba and Rizet (1960), who studied heteroallelic recombination of spore color markers of *A. immersus*.

A collection of pale spore mutants of series 46 behaved as closely linked alleles of a single gene in intercrosses with one another. Asci from heteroallelic crosses consisted primarily of eight mutant spores. Depending upon the heteroalleles a certain fraction of asci also exhibited $2+ : 6-$ segregation. The frequency of dark (wild-type) ascospores observed in the two point crosses could be used to order the mutant sites (Fig. 2). When the mutant members of $2+ : 6-$ tetrads were examined by backcross analysis, it was found that they observed the following rules: gene conversion rather than crossing-over was responsible for the wild-type ascospores, and the conversion of mutant to wild type always involved the allele closest to the right (arbitrary) end of the map.

Strict polarity and absence of reciprocal recombinants has proven to be the exception rather than the rule for other loci of *Ascobolus* and of other species. Reciprocal recombinants are rare but nonetheless present and polarity is usually less extreme (reviewed by Holliday, 1964, 1968, 1974).

### 4. Fine Structure Mapping

Though the analysis of tetrads containing wild-type recombinants derived from heteroallelic crosses indicates that recombination of alleles is most often due to gene conversion rather than classical reciprocal recombination, the frequency of intragenic recombinants observed in two-point crosses ($a$1-1 $\times$ $a$1-2) can be employed to construct fine structure maps. The ordering of sites, however, is frequently uncertain due to inconsistencies in the data (Esposito M., 1968), and indicates that the frequency of heteroallelic recombination is not strictly a function of the physical distance separating the sites. The most serious drawback of fine structure mapping based upon recombination values from two-point crosses is that the order of sites that best fits the data can be incorrect. This has been recently demonstrated in *S. cerevisiae*, in the case of mutations of the

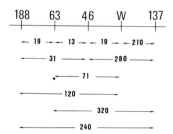

**Fig. 2.** A fine structure map of five mutants of series 46 of *Ascobolus immersus*. Values shown are the number of wild–type (dark) ascospores per $10^5$ ascospores resulting from heteroallelic crosses involving the two loci delimited by the arrows (Lissouba and Rizet, 1960).

structural gene $CYC1$, for iso-1-cytochrome $c$ (Moore and Sherman, 1975). The order of sites within the gene was unambiguously determined by amino acid sequence analysis of altered polypeptides. Five methods of fine structure mapping, including meiotic mapping as well as spontaneous and induced mitotic recombination (X-rays, near UV, and UV), either failed to order certain sites correctly, or did not accurately reflect the relative physical distances between the sites. As in prokaryotes, ordering of intragenic mutant sites in eukaryotes is most reliably performed by deletion mapping in which a qualitative rather than a quantitative response determines the order of the mutant sites. This method is least influenced by the effects of a particular mutation upon the probability of recombination in its vicinity (Sherman *et al.*, 1975).

### 5. Coconversion

Although fine structure maps constructed employing recombination data from two-point crosses usually contain some ambiguities, it appears that the recombination frequencies generally reflect the physical distance separating sites. When genetic analysis is restricted to tetrads containing only prototrophic recombinant ascospores, one observes that the prototrophic recombinant is generated by aberrant segregation at one of the sites, ie., one site segregating $3+ : 1a$ 1-1 and the other $2+ : 2a$ 1-2 (see Table IV, Single site conversion). How can gene conversion be sensitive to the distance separating mutant sites?

Unselected tetrad analysis of two-point and multipoint heteroallelic crosses involving mutants of the $arg4$ locus of *S. cerevisiae* has been most instructive in answering this question. Fogel and Mortimer (1969) and Fogel *et al.* (1971), have demonstrated that during conversion a segment of DNA base pairs of one chromatid, rather than a single base pair, is replaced by a homologous sequence of DNA from a nonsister chromatid. This conclusion was arrived at by the discovery that asci from heteroallelic crosses exhibiting $0+ : 4a$ with respect to phenotype, can be shown by genetic analysis to consist of two types of segregation patterns: asci in which both mutant sites segregated $2+ : 2a$ in the parental configuration, and rare asci that reflect coconversion in which the two sites of one chromatid converted simultaneously (Table IV). The coconversion asci contain no intragenic recombinants but nonetheless reflect conversional events and exhibit the usual association with reciprocal exchange of bracketing outside markers.

The closer a pair of mutant sites are to one another, the greater is the probability that they will be converted simultaneously. Since wild-type recombinants due to conversion arise only when one site segregates $3+ : 1a$ while the other segregates $2+ : 2a$, the distance between mutant

**TABLE IV**

**Evidence for Coconversion during Meiosis at the *arg*4 and *leu*1 Loci of *Saccharomyces cerevisiae*[a]**

| Tetrad classes | | Heteroallelic hybrids | | | |
|---|---|---|---|---|---|
| Sites | | $\dfrac{arg\,4\text{-}4}{arg\,4\text{-}17}$ | $\dfrac{arg\,4\text{-}1}{arg\,4\text{-}2}$ | $\dfrac{arg\,4\text{-}2}{arg\,4\text{-}17}$ | $\dfrac{leu\,1\text{-}1}{leu\,1\text{-}12}$ |
| *a*1-1 | *a*1-2 | | | | |
| **Single site** | | | | | |
| **conversion** | | | | | |
| − | + | | | | |
| + | + | 3 | 3 | 1 | 4 |
| + | − | | | | |
| + | − | | | | |
| − | + | | | | |
| − | + | 5 | 3 | 3 | 6 |
| − | − | | | | |
| + | − | | | | |
| − | + | | | | |
| − | + | 18 | 10 | 3 | 0 |
| + | + | | | | |
| + | − | | | | |
| − | + | | | | |
| − | − | 20 | 11 | 2 | 0 |
| + | − | | | | |
| + | − | | | | |
| **Coconversion** | | | | | |
| − | + | | | | |
| + | − | 2 | 13 | 14 | 1 |
| + | − | | | | |
| + | − | | | | |
| − | + | | | | |
| − | + | 1 | 10 | 13 | 1 |
| − | + | | | | |
| + | − | | | | |
| **Reciprocal** | | | | | |
| **exchange** | | | | | |
| − | + | | | | |
| + | + | 9 | 5 | 0 | 0 |
| − | − | | | | |
| + | − | | | | |
| Anomalous tetrads | | 1 | 1 | 0 | 0 |
| Total tetrads analyzed | | 697 | 502 | 544 | 544 |

[a] Data of Fogel and Mortimer (1969).

sites becomes important in determining the frequency of wild-type (as well as double mutant) recombinants arising by conversion. The average length of coconversion segments in yeast was estimated to be ca. 1000 DNA-base pairs, based upon X-ray fine structure mapping calibrated with reference to the X-ray response of *cyc*1 heteroalleles, whose physical distance from one another can be inferred from amino acid sequencing of appropriate revertant polypeptides. Coconversion has been demonstrated in other ascomycetes and in *Drosophila* (reviewed by Whitehouse 1974).

## B. Mitotic Gene Conversion

The appearance in mitotic cell populations of individuals exhibiting segregation for recessive markers originally in heterozygous condition was first demonstrated to occur in *Drosophila melanogaster* by Stern (1936). He observed somatic segregation in females heterozygous for the recessive sex-linked markers *y* (yellow body color) and *sn* (singed bristles) in repulsion.* Stern noted rare instances of twin spots consisting of patches of yellow tissue with normal bristles adjacent to tissue of wild-type color bearing singed bristles. He interpreted twin spots to be the clonal descendants of a diploid cell that, had undergone crossing-over at the four strand stage (G2), between the centromere proximal gene *sn* and the centromere, followed by mitotic disjunction of chromosomes (see Fig. 3).

Stern's hypothesis has been widely accepted as it has not been obvious how reciprocal recombination between unreplicated chromosomes in G1 can yield sectored clones or twin spots (i.e., AB/ab→Ab/aB). However in view of recent evidence that DNA molecules can participate in half-chiasma formation (see following section: Models of Recombination), it is worthwhile to note that the formation of Holliday structures in G1 and their subsequent replication can also lead to mitotic exchange and segregation of genetic markers (Fig. 3). This mechanism, like that of chromatid exchange in G2, yields at mitosis a diploid cell containing two parental and two recombinant chromatids, which depending upon the orientation of chromosomes during mitotic segregation, can result in daughter cells homozygous for markers distal to the site of exchange. Precise definition of the stage(s) in which spontaneous mitotic recombination occurs awaits further experimentation.

Genetic analysis of diploid cell populations of *Aspergillus nidulans* (see Pontecorvo, 1958, for review), revealed that mitotic segregation also occurs by nondisjunction of chromosomes at mitosis. This leads initially to the production of $2n + 1$ and $2n - 1$ segregants in which the latter exhibit

---

* That is, each recessive marker is on a different homologous chromosome.

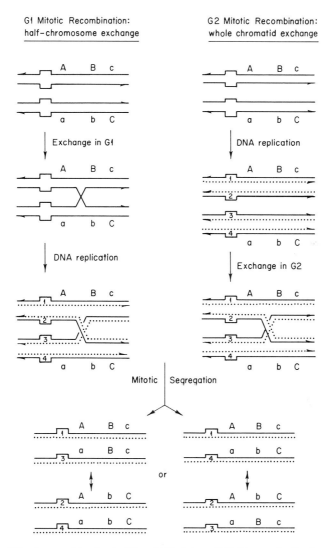

**Fig. 3.** Mitotic recombination in G1 and G2 of the cell division cycle. Reciprocal recombination in mitotic cells may result from whole chromatid exchange between non-sister chromatids in G2 and/or from half-chromosome exchange in G1 involving breakage and reunion of DNA chains of the same chemical polarity and subsequent replication of the Holliday structure. In each instance when chromatids 1 and 3 segregate from 2 and 4, the resultant cells are homozygous for markers distal to the point of exchange; when chromatids 1 and 4 segregate from 2 and 3, the cells remain heterozygous for markers distal to the point of exchange.

the recessive characters for all markers present on the monosomic chromosome. An initial event of mitotic nondisjunction leads to haploidization $(2n - 1 \rightarrow n)$ and diploidization $(2n - 1 \rightarrow 2n)$, owing to the selective advantage of euploid segregants (Käfer, 1961). The fusion of nuclei of unlike genotype, usually haploid, to give diploid nuclei or nuclei of higher ploidy, that subsequently undergo mitotic exchange and nondisjunction, is usually referred to as a parasexual cycle. The parasexual cycle has been exploited to assign genetic markers to linkage groups in a number of fungi, and has been applied to mapping the human genome with somatic cell hybrids (Ruddle and Creagan, 1975).

The genetic events leading to segregation and recombination in diploid cells are not restricted to reciprocal exchange and nondisjunction. Gene conversion also occurs during mitosis and has been demonstrated in both heterozygotes and heteroallelic diploids. Mitotic gene conversion has been most extensively studied in *S. cerevisiae*, which possesses a stable diploid phase as well as a stable haploid phase. Spontaneous gene conversion during mitosis of heteroallelic diploids was first reported by Roman (1956), who demonstrated that recombinant prototrophic diploid cells rarely harbor the double mutant chromosome ($a$1-1, $a$1-2), expected from reciprocal crossing-over, but instead are heterozygous for one or the other input parental allele. Subsequently a variety of agents including UV light, X-rays, ethylmethanesulfonate, 5-fluorodeoxyuridine, mitomycin-c, and nitrosoguanidine, have been shown to induce mitotic segregation for markers in heterozygous condition, due to either gene conversion or reciprocal exchange rather than mutation (James and Lee-Whiting 1955; Roman and Jacob, 1958; Esposito and Holliday, 1964; Holliday, 1964; Morpugo, 1965; Zimmermann and Schwaier, 1967; Esposito R., 1968; Nakai and Mortimer, 1969; Wildenberg, 1970; Roman, 1971, 1973).

Mitotic and meiotic gene conversion differ primarily in two respects: (1) there is no polarity in mitotic conversion for alleles that exhibit polarity in meiotic conversion (Hurst and Fogel, 1964; Wildenberg, 1970), and (2) mitotic conversion is infrequently accompanied by a reciprocal exchange in the vicinity of the conversional event (Roman, 1956). The stage(s) during the cell cycle in which spontaneous gene conversion occurs, as previously noted for reciprocal exchange, is not yet clear.

## C. Models of Genetic Recombination

Molecular models for gene conversion and reciprocal crossing-over in eukaryotes consistent with semiconservative DNA synthesis were first proposed by Whitehouse (1963) and Holliday (1964). The basic insight of both models was the idea that if two nonsister chromatids of a heterozygote $(A/a)$ undergo an exchange of single polynucleotide chains, a heteroduplex region consisting of a pair of mismatched base pairs $(A = a)$ is established on both chromatids (Fig. 4). If neither mismatch is repaired to the homoduplex condition before the completion of meiosis, $4A : 4a^a$ postmeiotic segregation results. Repair of both mismatches in the same direction results in $6A : 2a$ or $2A : 6a$ segregation depending upon whether

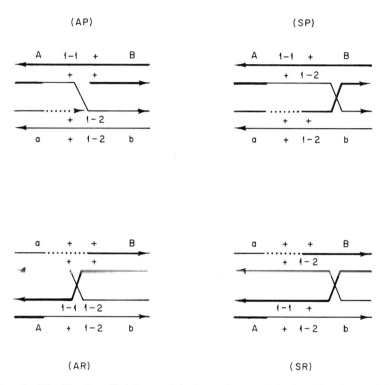

**Fig. 4.** The Meselson–Radding model of genetic recombination (Meselson and Radding, 1975). Recombination is initiated by uptake into the $A\ m1 - 1 + B$ homologue of a single strand, generated by DNA synthesis and strand displacement at the site of a nick on the $a + m1 - 2\ b$ homologue. A heteroduplex region is formed on only one chromatid and the bracketing markers remain in parental configuration, *AP*. Completion of recombination at this stage involves repair of heteroduplex regions by excision and resynthesis, closure of gaps by DNA polymerase and DNA ligase, and scission of the cross connection between the helices by nuclease action, etc. The *AP* structure can be reversibly isomerized to *AR* by rotation of the arms of the helices yielding the Holliday structure with a heteroduplex region restricted to one chromatid and bracketing markers in recombinant configuration following severance of the cross connections between the helices *AR*. Branch migration converts *AP* to *SP* in which there is a heteroduplex region on both chromatids, the *symmetric phase*. Branch migration similarly converts *AR* to *SR* with heteroduplex regions on both chromatids and outside markers in the recombinant array. *SP* can also reversibly isomerize yielding *SR*.

repair proceeded $A = a \to A = A$ or $A = a \to a = a$. Repair of one heteroduplex but not the other results in $5A : 3a$ or $3A : 5a$ postmeiotic segregation, again depending upon the direction of repair. In the case of the Holliday model, the heteroduplexes are established by an exchange of polynucleotide chains of like polarity yielding a half chiasma which can be resolved with or without recombination of heterozygous markers bracketing the site of conversion (i.e., heteroduplex formation and repair). The Holliday structure and related configurations of DNA molecules participating in genetic recombination have been visualized by electron microscopy (Rush and Warner, 1968; Broker and Lehman, 1971; Doniger *et al.*, 1973; Benbow *et al.*, 1975; Valenzuela and Inman, 1975; Thompson *et al.*, 1976; Potter and Dressler, 1976).

Several authors have recently formulated models of genetic recombination reflecting current knowledge in the area of DNA synthesis and repair of heteroduplex regions, and the physical properties of the Holliday structure and its derivatives (Sobell, 1972, 1975; Meselson and Radding, 1975; Wagner and Radman, 1975). The impact of the original Holliday model is obvious; the current issue is not whether the Holliday structure is formed, but how it is formed.

The general model for recombination presented by Meselson and Radding (1975), which accounts for the overall features of recombination in prokaryotic and eukaryotic systems as well as species specific variations in the properties of conversion in fungi, is illustrated in Fig. 4. The diagram depicts the interaction of two nonsister chromatids of DNA helices of a heteroallelic hybrid ($m1$-$1/m1$-$2$), heterozygous for markers flanking the heteroallelic pair. Recombination is initiated by DNA synthesis and strand displacement at the site of a single strand nick on the $a$ $m1$-$2$ $b$ homologue. The displaced strand induces a single strand break in the $A$ $m1$-$1$ $B$ molecule and pairs with the complementary region exposed by exonuclease activity at the site of the induced break. Transfer of a single strand from one duplex to the other thus involves concerted DNA polymerase and exonuclease activity. In this model the initiation of recombination is asymmetric, involving at first a donor and recipient chromatid and the formation of only one heteroduplex region. Repair of the heteroduplex region of $m1$-$1/+$ of Fig. 4 ($AP$) to $+/+$ by excision and resynthesis using the donor strand as template results in $6+ : 2m$ segregation of the $m1$-$1$ site and formation of a prototrophic heteroallelic recombinant as well. Had strand transfer proceeded further, both mutant sites would be in the heteroduplex. In this circumstance coconversion can result by excision and resynthesis at both sites using the donor DNA strand as a template.

The structure shown in Fig. 4 ($AP$) can isomerize to yield reciprocal recombination of flanking markers as indicated in Fig. 4 ($AR$). Conversion and recombination involving asymmetric strand transfer accommodates observations made in mitotic and meiotic yeast cells, and in *Ascobolus*, suggesting that a heteroduplex is sometimes formed on only one chromatid (Fogel and Mortimer, 1970; Roman, 1971; Stadler and Towe, 1971). Furthermore, the infrequent association of both spontaneous and UV-induced mitotic gene conversion in yeast with reciprocal recombination (Roman, 1956; Roman and Jacob, 1958) may reflect frequent failure to isomerize *AP* to *AR* in mitotic cells in contrast to meiotic cells. The model may also explain the observation that gene conversion events associated with reciprocal exchange of outside markers interfere with reciprocal recombination but not conversion in a neighboring interval (Stadler, 1959; Fogel and Hurst, 1967; Mortimer and Fogel, 1974). It is conceivable that two adjacent regions cannot undergo the required isomerizations simultaneously, since these involve rotations of the arms of the Holliday structure.

The next stage of the Meselson-Radding model involves entry into the symmetric phase, in which there is a heteroduplex region on both chromatids. This requires migration of the half chiasma driven by rotary diffusion and may proceed from *AP* to *SP*, or *AR* to *SR*. The primary consequence of a heteroduplex region on both chromatids is the occurrence of $4+ : 4m''$ postmeiotic segregation. It can be seen in Fig. 4 (*SP*) that site $m$1-2 will exhibit $4+ : 4m''$ postmeiotic segregation in the absence of repair of the heteroduplexes of both strands. In this particular instance a prototrophic strand is also generated. In the symmetric phase $6 + : 2m$, $2+ : 6m$, $5+ : 3m$, and $3+ : 5m$ aberrant segregations can also occur depending upon whether one or both heteroduplex regions are repaired and the direction of repair. The structure of *SP* can be rearranged to give *SR* in which the outside markers are recombined, satisfying the frequent association of $4+ : 4m''$ postmeiotic segregation and exchange of flanking markers. In addition to the features already described, the model can also account for polarity. Polarity is most easily explained as evidence of preferred sites for the initiation of strand transfer. Markers nearer to this sequence would have a greater probability of being included in a heteroduplex than markers further away.

## D. Genetic Control of Recombination

Numerous mutations affecting recombination during vegetative cell division and/or meiosis have been isolated by a variety of techniques (reviewed by Baker *et al.*, 1976). Following the report of Clark and Margulies (1965) that certain recombination deficient mutants of *Escherichia coli* are more sensitive to killing by ultraviolet light than the wild type, radiation and mutagen-sensitive mutants of eukaryotic cells have been isolated and tested for their effects on intragenic and intergenic recombination, and mutability. Some mutants affect both spontaneous and induced recombination, while others affect only the frequency of induced exchange. The former illustrate that certain gene products participate in both spontaneous recombination as well as repair of mutagen damage, while the latter indicate that induced recombination depends upon some functions that are not involved in spontaneous recombination. Furthermore, certain mutants enhance or abolish heteroallelic recombination without affecting intergenic exchange. These lend added credibility to models of recombination, such as the one described above, in which heteroallelic recombination results primarily from repair of mismatched base pairs, while intergenic exchange results primarily from the isomerization of the arms of the Holliday structure. In the case of the smut fungus, *Ustilago maydis*, a mutant defective only in heteroallelic recombination has been shown to lack the activity of a nuclease which specifically inactivates transforming DNA containing heteroduplex regions (Ahmad *et al.*, 1975). Accordingly, absence of heteroallelic recombination in the mutant is very likely due to absence of repair of mismatched base pairs.

In addition to gene products required for recombination anywhere in the genome, there are genes that modulate conversion and crossing-over in restricted regions. This has been demonstrated by studies performed in *N. crassa,* by Catcheside and his collaborators (cf. Catcheside, 1974: Baker *et al.*, 1976). The regulatory system discovered in *Neurospora* involves (1) dominant trans-acting mutations ($rec^+$) that are unlinked to the specific genetic regions in which they repress recombination when a recognition site ($con^+$) is present near the regulated region, and (2) cis-acting mutations ($cog^+$) that enhance recombination in their vicinity.

The phenotypes of other mutants indicate that genetic recombination in eukaryotes, as in bacteria and viruses (Clark, 1974), is a multistep process involving enzymes contributed by several pathways of DNA metabolism, namely DNA replication (e.g., polymerases and ligases), excision of mutagen damage or base pair mismatches (e.g., endonucleases and exonucleases), and repair replication following excision. Many mutants selected initially on the basis of effects on spontaneous mutability, recombination, ability to complete meiosis, or sensitivity to mutagens, exhibit pleiotrophic effects in one or more of these categories. The specific effects of such mutants with respect to fidelity of conversion, coconversion, postmeiotic segregation, as well as intergenic exchange, will hopefully lead to fruitful biochemical characterization of their defects.

It seems reasonable to conclude, in retrospect, that research directed toward an explanation of rare conversional exceptions has in fact culminated in a better understanding of orthodox reciprocal recombination. The phenomenon of paramutation, to which we now turn, is evidence for another feature of DNA metabolism of eukaryotes whose nature and significance are not yet as well understood as gene conversion.


## III. PARAMUTATION

Paramutation, like gene conversion, represents an apparent exception to the law of segregation. The phenomenon of paramutation involves the heritable alteration at high frequency of a responsive allele, the paramutable allele, in the presence of an allele that stimulates genetic alterations of responsive partners, the paramutagenic allele. Paramutation has been found at two loci of *Zea mays*, *R* and *B*, and occurs in other plants as well (reviewed by Brink, 1973). In this section we will describe paramutation at the extensively studied *R* locus of maize.

Paramutation at the *R* locus is a process unlike spontaneous mutation and gene conversion in several respects. Paramutation occurs at very high frequencies approaching 100% in the progeny of cells that can be assayed. The allelic alterations observed are directed, in that a particular phenotypic class of mutants is generated rather than the full range of known variants found at the *R* locus. Paramutable alleles are also inherently unstable and paramutate even when opposite a complete deletion of the *R* locus. When paramutation occurs in the presence of a paramutagenic allele, the paramutagenic allele remains unaltered. Paramutation thus provides evidence for the existence in eukaryotes of unstable DNA sequences that can be further provoked to change in the presence of another DNA sequence located on a homologous chromosome.

Although the evidence for paramutation is clear in the sense that it is different from ordinary gene mutation and intragenic recombination, it has not yet been associated with a specific aspect of DNA metabolism, e.g., replication, repair, or recombination, and the mechanism has remained obscure.


## A. Induction of Paramutation

Paramutation at the *R* locus of maize was discovered by Brink (1956). The *R* locus governs the formation of anthocyanin in the endosperm of the seed and other plant parts. As in the case of controlling elements, to be discussed later, the manifestation of paramutation is most conveniently studied by examining the phenotypes of kernels (the petri plate of the maize geneticist). To avoid confusion, it is useful to recall the life cycle of maize. The adult plant is diploid and monoecious. The formation of the seed involves double fertilization yielding a diploid $2N$ embryo and triploid $3N$ endosperm. Triploid endosperm results from the fusion of two haploid egg nuclei with one of the gametic nuclei of the pollen tube, while the diploid embryo results from the fusion of one haploid egg nucleus with the other haploid gametic nucleus of the pollen tube. The egg nuclei that participate in the fertilizations are identical in genotype, having arisen by mitosis from a single haploid meiotic product. The gametic nuclei of the pollen tube are likewise alike in genotype for the same reason. The embryo and endosperm are thus identical in genotype except for the gene dosage of the maternal contribution. The phenotype of the endosperm can be employed to infer the genotype of the embryo when genetic markers affecting endosperm traits such as the deposition of anthocyanin are involved. The mature seed consists of an outer seed coat, the pericarp (diploid tissue of seed parent origin), the aleurone layer of the endosperm ($3N$), the endosperm ($3N$), and the diploid embryo.

Paramutation at the $R$ locus is readily detectable in the aleurone layer of the endosperm. The first report of its occurrence involved the interaction of two alleles, $R^r$ which results in pigmentation of the plant and darkly mottled aleurone, and $R^{st}$ which results in fine stippling of the aleurone and pigmentation of the plant (Brink, 1956). Plants of genotype $R^r R^{st}$ were crossed with a plant of genotype $r^g r^g$, containing no anthocyanin in the aleurone and plant, as the seed parent. Since $r^g$ is recessive to both $R^r$ and $R^{st}$, half of the seeds should have been $R^r r^g r^g$ and darkly mottled, and the other half $R^{st} r^g r^g$ and finely stippled. Approximately 50% of the kernels (Table V) were finely stippled, but the remainder exhibited a readily observable and heritable dimunition of aleurone pigmentation (see Fig. 5). Such paramutated $R^r$ alleles are designated $R^{r'}$. Virtually all of the $R^r$

**TABLE V**

**Evidence for Paramutation of $R^r$ by $R^{st}$ in $R^r R^{st}$ Hybrids**[a]

| Crosses | Endosperm genotypes | |
|---|---|---|
| Seed parent × pollen parent | Expected | Observed |
| $r^g r^g \times R^r R^r$ | All $R^r r^g r^g$ | Observed = expected |
| $r^g r^g \times R^{st} R^{st}$ | All $R^{st} r^g r^g$ | Observed = expected |
| $r^g r^g \times R^r R^{st}$ | $\frac{1}{2}\,R^r r^g r^g$ and $\frac{1}{2}\,R^{st} r^g r^g$ | Observed = expected $\frac{1}{2}\,R^{r'} r^g r^g$ and $\frac{1}{2}\,R^{st} r^g r^g$ |

[a] See Brink *et al.* (1968).

**Fig. 5.** Paramutation of $R^r$ in maize. Top left: kernels from the mating of $r^g r^g ♀ \times R^r R^r ♂$. Top right: kernels from the mating of $r^g r^g ♀ \times R^{st} R^{st} ♂$. Bottom left: paramutated $R^{r'} r^g r^g$ kernels from the mating of $r^g r^g ♀ \times R^r R^{st} ♂$. Bottom right: $R^{st} r^g r^g$ kernels from the mating of $r^g r^g ♀ \times R^r R^{st} ♂$. Adapted from Brink *et al.*, 1968.

alleles emerging from $R^r R^{st}$ plants are changed to $R^{r'}$. The change from $R^r$ to $R^{r'}$ is the only one observed. For example, one does not detect generation of $R^{sc}$ alleles, which result in full pigmentation of the aleurone as well as pigmentation of certain plant parts (sc = self color). $R^{st}$ alleles emerging from the $R^r R^{st}$ hybrid retain their paramutagenic ability to stimulate the $R^r$ to $R^{r'}$ alteration. $R^{r'}$ alleles also become paramutagenic but less so than $R^{st}$. The response to $R^{st}$ is characteristic of most but not all $R^r$ alleles. Other paramutagenic $R$ alleles have also been described (Brink and Weyers, 1957; Brown and Brink, 1960).

## B. Paramutation during Mitosis and Meiosis

Paramutation of $R^r$ to $R^{r'}$ in response to the presence of $R^{st}$ appears to be an event that occurs chiefly during the mitotic cell divisions that establish the adult $R^rR^{st}$ plant. This was convincingly demonstrated by Sastry *et al.* (1965), who sampled the pollen from different tassel branches of $R^rR^{st}$ plants. They observed that several plants were mosaics with respect to the frequency of $R^{r'}$ bearing pollen in different parts of the tassel. Paramutation during meiosis is not ruled out but the frequency of its occurrence at meiosis must be sufficiently low so as not to obscure the detection of somatic sectors. In the case of paramutation of the *B* locus of maize, however, paramutation appears to occur predominantly during meiosis (Coe, 1966).

## C. Evidence for a Discrete Paramutagenic Sequence

Two separate lines of evidence indicate that the paramutagenic potential of $R^{st}$ is due to a DNA sequence different from that determining the stippled deposition of anthocyanin in the aleurone. Plants homozygous for $R^{st}$ give rise spontaneously to $R^{sc}$ mutants at a frequency of approximately $4/10^3$ gametes (Ashman, 1960; Kermicle, 1970). Some $R^{sc}$ mutants derived from $R^{st}$ retain paramutagenicity whereas others do not, indicating that the genetic determinants for stippling and paramutagenicity are separable by mutation though capable of simultaneous mutation (McWhirter and Brink, 1962). A critical experiment indicating the existence of a separate paramutagenic segment in $R^{st}$ was performed by Ashman (1965a,b), who recovered rare heteroallelic recombinants from an $R^rR^{st}$ hybrid that were reciprocally recombined for markers flanking the *R* locus and exhibited intermediate levels of paramutagenicity. These recombinants were recognized by their nearly colorless aleurone, neither an $R^r$ nor an $R^{st}$ phenotype. Subsequent studies indicate that the paramutagenic sequence is closely linked to the site conferring nearly colorless aleurone (Ashman, 1970; Satyanarayana and Kermicle, 1973a,b).

## D. The Genetic Fine Structure of the *R* Region

In the preceding introductory summary of the properties of paramutation, we have not dealt with the evidence indicating that the *R* region is a complex locus in order to avoid confounding observation and theory. It is useful, however, to recall that the *R* gene is a tandem repeat, since this may account for some of the similarities in behavior of paramutable *R* alleles and *bobbed* mutants of *Drosophila melanogaster*. The *bobbed* loci of the X and Y chromosomes of *Drosophila* together specify approximately 150 copies of the genes encoding ribosomal RNA (18 and 28 S). Deficiencies of various sizes in this region confer the bobbed bristle phenotype and are capable of reversion to the wild-type number of gene copies in both somatic tissue, termed gene compensation, and in germinal tissue, termed gene magnification (Ritossa, 1968; Atwood, 1969; Ritossa and Scala, 1969; Henderson and Ritossa, 1970; Tartof, 1971). The fact that the $R$ gene involves tandem repeats and that paramutation proceeds in hemizygotes[*], suggests that a process involving sister chromatid crossing-over may be involved (Brink, 1973), as postulated in the case of gene magnification in *Drosophila* (Tartof, 1973). Although we do not wish to suggest that paramutation and gene magnification are the same event, they share an outstanding attribute in common in that mutant alleles of a tandemly repetitious region undergo directed genetic alteration in hemizygotes (cf. Concluding Remarks).

The evidence that the standard $R^r$ allele governing anthocyanin deposition in both the seed and the plant is a tandem repeat was first obtained by Stadler (1951). Among the progeny of $R^r R^r$ homozygotes, mutations resulting in colorless seeds and red anthers, $r^r$, were discovered which were frequently associated with recombination of flanking heterozygous markers. Stadler and Nuffer (1953), thus reasoned that the $R^r$ gene consisted of two regions, one controlling seed pigmentation ($S$) and one controlling plant pigmentation ($P$), of sufficient homology to undergo unequal crossing-over at meiosis, generating variants deleted in $S$ and $P$, e.g., $r^r$ and $R^g$ mutants, respectively. Genetic evidence indicates that some $r^r$ alleles are in fact deletions for $S$ while others are mutant in the $S$ region, and similarly that some $R^g$ mutants are deletions of $P$ while others are mutant in the $P$ region. $R^g$ mutants derived from $R^r$, that have lost the $P$ region, are still paramutable (Bray and Brink, 1966). Since paramutable $R^r$ alleles become somewhat paramutagenic themselves after passage through $R^r R^{st}$ heterozygotes, one can determine whether $r^r$ mutants deleted for $S$ similarly remain paramutable, by whether they become paramutagenic after passage through $r^r R^{st}$ heterozygotes. The results of this test indicate that paramutability is lost concomitantly with loss of the $S$ region. The paramutable property therefore resides near or within the $S$ region (Brown, 1966).

Comparisons between the processes of gene conversion and paramutation will be deferred until the section, Concluding Remarks. We will now turn to the third and final exception considered here, controlling elements.

## IV. CONTROLLING ELEMENTS

Like paramutation, controlling elements affect the pattern of gene expression during development of the seed and other plant parts, but repre-

[*] An organism containing a gene in single, rather than double dose.

sent a distinctly different phenomenon. The extensive studies in maize by McClintock (1956) of genetic determinants inhibiting gene expression when near or within sensitive loci and capable of movement to other regions of the genome (i.e., transposition) added a new dimension to the transmission genetics (see Chapter 1) of eukaryotes. The properties of controlling elements in maize, other plants, and *Drosophila*, have been the subject of recent reviews (Green, 1973; Fincham and Sastry, 1974). In this discussion we will present certain aspects of the maize results that define the basic features of controlling elements.

## A. Autonomous and Nonautonomous Systems

Two types of controlling elements have been recognized: autonomous or one-element systems and nonautonomous or two-element systems. In the instance of autonomous systems the genetic data lead one to the view that a single element or DNA sequence resides at or within the locus of a wild-type gene and prevents normal expression. Transposition of the inhibiting segment from the affected locus, an autonomous function of the controlling element, restores wild-type gene expression. As in the case of paramutation, the behavior of controlling elements can be inferred by changes in the pigmentation of the seed and plant. The properties of one-element systems are illustrated below by the behavior of *Mp*, modulator of *P*, discovered by Brink and Nilan (1952). *Mp* is responsible for the autonomous instability of $P^{vv}$, an allele of the *P* gene, that causes variegated pigmentation of the pericarp and cob (Emerson, 1917).

In the case of two-element systems the inhibiting sequence at the locus is incapable of directing its own transposition. Transposition of the controlling element at the affected locus requires the function of a second element in the genome. The presence of the second element can be recognized by the instability of formerly stable mutant alleles. The inhibitory sequence which receives the signal for transposition has been referred to as the *operator* (McClintock, 1961), and more recently as the *receptor* (Fincham and Sastry, 1974), while the element fostering transposition is referred to as the *regulator* (McClintock, 1961). The element bearing the regulator can also direct its own transposition and can be shown to harbor a receptor sequence, as will be discussed later.

Regulators and chromosomally distant receptors can interact both with respect to transposition of the receptor at the affected genetic locus, as well as regulation of gene expression of the locus bearing the receptor. This feature of controlling elements is illustrated by the behavior of *Spm*, suppressor–mutator, a regulatory element discovered by McClintock (1953). Nonautonomous mutable alleles responsive to *Spm* exhibit only slight inhibition of wild-type activity in the absence of *Spm*. When *Spm* is present most cells exhibit the mutant phenotype, the suppressor aspect of *Spm*, while in other cells the presence of *Spm* results in transposition of the distant receptor and restoration of wild-type gene expression, the mutator aspect of *Spm*.

The behavior of a two-element system is illustrated by the case of the regulator *Ac*, activator, and one of its receptors *Ds*, dissociation. The latter receptor becomes the site of frequent chromosomal breaks during transposition mediated by *Ac*, providing cytological evidence of the chromosomal residence of *Ds* both before and after transposition (McClintock, 1951).

## B. Specificity of the Receptor–Regulator Relationship

Three major regulators have been distinguished in maize primarily due to the specificity of the regulator–receptor interaction: *Dt*, dotted (Rhoades, 1938); *Ac*, activator (McClintock, 1950); and *Spm*, suppressor–mutator (McClintock, 1953). Nonautonomous mutable alleles including alleles of the same locus, respond to the presence of only one of these regulators. Several autonomous mutable alleles bearing either *Spm* or *Ac* at the site of the mutable locus have given rise to derivative nonautonomous mutable alleles. These derivatives respond only to un-linked regulators of the type which were previously present at the locus, e.g., *Spm* or *Ac*. Apparently elements such as *Ac* and *Spm* encode both a regulator sequence required for transposition and a receptor sequence which can be retained when the regulatory sequence is lost or be-comes inactive. Both *Ac* and *Spm* foster their transposition as well as that of distant receptors based on mapping data (McClintock; 1956).

## C. One-Element Systems: Transposition of Modulator

The wild-type $P^{rr}$ gene of maize specifies red pigmentation of the cob and pericarp. In 1917, Emerson described a mitotically unstable allele of the $P$ gene, designated $P^{vv}$, which results in medium variegation (red sectors on a white background) of the cob and pericarp. The $P^{vv}$ gene mutates at low frequency during mitosis yielding cobs bearing a few per-cent of red wild-type kernels and a novel mutant phenotype called light variegated due to the lower number of red sectors of the pericarp. Brink and Nilan (1952), noted that some but not all red kernels occurred as twin spots in conjunction with a nearly equal number of light variegated ker-nels. Although the pericarp is diploid tissue of maternal origin, the transmission of the red and light variegated phenotypes could be studied, since the female gametic tissue was also part of the sectors. Genetic analysis of the members of twin sectors led to the conclusion that the $P^{vv}$ allele is a complex consisting of the wild-type $P^{rr}$ allele and a tightly linked modulator of its activity, $Mp$, capable of transposition. This conclusion was based on the observation that the light variegated members of twin sectors owed their phenotype to the presence of an additional copy of $Mp$. It appeared therefore, that the production of the cells giving red kernels was due to the migration of the modulator of $P^{rr}$ expression from the site of the $P$ gene to another site on the same chromosome, the other homolog, or a different chromosome entirely.

When migration of $Mp$ occurs after duplication of the $P$ bearing chromosome this can lead to the formation of red $P^{rr}/P^{vv}$ and light variegated $P^{vv}/P^{vv}$ + $Mp$ sectors. Other types of sectors, e.g., red $P^{vv}/P^{rr}$ + $Mp$ and light variegated $P^{vv}/P^{vv}$ + $Mp$, were also encountered, suggesting that $Mp$ elements can migrate before duplication of the $P$ gene bearing chromosome is completed (Greenblatt and Brink, 1962; Greenblatt, 1968). The interpretation of $P^{vv}$ as a $P^{rr}Mp$ complex was unambiguously proven by reconstitution of the $P^{vv}$ allele from separate $P^{rr}$ and $Mp$ components (Orton, 1966; Orton and Brink, 1966). Furthermore, the modulator, $Mp$, that modifies the phenotype of the $P^{vv}$ allele and that transposes from $P^{vv}$ generating $P^{rr}$ was shown to be a controlling element identical to $Ac$ with respect to the family of nonautonomous mutable alleles that respond to it, and its ability to provoke chromosomal breaks in stocks carrying the $Ds$ receptor (Barclay and Brink, 1954).

### D. Two-Element Systems: Activator–Dissociation and Suppressor–Mutator

McClintock's extensive analysis of controlling elements gained impetus from the appearance of unstable mutant alleles among the progeny of plants subjected to extensive breakage and rejoining of the short arm of chromosome 9. The extensive repatterning of the short arm of chromosome 9 was provoked by a bridge–breakage–fusion cycle initiated during meiosis of structural heterozygotes* consisting of a chromosome 9 with an inverted tandem duplication of the short arm and a normal homologue (McClintock, 1951). Most of the unstable mutations recovered were alleles of loci known to be located on the short arm of 9. The most important

---

* Structural heterozygotes contain a pair of homologues that differ in microscopically detectable ways resulting, e.g., from an inversion in one and not the other homologue.

variant discovered was a locus called dissociation, $Ds$, which became the preferred site for chromosomal breakage in structural homozygotes in the presence of a second element referred to as activator, $Ac$. It became obvious in later studies that $Ds$ was in fact a receptor and transposed in the presence of $Ac$, the regulator. A mutable allele of the $C_1$ locus involved in anthocyanin pigmentation of the aleurone, $c^{m-1}$, was recovered in a plant bearing $Ac$ and $Ds$. The appearance of this nonautonomous mutable allele was accompanied by a change in the site of $Ds$ instigated breaks from the $Ds$ locus to the vicinity of the cytological location of the $C_1$ gene (McClintock, 1949, 1951). This type of change in the position of $Ds$ accompanied by the appearance of other mutable alleles on the short arm of chromosome 9 led to the conclusion that the mutable alleles were generated by $Ac$ mediated transposition of $Ds$ from its original locus to the site of the affected gene (Fig. 6). The fact that $Ds$ is a preferred region for chromosomal breakage supports the conclusion from other studies that restoration of wild-type activity as a result of the loss of the receptor is the basis of two-element systems and the source of the wild–type sectors generated in the presence of $Ac$.

## E. Controlling Elements and Modern Biology

The nonautonomous mutable alleles recovered among descendants of plants involved in the bridge–breakage–fusion cycle included several mutants that did not respond to $Ac$, but responded instead to another element, $Spm$, whose dual phenotype has already been described. $Spm$ is a *bona fide* trans-acting repressor of wild-type genes bearing a responsive receptor sequence. As stated by McClintock (1961), $Spm$ and other controlling elements may represent an observable aberration of one of the systems involved in the coordination of gene expression during development. Certainly controlling elements provide evidence that there are DNA sequences capable of controlling the expression of a wide variety of gene

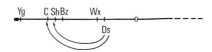

**Fig. 6.** Transposition of $Ds$ (dissociation) from its original locus to different positions on the short arm of chromosome 9, as described by McClintock (1956). Gene symbols are $Bz$, bronze plant; $C$, anthocyanin pigmentation; $Sh$, shrunken endosperm; $Wx$, waxy pollen; and $Yg$, yellow plant.

loci and responsive to specific regulators. The existence of controlling elements therefore adds to the credibility of models for development such as proposed by Britten and Davidson (1969), which involve major controlling genes and unique structural DNA sequences preceded by members of various multiple copy DNA families that participate in the regulation of transcription of the structural genes. The peripatetic nature of controlling elements also suggests that short range transpositions might be the mechanism whereby DNA regions encoding variable and constant immunoglobin chains are fused and triggered for transcription and translation (Hood, 1973).

## V. CONCLUDING REMARKS

In the preceding sections we have stressed the distinct nature of gene conversion, paramutation, and controlling elements. The three processes nonetheless do bear a relationship to one another since each involves recombination or recombination-like events. Conversion in both mitosis and meiosis is an aspect of genetic recombination between homologous chromosomes that recombines heteroalleles but does not generate novel alleles. Although the recombination processes associated with paramutation and transposition of controlling elements appear to differ from one another and ordinary gene conversion and exchange in several respects, there are similarities to aspects of recombination observed in other systems.

Paramutation of $R^r$ alleles as discussed above occurs in $R^r$ homozygotes and hemizygotes and is enhanced by the presence of a paramutagenic allele. How can these properties be analogized to events in other systems? We begin by noting that recombination can occur between incompletely homologous DNA regions by mechanisms resembling gene conversion, i.e., interspecific transformation in bacteria (Shaeffer and Ritz, 1955; Ravin, 1961), and that there are DNA sequences in both eukaryotes, e.g., $cog^+$ sites (Catcheside, 1974), and in prokaryotes, e.g., $chi$ sites (Stahl $et$ $al.$, 1975), that stimulate recombination in their vicinity. Since the $R$ gene is a complex locus that may have arisen by gene duplication, one may imagine that a paramutable $R$ allele contains at least two incompletely homologous, but nevertheless sufficiently similar, regions so that recombination between them is possible. Paramutation in $R^r$ homozygotes and hemizygotes may therefore be due to a process of *nonhomologous gene conversion* resembling heterospecific transformation between these different regions. The DNA sequence conferring the $R^{r'}$ paramutant phenotype can thus be generated by repair of a heteroduplex formed in the $S$ region using the donor strand from another region as template. This event can occur between both sister and nonsister chromatids, and thereby in both homozygotes and hemizygotes, since it is an interaction between different regions of the same gene.

One can further postulate that paramutagenic alleles instigate nonhomologous recombination in regions containing paramutable alleles because the paramutagenic alleles harbor DNA sequences resembling $cog^+$ or *chi* sites, which they donate to chromosomes bearing paramutable alleles stimulating exchange in the latter. This role of the paramutagenic allele must be consistent with the observations that paramutagenic alleles emerge unchanged in phenotype and paramutagenicity from association with paramutable alleles, while paramutants acquire various degrees of paramutagenicity. The sequence of events may therefore involve: (1) asymmetric strand transfer (and resynthesis in the donor molecule) of a variable length of the paramutagenic sequence to a recipient DNA molecule containing the paramutable allele, (2) displacement of a portion of the recipient molecule by the sequence bearing the recombinational hot spot(s), and (3) stimulation of nonhomologous recombination in the chromosome bearing the paramutable allele (e.g., within the same chromatid or between sister chromatids), immediately or in subsequent cell divisions. In summary, directed mutation of $R^r$ to $R^{r'}$ could result from nonhomologous exchange provoked by incorporation of all or part of a conversional hotspot into a region containing the paramutable allele. Whether nonhomologous gene conversion within complex genes or other loci containing partially homologous sequences can be invoked to explain other cases of directed mutation or directed replacement of DNA sequences remains to be seen.

The recombinational and mutational events that occur during transposition of controlling elements also simulate types of recombination observed in other systems (cf. Fincham and Sastry, 1974). Receptor elements like those of $Ac$, which inhibit wild-type gene expression when they integrate, may do so by having disrupted the DNA sequence of a structural gene by insertion into it. DNA viruses and plasmids of *E. coli*, which exhibit nonhomologous recombination of this type have been described, and serve as molecular models for controlling elements that invariably inhibit wild-type gene expression. Receptor elements such as those of $Spm$, which do not abolish wild-type gene activity, may insert in a manner more analogous to integration of bacteriophage $\lambda$ which normally occurs in a restricted region of the *E. coli* genome. Since both paramutation and controlling elements can alter the pattern of wild-type gene expression in somatic cell lineages, we wonder whether recombination-like events play an over-looked role during development.

## REFERENCES

Ahmad, A., Holloman, W. K., and Holliday, R. (1975). Nuclease that preferentially inactivates DNA containing mismatched bases. *Nature (London)* **258**, 54–56.

Ashman, R. B. (1960). Stippled aleurone in maize. *Genetics* **45**, 19–34.

Ashman, R. B. (1965a). Mutants from maize plants heterozygous $R^r R^{st}$ and their association with crossing-over. *Genetics* **51**, 305–312.

Ashman, R. B. (1965b). Paramutagenic action of mutants from maize plants heterozygous $R^r$ $R^{st}$. *Genetics* 52, 835–841.

Ashman, R. B. (1970). The compound structure of the $R^{st}$ allele in maize. *Genetics* **64**, 239–245.

Atwood, K. C. (1969). Some aspects of the *bobbed* problem in *Drosophila*. *Genetics* **61**, Suppl. 1, Part 2, 319–327.

Baker, B. S., Carpenter, A. T. C., Esposito, M. S., Esposito, R. E., and Sandler, L. (1976). The genetic control of meiosis. *Annu. Rev. Genet.* **10**, 53–134.

Baker, W. K. (1968). Position-effect variegation. *Adv. Genet.* **14**, 133–169.

Barclay, P. C., and Brink, R. A. (1954). The relation between modulator and activator in maize. *Proc. Natl. Acad. Sci. U.S.A.* **40**, 1118–1126.

Bateson, W. ed, (1928). "William Bateson FRS Naturalist; His Essays and Addresses Together with a Short Account of his Life." Cambridge, Univ. Press, London and New York.

Benbow, R., Zuccarelli, A., and Sinsheimer, R. (1975). Recombinant DNA molecules of bacteriophage $\phi$X174. *Proc. Natl. Acad. Sci. U.S.A.* **72**, 235–239.

Bray, R. A., and Brink, R. A. (1966). The metastable nature of paramutable R alleles in maize. I. Heritable enhancement in level of standard $R^r$ action. *Genetics* **54**, 433–439.

Brink, R. A. (1956). A genetic change associated with the R locus in maize which is directed and potentially reversible. *Genetics* **41**, 872–889.

Brink, R. A. (1973). Paramutation. *Annu. Rev. Genet.* **7**, 129–152.

Brink, R. A., and Nilan, R. A. (1952). The relation between light variegated and medium variegated pericarp in maize. *Genetics* **37**, 519–544.

Brink, R. A., and Weyers, W. H. (1957). Invariable genetic change in maize plants heterozygous for marbled aleurone. *Proc. Natl. Acad. Sci. U.S.A.* **43**, 1053–1060.

Brink, R. A., Styles, E. D., and Axtell, J. D. (1968). Paramutation: Directed genetic change. *Science* **159**, 161–170.

Britten, R. J., and Davidson, E. H. (1969). Gene regulation for higher cells: A theory. *Science* **165**, 349–357.

Broker, T. R., and Lehman, I. R. (1971). Branched DNA molecules: Intermediates in T4 recombination. *J. Mol. Biol.* **60**, 131–149.

Brown, D. F. (1966). Paramutability of $R^g$ and $r^r$ mutant genes derived from an $R^r$ allele in maize. *Genetics* **54**, 899–910.

Brown, D. F., and Brink, R. A. (1960). Paramutagenic action of paramutant $R^r$ and $R^g$ alleles in maize. *Genetics* **45**, 1313–1316.

Capecchi, M. R., Hughes, S. H., and Wahl, G. M. (1975). Yeast supersuppressors are altered tRNA's capable of translating a nonsense codon *in vitro*. *Cell* **6**, 269–277.

Carlson, P. S. (1971). A genetic analysis of the rudimentary locus of *Drosophila melanogaster*. *Genet. Res.* **17**, 53–81.

Catcheside, D. G. (1974). Fungal genetics. *Annu. Rev. Genet.* **8**, 279–300.

Chovnick, A., Ballantyne, G. H., Baillie, D. L., and Holm, D. G. (1970). Gene conversion in higher organisms: Half-tetrad analysis of recombination within the rosy cistron of *Drosophila melanogaster*. *Genetics* **66**, 315–329.

Clark, A. J. (1974). Process toward a metabolic interpretation of genetic recombination of *Escherichia coli* and bacteriophage lambda. *Genetics* **78**, 259–271.

Clark, A. J., and Margulies, A. D. (1965). Isolation and characterization of recombination-deficient mutants of *Escherichia coli K12*. *Proc. Natl. Acad. Sci. U.S.A.* **53**, 451–459.

Coe, E. H., Jr. (1966). The properties, origin, and mechanism of conversion–type inheritance at the *B* locus in maize. *Genetics* **53**, 1035–1063.

Crick, F. H. C. (1966). Codon-anticodon pairing: The wobble hypothesis. *J. Mol. Biol.* **19**, 548–555.

Demerec, M. (1963). Selfer mutants of *Salmonella typhimurium*. *Genetics* **48**, 1519–1531.

Doniger, J., Warner, R. C., and Tessman, I. (1973). Role of circular dimer DNA in the primary recombination mechanism of bacteriophage S13. *Nature (London), New Biol.* **242**, 9–12.

Emerson, R. A. (1917). Genetical studies of variegated pericarp in maize. *Genetics* **2**, 1–35.

Esposito, M. S. (1968). X-Ray and meiotic fine structure mapping of the adenine-8 locus in *Saccharomyces cerevisiae*. *Genetics* **58**, 507–527.

Esposito, M. S. (1971). Postmeiotic segregation in *Saccharomyces*. *Mol. Gen. Genet.* **111**, 297–299.

Esposito, R. E. (1968). Genetic recombination in synchronized cultures of *Saccharomyces cerevisiae*. *genetics* **59**, 191–210.

Esposito, R. E., and Holliday, R. (1964). The effect of 5-fluorodeoxyuridine on genetic replication and mitotic crossing over in synchronized cultures of *Umstilago Maydis*. *Genetics* **50**, 1009–1017.

Fincham, J. R. S., and Sastry, G. R. K. (1974). Controlling elements in maize. *Annu. Rev. Genet.* **8**, 15–50.

Fogel, S., and Hurst, D. D. (1967). Meiotic gene conversion in yeast tetrads and the theory of recombination. *Genetics* **57**, 455–481.

Fogel, S., and Mortimer, R. K. (1969). Informational transfer in meiotic gene conversion. *Proc. Natl. Acad. Sci. U.S.A.* **62**, 96–103.

Fogel, S., and Mortimer, R. K. (1970). Fidelity of meiotic gene conversion in yeast. *Mol. Gen. Genet.* **109**, 177–185.

Fogel, S., Hurst, D. D., and Mortimer, R. K. (1971). Gene conversion in unselected tetrads from multipoint crosses. *Stadler Genet. Symp.* **1** and **2**, 89–110.

Green, M. M. (1973). Some observations and comments on mutable and mutator genes in *Drosophila*. *Genetics* **73**, Suppl., 187–194.

Greenblatt, I. M. (1968). The mechanism of modulator transposition in maize. *Genetics* **58**, 585–597.

Greenblatt, I. M., and Brink, R. A. (1962). Twin mutations in medium variegated pericarp maize. *Genetics* **47**, 489–501.

Hawthorne, D. C. (1963). Directed mutation of the mating type alleles as an explanation of homothallism in yeast. *Genet. Today, Proc. Int. Congr., 11th, 1963* Vol. 1, pp. 34–35.

Henderson, A., and Ritossa, F. (1970). On the inheritance of rDNA of magnified bobbed loci in *D. melanogaster*. *Genetics* **66**, 463–473.

Hexter, W. M. (1963). Nonreciprocal events at the garnet locus in *Drosophila melanogaster*. *Proc. Natl. Acad. Sci. U.S.A.* **50**, 372–379.

Holliday, R. (1964). A mechanism for gene conversion in fungi. *Genet. Res.* **5**, 282–304.

Holliday R. (1965). Induced mitotic crossing-over in relation to genetic replication in synchronously dividing cells of *Ustilago maydis*. *Genet. Res.* **6**, 104–120.

Holliday, R. (1968). Genetic recombination in fungi. *In* "Replication and Recombination of Genetic Material" (W. J. Peacock and R. D. Brock, eds.), pp. 157–174. Aust. Acad. Sci., Canberra.

Holliday, R. (1974). Molecular aspects of genetic exchange and gene conversion. *Genetics* **78**, 273–287.

Hood, L. (1973). The genetics, evolution and expression of antibody molecules. *Stadler Genet. Symp.* **5**, 74–142.

Hurst, D. D., and Fogel, S. (1964). Mitotic recombination and heteroallelic repair in *Saccharomyces cerevisiae*. *Genetics* **50**, 435–458.

Hurst, D. D., Fogel, S., and Mortimer, R. K. (1972). Conversion-associated recombination in yeast. *Proc. Natl. Acad. Sci. U.S.A.* **69**, 101–105.

James, A. P., and Lee-Whiting, B. (1955). Radiation-induced genetic segregations in vegetative cells of diploid yeast. *Genetics* **40**, 826–831.

Kafer, E. (1961). The processes of spontaneous recombination in vegetative nuclei of *Aspergillus nidulans*. *Genetics* **46**, 1581–1609.

Kermicle, J. L. (1970). Somatic and meiotic instability of R-stippled, an aleurone spotting factor in maize. *Genetics* **64**, 247–258.

Kitani, Y., Olive, L. S., and El-Ani, A. S. (1962). Genetics of *Sordaria fimicola*. V. Aberrant segregation at the G locus. *Am. J. Bot.* **49**, 697–706.

Kruszewska, A., and Gajewski, W. (1967). Recombination within the Y locus in *Ascobolus immersus*. *Genet. Res.* **9**, 159–177.

Leblon, G. (1972a). Mechanism of gene conversion in *Ascobolus immersus*. I. Existence of a correlation between the origin of mutants induced by different mutagens and their conversion spectrum. *Mol. Gen. Genet.* **115**, 36–48.

Leblon, G. (1972b). Mechanism of gene conversion in *Ascobolus immersus*. II. The relationships between the genetic alterations in b, or b₂ mutants and their conversion spectrum. *Mol. Gen. Genet.* **116**, 322–335.

Leblon, G., and Rossignol, J. L. (1973). Mechanism of gene conversion in *Ascobolus immersus*. III. The interaction of heteroalleles in the conversion process. *Mol. Gen. Genet.* **122**, 165–182.

Lindegren, C. (1953). Gene conversion in *Saccharomyces*. *J. Genet.* **51**, 625–627.

Lissouba, P., and Rizet, G. (1960). Sur l'existence d'une unité génétique polarisée ne subissant que des échanges non réciproques. *C. R. Hebd. Seances Acad. Sci.* **250**, 3408–3411.

Lissouba, P., Mousseau, J., Rizet, G., and Rossignol, J. L. (1962). Fine structure of genes in the ascomycete *Ascobolus immersus*. *Adv. Genet.* **11**, 343–380.

McClintock, B. (1949). Mutable loci in maize. *Carnegie Inst. Washington, Yearb.* **48**, 142–154.

McClintock, B. (1950). The origin and behavior of mutable loci in maize. *Proc. Natl. Acad. Sci. U.S.A.* **36**, 344–345.

McClintock, B. (1951). Chromosome organization and genic expression. *Cold Spring Harbor Symp. Quant. Biol.* **16**, 13–47.

McClintock, B. (1953). Mutation in maize. *Carnegie Inst. Washington, Yearb.* **52**, 227–233.

McClintock, B. (1956). Controlling elements and the gene. *Cold Spring Harbor Symp. Quant. Biol.* **21**, 197–216.

McClintock, B. (1961). Some parallels between gene control systems in maize and in bacteria. *Am. Nat.* **95**, 265–277.

McWhirter, K. S., and Brink, R. A. (1962). Continuous variation in level of paramutation at the R locus in maize. *Genetics* **47**, 1053–1074.

Meselson, M. S., and Radding, C. M. (1975). A general model for genetic recombination. *Proc. Natl. Acad. Sci. U.S.A.* **72**, 358–361.

Mitchell, M. B. (1955). Aberrant recombination of pyridoxine mutants of *Neurospora*. *Proc. Natl. Acad. Sci. U.S.A.* **41**, 215–220.

Moore, C. W., and Sherman, F. (1975). Role of DNA sequences in genetic recombination in the iso-1-cytochrome c gene of yeast. I. Discrepancies between physical distances and genetic distances determined by five mapping procedures. *Genetics* **79**, 397–418.

Morpurgo, G. (1963). Induction of mitotic crossing over in *Aspergillus nidulans* by bifunctional alkylating agents. *Genetics* **48**, 1259–1263.

Mortimer, R. K., and Fogel, S. (1974). Genetical interference and gene conversion. *In* Mechanisms of Recombination" (R. F. Grell, ed.), p. 263–275. Plenum, New York.

Nakai, S., and Mortimer, R. K. (1969). Studies of the genetic mechanism of radiation-induced mitotic segregation in yeast. *Mol. Gen. Genet.* **103**, 329–338.

Oliver, C. P. (1940). A reversion to wild-type associated with crossing-over in *Drosophila melanogaster. Proc. Natl. Acad. Sci. U.S.A.* **26**, 452–454.

Orton, E. R. (1966). Frequency of reconstitution of the variegated pericarp allele in maize. *Genetics* **53**, 17–25.

Orton, E. R., and Brink, R. A. (1966). Reconstitution of the variegated pericarp allele in maize by transposition of modulator back to the P locus. *Genetics* **53**, 7–16.

Pontecorvo, G. (1958). "Trends in Genetic Analysis." Columbia Univ. Press, New York.

Potter, H., and Dressler, D. (1976). On the mechanism of genetic recombination: Electron microscopic observation of recombination intermediates. *Proc. Natl. Acad. Sci. U.S.A.* **73**, 3000–3004.

Radding, D. M. (1973). Molecular mechanisms in genetic recombination. *Annu. Rev. Genet.* **7**, 87–111.

Ravin, A. (1961). The genetics of transformation. *Adv. Genet.* **10**, 61–163.

Rhoades, M. M. (1938). Effect of the Dt gene on the mutability of the $a_1$ allele in maize. *Genetics* **23**, 377–397.

Ritossa, F. M. (1968). Unstable redundancy of genes for ribosomal RNA. *Proc. Natl. Acad. Sci. U.S.A.* **60**, 509–516.

Ritossa, F. M., and Scala, G. (1969). Equilibrium variations in the redundancy of rDNA in *Drosophila melanogaster. Genetics* **61**, Suppl. 1, Part 2, 305–317.

Roman, H. L. (1956). Studies of gene mutation in *Saccharomyces. Cold Spring Harbor Symp. Quant. Biol.* **21**, 175–185.

Roman, H. L. (1963). Genic conversion in fungi. *In* "Methodology in Basic Genetics" (W. J. Burdette, ed.), p. 209–246. Holden-Day, San Francisco, California.

Roman, H. L. (1971). Induced recombination in mitotoc diploid cells of *Saccharomyces. Genet. Lect.* **2**, 43–59.

Roman, H. L. (1973). Studies of recombination in yeast. *Stadler Genet. Symp.* **5**, 35–48.

Roman, H. L., and Jacob, F. (1958). A comparison of spontaneous and ultraviolet-induced allelic recombination with reference to the recombination of outside markers. *Cold Spring Harbor Symp. Quant. Biol.* **23**, 155–160.

Roper, J. A. (1950). Search for linkage between genes determining a vitamin requirement. *Nature (London)* **166**, 956–957.

Ruddle, F., and Creagan, R. P. (1975). Parasexual approaches to the genetics of man. *Annu. Rev. Genet.* **9**, 407–486.

Rush, M. G., and Warner, R. C. (1968). Molecular recombination in a circular genome— $\phi$X174 and S13. *Cold Spring Harbor Symp. Quant. Biol.* **33**, 459–466.

Sastry, G. R. K., Cooper, H. B., Jr., and Brink, R. A. (1965). Paramutation and somatic mosaicism in maize. *Genetics* **52**, 407–424.

Satyanarayana, K. V., and Kermicle, J. L. (1973a). The R-stippled allele in maize. I. Relation of changes in paramutagenic action to recombinant origin of self-colored mutations. Cited in Brink (1973).

Satyanarayana, K. V., and Kermicle, J. C. (1973b). The R-stippled allele in maize. II. Arrangement of the pigmenting and paramutagenic components. Cited in Brink (1973).

Shaeffer, P., and Ritz, E. (1955). Transfert interspécifique d'un caractère héréditaire chez des bactéries du genre *Hemophilus. C. R. Hebd. Seances Acad. Sci.* **240**, 1491–1493.

Sherman, F., Jackson, M., Liebman, S. W., Schweingruber, A. M., and Stewart, J. W. (1975). A deletion map of *cycl* mutants and its correspondence to mutationally altered iso-l-cytochromes *c* of yeast. *Genetics* **81**, 51–73.

Sobell, H. M. (1972). Molecular mechanism for genetic recombination. *Proc. Natl. Acad. Sci. U.S.A.* **69**, 2483–2487.

Sobell, H. M. (1975). A mechanism to activate branch migration between homologous DNA molecules in genetic recombination. *Proc. Natl. Acad. Sci. U.S.A.* **72**, 279–283.

Stadler, D. R. (1959). The relationship of gene conversion to crossing-over *Neurospora. Proc. Natl. Acad. Sci. U.S.A.* **45**, 1625–1629.

Stadler, D. R. (1973). The mechanism of intragenic recombination. *Annu. Rev. Genet.* **7**, 113–127.

Stadler, D. R., and Towe, A. M. (1963). Recombination of allelic cysteine mutants in *Neurospora. Genetics* **48**, 1323–1344.

Stadler, D. R., and Towe, A. M. (1971). Evidence for meiotic recombination in *Ascobolus* involving only one member of a tetrad. *Genetics* **68**, 401–413.

Stadler, L. J. (1951). Spontaneous mutation in maize. *Cold Spring Harbor Symp. Quant. Biol.* **16**, 49–63.

Stadler, L. J., and Nuffer, M. G. (1953). Problems of gene structure. II. Separation of R$^r$ elements (S) and (P) by unequal crossing-over. *Science* **117**, 471–472.

Stahl, F. W., Crasemann, J. M., and Stahl, M. M. (1975). Rec-mediated recombinational hot spot activity in bacteriophage lambda. *J. Mol. Biol.* **94**, 203–212.

Stern, C. (1936). Somatic crossing-over and segregation in *Drosophila melanogaster. Genetics* **21**, 625–730.

Tartof, K. D. (1971). Increasing the multiplicity of ribosomal RNA genes in *Drosophila melanogaster. Science* **171**, 294–297.

Tartof, K. D. (1973). Unequal mitotic sister chromatid exchange and disproportionate replication as mechanisms regulating ribosomal RNA gene redundancy. *Cold Spring Harbor Symp. Quant. Biol.* **38**, 491–500.

Thompson, B., Escarmis, C., Parker, B., Slater, W., Doniger, J., Tessman, I., and Warner, R. (1975). Figure-8 configuration of dimers of S13 and $\phi$X174 replicative form DNA. *J. Mol. Biol.* **91**, 409–419.

Thompson, B. J., Carmien, M. N., and Warner, R. C. (1976). Kinetics of branch migration in double-stranded DNA. *Proc. Natl. Acad. Sci. U.S.A.* **73**, 2299–2303.

Valenzuela, M. S., and Inman, R. B. (1975). Visualization of a novel junction in bacteriophage λ DNA. *Proc. Natl. Acad. Sci. U.S.A.* **72**, 3024–3028.

Wagner, R. E., Jr., and Radman, M. (1975). A mechanism for initiation of genetic recombination. *Proc. Natl. Acad. Sci. U.S.A.* **72**, 3619–3622.

Whitehouse, H. L. K. (1963). A theory of crossing-over by means of hybrid deoxyribonucleic acid *Nature (London)* **199**, 1034–1040.

Whitehouse, H. L. K. (1974). Advances in recombination research. *Genetics* **78**, 237–245.

Wildenberg, J. (1970). The relation of mitotic recombination to DNA replication in yeast pedigrees. *Genetics* **66**, 291–304.

Winkler, H. (1930). "Die Konversion der Gene." Fischer, Jena.

Zimmering, S., Sandler, L., and Nicoletti, B. (1970). Mechanisms of meiotic drive. *Annu. Rev. Genet.* **4**, 409–436.

Zimmermann, F. K. (1968). Enzyme studies on the products of mitotic gene conversion in *Saccharomyces cerevisiae. Mol. Gen. Genet.* **101**, 171–184.

Zimmermann, F. K., and Schwaier, R. (1967). Induction of mitotic gene conversion with nitrous acid, 1-methyl-3-nitro-1-nitrosoguanidine and other alkylating agents in *Saccharomyces cerevisiae. Mol. Gen. Genet.* **100**, 63–76.

# 3

# The Onset of Meiosis

*Peter B. Moens*

## I. INTRODUCTION

This article describes some aspects of the early stages of the meiotic pathway that may be of interest to cell biologists. It enters in some detail, complementing the overview of meiosis by E. Garber in Chapter 6. The emphasis is on the wide variety of chromosome behavior during prophase of meiosis to emphasize that meiosis is a flexible process adapted to the reproductive pattern of the species and to the special interactions of each species with its environment. In my own work on meiosis, I have often found organisms with a meiotic pathway that proved to have differences from the expected sequence of events. The conclusion is that the meiotic process has a large number of variations on a basic theme and, on some occasions, the theme itself is lost!

## II. PREMEIOTIC CONDITIONS THAT AFFECT MEIOSIS

Although somatic cells and meiotic cells have obvious developmental differences, they do have metabolic processes in common and it can be expected that some conditions affecting the mitotic cell cycle can therefore also exert an influence on the meiotic process. The coincident disturbance of mitosis and meiosis is well illustrated in a series of mutant yeast strains which have defects in the mitotic cell cycle (Hartwell et al., 1974). Those strains which have a defect in mitotic DNA synthesis and in nuclear division (at 34°C but not at 25°C) are also defective in meiotic functions (Simchen, 1973).

In wheat, chromosome distribution in somatic and in meiotic cells is disturbed by the lack or excess of chromosome 5B (Feldman, 1966; Feldman et al., 1966). These coincident effects need not have a casual relationship such that homologous somatic associations are a necessary condition for homologous chromosome synapsis at meiosis (Palmer, 1971; Darvey and Driscoll, 1972; Darvey et al., 1973; Dvorak and Knott, 1973). Rather, a common factor, such as a nuclear envelope function, may be defective in both somatic and meiotic cells. A real carryover effect into meiosis of treatments prior to meiosis was demonstrated in wheat, where colchicine treatment of the last mitosis before meiosis reduces chiasma frequency at meiosis from 0.96 to 0.44 per pair of chromosome arms. No such reduction occurs in an isochromosome (2 homologous chromosomes attached to a single centromere). Presumably colchicine affects a precondition for meiosis that is necessary for proper synapsis. The isochromosome is not affected, possibly because the homologous arms are permanently associated with each other at the centromere (Driscoll and Darvey, 1970). Similarly McNelly-Ingle et al. (1966) demonstrated that growth and fertilization of the female parent (Neurospora, protoperthecial parent) at one temperature, and sexual maturation at another temperature produced recombination values intermediate between those obtained from cultures processed entirely at one temperature or the other.

In the latter two cases a meiotic effect was scored by a treatment just prior to meiosis. It is improper to assume from those experiments that chromosome pairing and recombination take place just prior to the onset of meiosis. The weight of the evidence is that pairing and recombination are meiotic functions, and it must be assumed that some necessary precondition for meiosis had been disturbed by the treatments just prior to meiosis. Similar problems in interpretations are encountered where treatments are given early in the meiotic pathway and the effects are scored at a later stage. For example, there is a longstanding debate over whether genetic exchange occurs during S phase or later during the pachytene stage of meiosis. It has been shown that temperature treatments during meiotic S phase do affect recombination rates (Grell and Day, 1974; Lu, 1974). Recombination and chiasma formation, however, can also be affected by later treatments (Lu, 1974; Peacock, 1970; Stern and Hotta, 1974), indicating that the process of genetic exchange is certainly not completed until pachytene.

## III. MEIOTIC DNA SYNTHESIS

The major DNA synthetic period of the meiotic developmental pathway has the unfortunate title of premeiotic S phase, an anachronism from the time that meiosis referred only to the segregation of chromosomes at the first and second division of the meiotic pathway. It is the present practice to include $G_1$, S, and $G_2$ phase as a part of the meiotic pathway. Consequently, the major DNA synthetic period preceding meiotic chromosome segregation can be called meiotic S phase (Rossen and Westergaard, 1966; Riley and Bennett, 1971).

Following the major S phase, additional periods of DNA synthesis occur during meiosis. In the oocytes of some animals, the DNA that codes for ribosomal RNA is amplified to produce many copies of these genes. In some amphibians and insects this amplification occurs extrachromosomally (Brown and Dawid, 1968; Cave and Allen, 1969; Miller and Beatty, 1969; Tocchini–Valentini et al., 1973). In a number of plants and animals small amounts of DNA are synthesized during zygotene and pachytene of meiotic prophase (Stern and Hotta, 1974; Flavell and Walker, 1973). An additional major DNA replication occurs during prophase of some parthenogenetic plants and animals (Koch et al., 1972). Tritiated thymidine incorporation has been reported in meiotic metaphase chromosomes of wheat and rye (Riley and Bennett, 1971; Flavell and Walker, 1973).

DNA replication of the meiotic pathway is itself quantitatively different from somatic cell DNA replication. In the lily, DNA replication during meiotic S phase is incomplete. Not until zygotene is the delayed portion replicated (the socalled Z-DNA) (Stern and Hotta, 1974). Unfortunately, few organisms have as well-synchronized zygotene cells as lily, and the chances of identifying Z-DNA in many other species are slim. Similarly, the repair DNA-synthesis that takes place during pachytene in pollen mother cells of lily (P-DNA) (Stern and Hotta, 1974) may be difficult to identify elsewhere.

In general, meiotic S phase takes longer than the S phase of somatic cells. The lengthy S phase is not the result of a slower rate of replication but is due to the reduction in the number of sites for the initiation of replication (Callan, 1972). The duration of the S phase in *Triturus* embryonic cells is 1–2 hours, while that of somatic cells is 2 days and that of meiotic cells is 9–10 days (Callan, 1972). It follows that the reduction in the number of initiation sites is not specific to meiosis but only most pronounced during meiotic DNA synthesis. The reduction in sites of initiation of DNA replication also occurs in the toad *Xenopus* (Callan, 1972). In *Drosophila* embryonic cells, Blumenthal *et al.* (1973) have shown a 4-minute S phase with an observed spacing of 10,000 to 20,000 bases between initiation sites and have calculated a distance of 3,100,000 bases between initiation sites in cultured somatic cells in which the S phase lasts 600 minutes. In male rats and mice, the S phase of successive spermatogonial generations becomes progressively longer as the spermatogonia approach meiosis (Hilscher, 1967; Monesi, 1962; Huckins, 1971).

It is not clear to what extent the characteristic mode of meiotic DNA replication is a necessary prerequisite for the meiosis-specific chromosomal events—pairing, genetic exchange, and segregation, that follow. It may be that the delayed replication of Z-DNA functions in

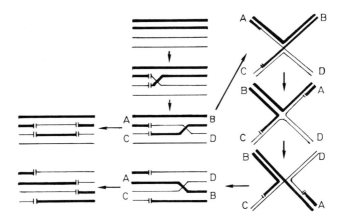

**Fig. 1.** Genetic exchange model (Holliday, 1964; Sigal and Alberts, 1972). Only two of four chromatids are diagrammed. The DNA strands of the one chromatid are drawn wide and the strands of the nonsister chromatid are drawn narrow. The nicks are indicated by gaps in the lines. At the sites of the initial nicks an exchange of strands is established. The exchange point can travel along the two DNA strands so that regions of hybrid DNA are formed. The crossed strands can be cut so that the chromatids become separated without signs of recombination between flanking markers, but gene conversion may occur in the hybrid region. Alternatively the strands can be flipped over so that the nonexchange strands become the crossed strands. A cut now results in a classic crossover event with recombination of flanking markers. Conversion can still occur through excision and repair of mismatches in the hybrid region (see Fig. 3).

synapsis and in the formation of hybrid DNA, and consequently in crossing-over and chiasma formation. The repair synthesis at pachytene, P-DNA, is probably required to seal the several gaps at the sites of strand exchange and at the sites of mismatches (Stern and Hotta, 1974) (Figs. 1, 3) and possibly to replace one strand in the single chromatid exchange (Stadler and Towe, 1971; Radding, 1973). Partial inhibitions of prophase DNA synthesis leads to chromosome fragmentation in lily (Ito *et al.*, 1967), to changes in recombination frequency in *Chlamydomonas reinhardi* (Chiu and Hasting, 1973), and to alterations in chromosome pairing and synaptonemal complex formation in lily (Roth and Ito, 1967). An unusual case exists in the yeast meiotic mutant *spo* 11, which has a normal S phase at 25°C but very little nuclear DNA synthesis at 34°C. Unlike most mutant strains that have a block in DNA synthesis and that fail to continue through meiosis, *spo* 11 at 34°C proceeds apparently normally through meiosis with its unreplicated chromosomes. The four products of meiosis are by necessity deficient in chromosomes and they degenerate eventually (Moens *et al.*, 1976). It follows that the kinetic functions of meiosis are not directly dependent on normal DNA replication.

## IV. MEIOTIC DNA REPAIR

The models of recombination at meiotic prophase imply that there are several single-stranded gaps along the length of the chromosome (Figs. 1, 3). The presence of such gaps can be verified by the appearance of relatively short segments of single-stranded DNA at the time of genetic exchange and by the presence of DNA repair processes at that time. Both of these have been documented in the lily by Stern and co-workers. They report that denatured DNA sediments at the 104 S region of a glycerol gradient during meiotic prophase, but that during pachytene a class of shorter strands appears at 62 S. The shorter segments were not observed at pachytene of an achiasmatic lily or in cells arrested at prepachytene with colchicine. Artificially introduced nicks caused the formation of segments of all different sizes. The data are interpreted to mean that endogenous nicking occurs during pachytene and that it is not randomly distributed (Hotta and Stern, 1974). The presence of repair enzymes and DNA replication have also been reported in lily pollen mother cells at pachytene (Stern and Hotta, 1974). Jacobson *et al.* (1975) estimate 70 to 200 single-strand scissions per meiotic cells in yeast.

The repair activities are similar to those carried out following radiation damage of somatic cells. In some cases impairment of repair mechanisms used by somatic cells also affects meiotic conversion and recombination, but there is no necessary correspondence between the two, indicating that some steps of the repair process are common to meiotic and mitotic cells while others are not (Holliday *et al.*, 1974). A striking case of coincident impairment has been reported in a human male who was deficient in chiasmata between the paired homologs of the spermatocyte nuclei and who was also found to have a reduced facility for chromosomal repair following radiation of lymphocyte cultures (Pearson *et al.*, 1970). No reduction of chiasmata was observed in a human male suffering from xeroderma pigmentosum, a defect in DNA repair mechanisms (Hulten *et al.*, 1974).

## V. Meiotic Prophase

Reproductive cells of different species or different sexes prepare for the meiotic divisions in a wide variety of ways and it is nearly impossible to generalize about the meiotic pathway. An exception can usually be found to every characteristic thought to be fundamental and necessary for meiosis. For example, pairing and segregation of homologous chromosomes seem basic elements of meiosis, yet male hymenoptera, such as ants, bees, wasps, and sawflies, have only one set of chromosomes, and they have evolved a meiosis without pairing and with a unipolar meiosis I spindle.

Similarly, apomictic reproduction in plants and animals is in some cases accomplished through a lack of chromosome reduction at meiosis I (Swanson, 1957, "various types of asexual reproduction that substitute for, and in many instances replace sexual reproduction"). Chromosome synapsis and genetic exchange have been abolished in males of some dipteran species (flies) and in the females of some lepidopteran species (butterflies, moths). The time course of meiotic prophase varies and can be greatly protracted, for example, in plants where the germ cells overwinter, in insects where egg development is conditional on nutritional status, and in mammalian oocytes.

It is no surprise, then, that chromosome behavior during meiotic prophase also varies among different species, and a single description fits only a few. For example, where in most organisms cell fusion (fertilization) occurs one or more cell generations prior to meiosis, in ascomycetous fungi, such as *Neotiella* (Rossen and Westergaard, 1966) and *Neurospora* (Singleton, 1953), meiotic S phase and early prophase occur while the two cells are still separate and fusion takes place in midprophase, followed immediately by chromosome pairing. In the mushroom *Coprinus* (and probably most other basidiomycetes) the two haploid nuclei are in a single cell at the onset of meiosis and they do not fuse until prophase (Lu, 1967).

The appearance of long unpaired chromosomes at leptotene followed by the synapsis of the homologs at zygotene, the classic image of chromosome behavior at meiotic prophase (Swanson, 1957; McLeish and Snoad, 1966), is in reality often difficult to observe microscopically. In many plants the prophase chromosomes are clumped in one region of the nucleus during early prophase, and this synizesis stage renders impractical the analysis of chromosome behavior with the light microscope. The Easter lily, often thought to have a "classic" prophase, may have a stage of complete chromosome condensation between interphase and leptotene. This stage has been reported in several of the common lily varieties and possibly in other plants (Walters, 1970, 1972) and in animals (Church, 1972). Stages such as leptotene and zygotene are not apparent in dipteran insects, and it seems that in this order of insects, which has homologous pairing in its somatic cells, chromosome associations are initiated between the condensed chromosomes during the telophase of the last mitosis prior to meiosis. In the face of this variability of chromosome behavior it is difficult to isolate a single basic mechanism which produces the fundamental meiotic functions.

## VI. THE SYNAPTONEMAL COMPLEX

The discovery of pairing structures within the meiotic bivalents by M. J. Moses (1956, 1958) introduced a unifying characteristic of meiotic prophase. The *synaptonemal complex* has been found, with few exceptions, among all sexually reproducing organisms examined with the electron microscope, and in 1972 the list included 61 genera from plants, animals, fungi, and protists (Westergaard and von Wettstein, 1972). Its fairly universal occurrence suggests that it plays a role in some fundamental aspects of meiosis, i.e., synapsis of homologous chromosomes and the genetic exchanges that occur while the complexes are present.

Evidence for the functions of the synaptonemal complex in chromosome synapsis is only circumstantial. In some organisms a dense axial core forms in each of the single chromosomes, and two of these cores can be seen to approach each other to a distance of 100 nm, where they become the parallel lateral elements of the synaptonemal complex (Fig. 2) (Moens, 1968, 1969; Rasmussen, 1976). *In vitro* exposure to cycloheximide by lily pollen mother cells in the process of chromosome pairing prevents further formation of synaptonemal complexes and of chro-

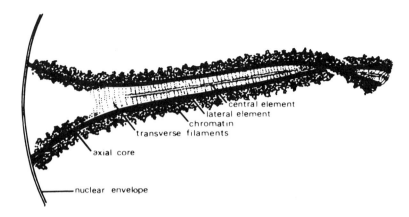

**Fig. 2.** A diagram of the synaptonemal complex and its terminology.

mosome pairing. After removal of cycloheximide, normal development continues (Roth and Parchman, 1971). As before, the complexes and pairing show coincident behavior. On the other hand in ♀ *Drosophila*, single cores undergoing pairing have not been reported, even though complexes have been extensively examined (Meyer, 1965; Carpenter, 1975; Rasmussen, 1973). Indications are that the complex may form after the homologs are paired. In some organisms other than Diptera, pairing of single cores is apparent near the nuclear envelope, where cores form first but in the interior of the nucleus synaptonemal complexes appear directly between portions of the paired chromosomes (Moens, 1973a). It is assumed that the fibrillar material of the central portion of the synaptonemal complex functions in the juxtapositioning of homologous DNA segments that are associated with the cores or lateral elements. DNA not associated with the complex cannot undergo genetic exchange at meiotic levels. In this way precise synapsis and the potential for genetic exchange can be related to the synaptonemal complex (Moens, 1973b).

Particularly in *Drosophila melanogaster*, the correlation between recombination and structure of synaptonemal complexes has been well-documented. The males, which do not have meiotic recombination, lack complexes (Meyer, 1965; Rasmussen, 1973), while the normal females have complexes; but homozygous c(3)G females, which have suppressed recombination, do not have complexes (Meyer, 1965; Smith and King, 1968; Rasmussen, 1975). Specialized areas of the complex have been designated as "recombination nodules" by Carpenter (1975) because their frequency and distribution correspond to the recombination pattern in the female.

## VII. CROSSOVER INTERFERENCE

Genetic exchange has two operationally different aspects (see Figs. 1, 3): (1) gene conversion, which is defined by nonreciprocal exchange, negative interference, and polarization, and which can be accounted for by the formation of hybrid DNA in one or two nonsister chromatids (Holliday, 1964; Sigal and Alberts, 1972; Radding, 1973), and (2) genetic recombina-

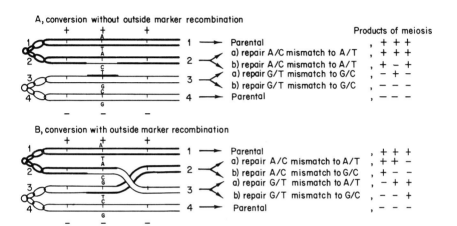

**Fig. 3.** Gene conversion and recombination (continuation of Fig. 1). Assume that a cross between two organisms of types $\frac{+++}{+++} \times \frac{---}{---}$ (3 closely linked genetic markers) resulted in an $F_1$ hybrid of type $\frac{+++}{---}$. At meiosis of the $F_1$ the homologous chromosomes are paired as shown in (A). Each chromosome consists of two chromatids and each chromatid consists of a helix of double stranded DNA. The processes described in Fig. 1 are shown to have caused a region of hybrid DNA in the center marker so that there is a mismatch of one base pair between the normal (+) and mutant (−) allele. Repair of the mismatches produces in the case of 2b and 3a an apparent double recombinant +−+ or −+−. Real double recombinants are virtually excluded from short genetic distances and their frequent appearance when monitoring intra-allelic exchange is referred to as negative interference. The process of conversion appears to be nonreciprocal when products 1, 2b, 3b and 4 are present in one tetrad (four products of meiosis found together, for example in a yeast ascus). Polarity in the process of conversion is observed when there are two mismatches within a given hybrid region. The site of initial breakage is to one side of the gene and branch migration proceeds from there. In (B) there is a classic crossover so that the outside markers show recombination. The frequency and distribution of such crossovers is regulated (positive interference, chiasma localization) and requires that segments of nonsister chromatids become continuous while these segments remain associated with their sister chromatids.

tion, which is characterized by an exchange between homologous, nonsister chromatids, reciprocity, and positive interference. (These characteristics are discussed more fully in Chapter 2 by Esposito and Esposito).

If the exchange strands of the hybrid DNA segment are cut, the result is a conversion event without reciprocal recombination of outside markers (Figs. 1, 3). If, at the site of a hybrid DNA segment the nonexchange strands are cut, then conversion events are accompanied by outside marker recombination. This constitutes a classic reciprocal crossover event for the outside markers. Hybrid DNA segments may occur frequently and independently along the length of the chromosome, but only in some of these do the nonsister chromatids become permanently connected to produce a reciprocal crossover. The frequency and distribution of such reciprocal events is regulated. E.g., one crossover may reduce the probability of another one from taking place nearby, that is, within a distance of 10 to 20 map units, which corresponds to several genes or $10^4$ or more nucleotides. In other cases the crossovers are localized in a given segment of a chromosome and the remainder is free from crossovers.

The regulation of the frequency and distribution of reciprocal crossovers may prove to be one of the functions of the synaptonemal complex. The systematic nonoccurrence of reciprocal exchange in competent chromosomes is not predictable from molecular models of conversion or exchange. Modifications of crossover distribution include positive interference, species–specific limitations on the numbers of crossovers per cell or per bivalent, the localization of crossovers to specific chromosome segments, and the abolishment of crossovers in meiotic cells of one or the other sex of members of one species. Unlike the strand exchange itself, these adaptations are often of recent evolutionary origin, and differences occur among the species of a single genus or family. A correlation between specialized crossover distribution and modified synaptonemal complexes has been reported for the boat lily *Rhoeo spathecea*, the grasshoppers *Choealtis conspersa, Chorthippus longicornis* and *Stethophyma grossum*, and in the fungi *Podospora anserina* and *Ascobolus immersus* (Moens, 1974). Similarly, the complexes of achiasmatic male scorpion–flies *Panorpa communis* differ at late stages from the synaptonemal complexes observed in the chiasmatic female (Welsch, 1973).

As a working hypothesis it is assumed that to establish a reciprocal crossover within the context of a synaptonemal complex, the lateral elements must break and rejoin and thereby come to lie along the crossover chromatids. The physical dimensions and properties of the lateral elements then regulate the frequency and distribution of reciprocal crossovers. For example, regions with very broad lateral elements or regions that do not form lateral elements can thus be exempted from crossing-over (Jones, 1973).

## VIII. CHROMOSOME REDUCTION

The essential function of meiosis is to reduce the chromosome number from $2n$ to $n$. It is normally accomplished through synapsis of homologous chromosomes and by the unusual behavior of the centromeres at the first meiotic division (kinetochores), which do not divide on this occasion, so that only one member of each set of paired homologues reaches one spindle pole. In general, the orientation of a given pair is independent of the other pairs, and, as a result, random assortment of chromosomes with respect to parental origin takes place. If $n$ is the number of chromosomes, then the number of possible gametes with different chromosome combinations is $2^n$. In addition to the genotypic variability introduced by chromosome segregation at meiosis, genetic exchange within a set of paired bivalents produces further genetic variability. Meiosis has probably evolved because of its value in producing genetic variability among the offspring of two parents (Williams and Mitton, 1973). At the same time pressure has existed to reduce variability where well-adapted genotypes had been produced and meiosis has been modified to reduce variability.

Genotypic variability among the offspring can be suppressed in several ways. Strikingly efficient is homothallism among fungi whereby a haploid product of meiosis of mating type $\alpha$ undergoes a cell division, and the new cell "mutates" from $\alpha$ to $a$, and the two cells immediately fuse (Oshima and Takano, 1970). The diploid product is homozygous at all genetic loci except the mating locus $a/\alpha$. Nearly as efficient are those cases where sexually reproducing organisms become parthenogenic, such that females produce only females. If parthenogenesis is accomplished through fusion of any two of the four products of one meiocyte (Swanson, 1957), it will take a number of generations before homozygosis is established. If one of the meiotic products doubles the chromosome number without a nuclear division, a homozygous diploid is established immediately (Templeton *et al.*, 1975).

In the case of male Hymenoptera, which are haploid, all the chromosomes go to one pole at meiosis I, and after meiosis II two sperm cells are produced (Hoage and Kessel, 1968). Because neither segregation of chromosomes nor crossing-over occurs, the spermatozoa of a given male are all genetically identical. The result is that the female offspring of a given cross are genetically unusually homogeneous. Among the Coccoidea, several forms of unusual meiotic segregation occur. In the mealybugs and the armored scales the male transmits only the chromosomes he received from his female parent. The paternal chromosomes are eliminated in a variety of ways (Brown, 1963).

Just prior to meiosis in triploid all-female fish, the chromosome number of the oogonium is raised to hexaploidy through chromosome duplication without nuclear division. Following the two meiotic divisions each egg is again a triploid. Sperm from a sympatric bisexual gonochoristic (bisexual) species stimulates the egg to develop but no fertilization occurs (Cimino, 1972). The same phenomenon has also been observed in triploid females of the salamander *Ambystoma jeffersonianum* (MacGregor and Uzzel, 1964). Since the endoduplicated chromosomes remain paired during meiotic prophase homologous chromosomes can not pair and segregation due to genetic exchange is suppressed.

Genetic variability may be curtailed in the boat lily *Rhoeo spathacea*, where each terminal chromosome segment is translocated to the next chromosome until all are linked together. The result is that at meiosis I chromosome set A goes to one pole and set B goes to the other (Wimber, 1968; Mertens, 1973). Set A can only be passed on through the egg and set B through the pollen. Thus all plants are always AB. A similar system has evolved among races of the evening primrose *Oenothera* (Cleland, 1973) and in *gibasis pulchella*.

These are just a few of the numerous variations on the theme of meiosis. It seems almost as though there are as many variations in the meiotic pathway as there are species. It would be even more apparent if not only the genetic-morphological characteristics were considered, but also the physiological aspects such as hormone regulation. Each habitat and each breeding regime seemingly has its own demands.

### ACKNOWLEDGMENTS

I thank Dr. Kathleen Church and my wife Marja Moens for their assistance in preparing the manuscript.

### REFERENCES

Blumenthal, A. B., Kreigstein, H. J., and Hogness, D. S. (1973). The units of DNA replication in *Drosophila melanogaster* chromosomes. *Cold Spring Harbor Symp. Quant. Biol.* **38,** 205–223.

Brown, D. D., and Dawid, I. B. (1968). Specific gene amplification in oocytes. *Science* **160,** 272–280.

Brown, S. W. (1963). The Comstockiella system of chromosome behavior in the armored scale insects (Coccoïdea: Diaspididae). *Chromosoma* **14**, 360–406.

Callan, H. B. (1972). Replication of DNA in the chromosomes of eukaryotes. *Proc. R. Soc. London, Ser. B* **181**, 19–41.

Carpenter, A. T. C. (1975). Electron microscopy of meiosis in *Drosophila melanogaster* females. II. The recombination module—a recombination-associated structure at pachytene? *Proc. Natl. Acad. Sci. U.S.A.* **72**, 3186–3189.

Cave, M. D., and Allen, E. R. (1969). Synthesis of nucleic acids associated with a DNA-containing body in oocytes of *Acheta. Exp. Cell Res.* **58**, 201–212.

Chiu, S. M., and Hasting, P. J. (1973). Premeiotic DNA synthesis and recombination in *Chlamydomonas reinhardi. Genetics* **73**, 29–43.

Church, K. (1972). Meiosis in the grasshopper. II. The preleptotene spiral stages during oogenesis and spermatogenesis in *Melanoplus femur-rubrum. Can. J. Genet. Cytol.* **14**, 397–401.

Cimino, M. C. (1972). Meiosis in triploid all-female fish (*Poeciliopsis*, Poeciliidae). *Science* **175**, 1484–1486.

Cleland, R. E. (1973). "Oenothera: Cytogenetics and Evolution." Academic Press, New York.

Darvey, N. L., and Driscoll, C. J. (1972). Evidence against somatic association in hexaploid wheat. *Chromosoma* **36**, 140–149.

Darvey, N. L., Driscoll, C. J., and Kaltsikes, P. J. (1973). Evidence against somatic and premeiotic association in hexaploid wheat. *Genetics* **74**, s57.

Driscoll, C. J., and Darvey, N. L. (1970). Chromosome pairing: Effect of colchicine on an isochromosome. *Science* **169**, 290–291.

Dvorak, J., and Knott, D. R. (1973). A study of somatic associations of wheat chromosomes. *Can. J. Genet. Cytol.* **15**, 411–416.

Feldman, M. (1966). The effect of chromosomes 5B, 5D and 5A on chromosome pairing in *Triticum aestivum. Proc. Natl. Acad. Sci. U.S.A.* **55**, 1447–1453.

Feldman, M., Mello-Sampayo, T., and Sears, F. R. (1966). Somatic association in *Triticum aestivum. Proc. Natl. Acad. Sci. U.S.A.* **56**, 1192–1199.

Flavell, R. B., and Walker, G. W. R. (1973). The occurrence and role of DNA synthesis during meiosis in wheat and rye. *Exp. Cell Res.* **77**, 15–24.

Grell, R. F., and Day, J. W. (1974). Intergenic recombination, DNA replication and synaptonemal complex formation in the *Drosophila* oocyte. *In* "Mechanisms of Recombination" (R. F. Grell, ed.), pp. 327–349. Plenum, New York.

Hartwell, L. H., Culotti, J., Pringle, J. R., and Reid, B. J. (1974). Genetic control of the cell division cycle in yeast. *Science* **183**, 46–51.

Hilscher, W. (1967). DNA synthesis: proliferation and regeneration of the spermatogonia in the rat. *Arch. Anat. Microsc. Morphol. Exp.* **56**, Suppl., 75–84.

Hoage, T. R., and Kessel, R. G. (1968). An electron microscope study of the process of differentiation during spermatogenesis in the honey bee (*Apis mellifera* L.) with special reference to centriole replication and elimination. *J. Ultrastruct. Res.* **24**, 6–32.

Holliday, R. (1964). A mechanism for gene conversion in fungi. *Genet. Res.* **5**, 282–304.

Holliday, R., Holloman, W. K., Banks, G. R., Unrau, P., and Pugh, J. E. (1974). Genetic and biochemical studies of recombination in *Ustilago maydis. In* "Mechanisms in Recombination" (R. F. Grell, ed.), pp. 239–262. Plenum, New York.

Hotta, Y., and Stern, H. (1974). DNA scission and repair during pachytene in *Lilium. Chromosoma* **46**, 279–296.

Huckins, C. (1971). The spermatogonial stem cell population in adult rats. II. A radioautographic analysis of their cell cycle properties. *Cell Tissue Kinet.* **4**, 313–318.

Hulten, M., de Weerd-Kastelein, E. A., Bootsma, D., Solari, A. J., Skakkebaek, N. E., and Swanbeck, G. (1974). Normal chiasma formation in a male with xeroderma pigmentosum. *Hereditas* **78**, 117–124.

Ito, M., Hotta, Y., and Stern, H. (1967). Studies of meiosis *in vitro*. II. Effect of inhibiting DNA synthesis during meiotic prophase on chromosome structure and behavior. *Dev. Biol.* **16**, 54–77.

Jacobson, G. K., Piñon, R., Esposito, R. E., and Esposito, M. S. (1975). Single strand scissions of chromosomal DNA during commitment to recombination at meiosis. *Proc. Natl. Acad. Sci. U.S.A.* **72**, 1887–1891.

Jones, G. H. (1973). Light and electron microscope studies of chromosome pairing in relation to chiasma localization in *Stethophyma grossum* (*Orthoptera : Acrididae*) *Chromosoma* **42**, 145–162.

Koch, P., Pynacker, L. P., and Kerke, J. (1972). DNA reduplication during meiotic prophase in the oocytes of *Carassius morosus* Br. (Insecta, Cheleutoptera). *Chromosoma* **36**, 313–321.

Lu, B. C. (1967). Meiosis in *Coprinus lagopus*: A comparative study with light and electron microscopy. *J. Cell Sci.* **2**, 529–536.

Lu, B. C. (1974). Genetic recombination in *Coprinus*. IV. A kinetic study of the temperature effect on recombination frequency. *Genetics* **78**, 661–677.

MacGregor, H. C., and Uzzel, T. M. (1964). Gynogenesis in salamanders related to *Ambystoma jeffersonianum*. *Science* **143**, 1043–1045.

McLeish, J., and Snoad, B. (1966). "Looking at Chromosomes." Macmillan, New York.

McNelly-Ingle, C. A., Lamb, B. C., and Frost, L. C. (1966). The effect of temperature on recombination frequency in *Neurospora crassa*. *Genet. Res.* **7**, 169–183.

Mertens, T. R. (1973). Meiotic chromosome behaviour in *Rhoeo spathecea*. *J. Hered.* **64**, 365–368.

Meyer, G. F. (1965). A possible correlation between the submicroscopic structure of meiotic chromosomes and crossing-over. *Electron Microsc., Proc. Eur. Reg. Conf., 3rd, 1964* Vol. B, pp. 461–462.

Miller, O. L., and Beatty, B. (1969). Visualization of nucleolar genes. *Science* **164**, 955–958.

Moens, P. B. (1968). The structure and function of the synaptonemal complex in *Lilium longiflorum* sporocytes. *Chromosoma* **23**, 418–451.

Moens, P. B. (1969). The fine structure of meiotic chromosome polarization and pairing in *Locusta migratoria* spermatocytes. *Chromosoma* **28**, 1–25.

Moens, P. B. (1973a). Quantitative electron microscopy of chromosome organization at meiotic prophase. *Cold Spring Harbor Symp. Quant. Biol.* **38**, 99–107.

Moens, P. B. (1973b). Mechanisms of chromosome synapsis at meiotic prophase. *Int. Rev. Cytol.* **35**, 117–133.

Moens, P. B. (1974). Coincidence of modified crossover distribution with modified synaptonemal complexes. *In* "Mechanisms in Recombination" (R. F. Grell, ed.), pp. 377–383. Plenum, New York.

Moens, P. B., Mowat, M., Esposito, M. S., and Esposito, R. E. (1977). Meiosis in a temperature sensitive DNA-synthesis mutant and in an apomictic yeast strain (*Saccharomyces cerevisiae*). *Phil. Trans. R. Soc. Lond. B.* 277, 351–358.

Monesi, V. (1962). Antoradiographic study of DNA synthesis and the cell cycle in spermatogonia and spermatocytes of mouse testis using tritiated thymidine *J. Cell Biol.* **14**, 1–18.

Moses, M. J. (1956). Chromosomal structures in crayfish spermatocytes. *J. Biophys. Biochem. Cytol.* **2**, 215–218.

Moses, M. J. (1958). The relation between axial complex of meiotic prophase chromosomes and chromosome pairing in a salamander (*Plethodon cinereus*). *J. Biophys. Biochem. Cytol.* **4**, 633–638.

Oshima, Y., and Takano, I. (1971). Mating types in *Saccharomyces*: Their convertibility and homothallism. *Genetics* **67**, 327–335.

Palmer, R. G. (1971). Cytological studies of a meiotic and normal maize with reference to premeiotic pairing. *Chromosoma* **35**, 233–246.

Peacock, W. J. (1970). Replication, recombination, and chiasmata in *Goniaea australiasia* (Orthroptera: Acrididae). *Genetics* **65**, 593–617.

Pearson, P. L., Ellis, J. D., and Evans, H. J. (1970). A gross reduction in chiasma formation during meiotic prophase and a defective DNA repair mechanism associated with a case of human male infertility. *Cytogenetics* **9**, 460–476.

Radding, C. M. (1973). Molecular mechanisms in recombination. *Annu. Rev. Genet.* **7**, 87–111.

Rasmussen, S. W. (1973). Ultrastructural studies of spermatogenesis in *Drosophila melanogaster* Meigen. *Z. Zellforsch. Mikrosk. Anat.* **140**, 125–144.

Rasmussen, S. W. (1975). Ultrastructural studies of meiosis in males and females of the c (3) G$^{17}$ mutant of *Drosophila melanogaster* Meigen. *C. R. Trav. Lab. Carlsberg* **40**, 163–173.

Rasmussen, S. W. (1976). The meiotic prophase in *Bombyx mori* females analyzed by three-dimensional reconstruction of synaptonemal complexes. *Chromosoma* **54**, 245–293.

Riley, R., and Bennett, M. D. (1971). Meiotic DNA synthesis. *Nature (London)* **230**, 182–185.

Rossen, J. M., and Westergaard, M. (1966). Studies on the mechanism of crossing-over. II. Meiosis and the time of meiotic chromosome replication in the ascomycete *Neottiella rutelans* (Fr) Dennis. *C. R. Trav. Lab. Carlsberg* **35**, 233–260.

Roth, T. F., and Ito, M. (1967). DNA-dependent formation of the synaptonemal complex at meiotic prophase. *J. Cell Biol.* **35**, 247–255.

Roth, T. F., and Parchman, L. G. (1971). Alterations of meiotic chromosomal pairing and synaptonemal complexes by cycloheximide. *Chromosoma* **35**, 9–27.

Sigal, N., and Alberts, B. (1972). Genetic recombination: The nature of crossed strand-exchange between two homologous DNA molecules. *J. Mol. Biol.* **71**, 789–793.

Simchen, G. (1973). Are mitotic functions required in meiosis? *Genetics* **76**, 745–753.

Singleton, J. R. (1953). Chromosome morphology and the chromosome cycle in the ascus of *Neurospora crassa*. *Am. J. Bot.* **40**, 124–144.

Smith, P. A., and King, R. C. (1968). Genetic control of synaptonemal complexes in *Drosophila melanogaster*. *Genetics* **60**, 335–351.

Stadler, D. R., and Towe, A. M. (1971). Evidence for meiotic recombination in *Ascobolus* involving only one member of a tetrad. *Genetics* **68**, 401–413.

Stern, H., and Hotta, Y. (1974). Biochemical controls of meiosis. *Annu. Rev. Genet.* **7**, 37–65.

Swanson, C. P. (1957). "Cytology and Cytogenetics," Prentice-Hall, Englewood Cliffs, New Jersey.

Templeton, A. R., Carson, H. L., and Sing, C. F. (1975). The population genetics of parthenogenetic strains of *Drosophila mercatorium*. II. The capacity for parthenogenesis in a natural, bisexual population. *Genetics* **82**, 527–542.

Tocchini-Valentini, G. P., Mahdavi, V., Brown, R., and Crippa, M. (1973). The synthesis of amplified ribosomal DNA. *Cold Spring Harbor Symp. Quant. Biol.* **38**, 551–558.

Walters, M. S. (1970). Evidence on the time of chromosome pairing from the preleptotene spiral stage in *Lilium longiflorum* "Croft." *Chromosoma* **29**, 375–418.

Walters, M. S. (1972). Preleptotene chromosome contraction in *Lilium longiflorum* "Croft." *Chromosoma* **39**, 311–332.

Welsch, B. (1973). Synaptonemal Complex und Chromosomenstructur in der achiasmatischen Spermatogenese von *Panorpa communis* (Mecoptera). *Chromosoma* **43**, 19–74.

Westergaard, M., and von Wettstein, D. (1972). The synaptonemal complex. *Annu. Rev. Genet.* **6**, 71–110.

Williams, G. C., and Mitton, J. B. (1973). Why reproduce sexually? *J. Theor. Biol.* **39**, 545–554.

Wimber, D. E. (1968). The nuclear cytology of bivalent and ring-forming Rhoeos and their hybrids. *Am. J. Bot.* **55**, 572–574.

# 4

# Chromosome Imprinting and the Differential Regulation of Homologous Chromosomes

*Spencer W. Brown and H. Sharat Chandra*

## I. INTRODUCTION

### A. Differential Regulation of Homologous Chromosomes

One of the principles of Mendelian genetics is that genes are equally transmitted and equally expressed regardless of parental origin. This principle has been confirmed not only by a multitude of studies in both plants and animals but also by the results of special experiments involving altered parental contributions to the zygote. For example, in *Drosophila* Bridges (1916) recovered XY sons which had received the Y chromosome from an XXY mother and the X chromosome from a normal XY father. This transmission is the reverse of the usual pattern in which the sons receive an X from the mother and a Y from the father. Yet the exceptional sons in Bridges' experiment were normal and fertile, indistinguishable from those produced in the typical way except by the use of marker genes on the X chromosome.

Exceptions to the Mendelian pattern did not become known until the pioneer studies of Metz (1938, review) and his colleagues on the genetic system of the genus *Sciara* of the Diptera, an order of insects including flies, gnats, midges, and mosquitoes. The results in *Sciara* were indeed startling. At meiosis in the male, the entire paternal set of chromosomes is eliminated; the male transmits only the set of chromosomes derived from his mother. Meiosis in the female is typical and genes from both parents are transmitted in accordance with Mendelian expectation. The *Sciara* male is thus equivalent to the male honeybee in regard to genetic transmission; both transmit only the maternal chromosome set to their offspring. The male honeybee comes from an unfertilized egg, has only maternal chromosomes and can, therefore, transmit only maternal chromosomes. The *Sciara* male not only comes from a typical zygote but the genes on chromosomes from the father are expressed during his development. At meiosis, however, the paternal and maternal chromosome sets show the differential behavior which leads to the elimination of the paternal set; only the maternal set is present in the sperm. Sex determination in *Sciara* is conventional; the females are XX, the males are XO; but these conditions are achieved by chromosome elimination in early embryogeny. Chromosome behavior at meiosis and in the embryos of *Sciara* will be described in detail in Section IV.

The behavior of the chromosomes of *Sciara* is in sharp contrast to that of *Drosophila*. The type of behavior shown by Metz and his co-workers for *Sciara* may be called *facultative* since the chromosomes of the male will be either eliminated at meiosis or transmitted via the sperm *on the basis of the prior circumstance of parental origin*. The type of behavior known for

*Drosophila* may be called *constitutive* since chromosome behavior and gene expression are normally solely dependent on genetic content and not influenced by the prior circumstance of parental origin. The terms, *facultative* and *constitutive,* were first applied to heterochromatin and heterochromatization by Brown and Nelson-Rees (1961) and Brown (1966) and the extension of their use to analogous chromosome systems not involving heterochromatization seems appropriate. Facultative systems are now known to regulate the behavior of the X chromosome of mammals, the X chromosome of *Sciara,* and entire sets of chromosomes in *Sciara* and certain coccid insects.

The genetic systems just mentioned are normal components of the life cycles of the organisms in which they occur and involve whole chromosomes or whole sets of chromosomes. In addition, there are three types of parental influence on genetic expression and/or transmission which are largely outside the scope of the present chapter and will be only briefly noted here.

In certain complex examples involving altered chromosomes and mutant loci of *Drosophila* and maize, a differential effect may be noted depending on the parental origin of the altered chromosome or gene (see Chapter 2). Such examples indicate that a wide-spread potential probably exists in the living world for the evolution of genetic systems geared to parental origin. The fact that such systems have not evolved more frequently indicates that they have selective advantage only under quite special circumstances.

In the second type of parental influence, cell organelles, such as the chloroplast, may appear to be transmitted to the offspring only from the mother. In certain of the more intensively studied cases, it is clear that both parents contribute organelles to the zygote and the paternal organelle is destroyed shortly thereafter (see Chapter 7).

For the third type of parental influence, it should be recalled that the cytoplasmic constituents of the egg are synthesized prior to the completion of meiosis and thus both sets of the mother's chromosomes are responsible for the cytoplasmic composition of the egg. Thus, it is not surprising that structural characteristics appearing very early in development should be under the control of the genes of the mother rather than those of the zygote. The classic example of such maternal control was reported early in the history of genetics in Sturtevant's (1923) analysis of the results of certain crosses made with a species of snail of the genus *Limnaea*, which is hermaphroditic. The shell of the snail may be coiled either to the right (dextral) or to the left (sinistral). This characteristic is controlled by a single pair of Mendelizing alleles with that for dextral dominant over that for sinistral. The direction of coiling, already evident in the plane of the second cleavage division, is determined by the genotype of the mother rather than that of the zygote. In the mother, the expression of the genes is orthodox; the dextral allele is dominant to the sinistral allele regardless of parental origin of the alleles. There is no differential expression of homologs in this case but there is a lag of one generation before the results of gene expression become visible. Thus, a heterozygous dextral/sinistral snail mated as egg parent to a homozygous sinistral snail as sperm parent will produce all dextral offspring; half the offspring will be homozygous sinistral and will produce only sinistral young even though dextrally coiled themselves.

There are other reports of maternal influence or determination but these will not be discussed and attention will be returned to the primary subject of this review.

## B. Chromosome Imprinting and Chromosome Regulation

As noted in the preceding section, paternal chromosomes behave differently from maternal chromosomes in males of *Sciara*. For some time it was believed that the paternal chromosomes had been preconditioned by passage through the father, and Crouse (1960a) introduced the term "imprinting" to denote such preconditioning. Imprinting may be defined as the molecular event, or set of events, which determines that a chromosome will later behave differently from a homologous chromosome in the same nucleus. The nature of the molecular change is not known but certain ideas on this subject are of considerable current interest (Section V).

Some evidence is now available from both coccids and mammals on the time and place in the life cycle at which imprinting occurs. This evidence is complex and will be described in detail in Sections II and III, but its significance will be briefly noted now. Studies of certain parthenogenetic coccids and human ovarian teratomas* indicate that imprinting probably occurs in the oocyte (Chandra and Brown, 1975). In both cases development proceeded parthenogenetically from just an oocyte. But the differential behavior of homologs which appeared later in development indicated that typical imprinting must have occurred. In both cases, two haploid chromosome sets were briefly separated before recombining to form a diploid zygote substitute. This brief spatial separation would permit the oocyte to imprint chromosomes in one of the haploid nuclei without affect-

---

* Human ovarian teratomas are differentiated but disorganized assemblages of tissues occurring in ovaries. Evidence points to their parthenogenetic origin on union of the nucleus of polar body II with the egg nucleus (Linder 1969; Linder and Power, 1970; see Section II); they could thus be considered to be grossly abnormal embryos.

ing homologous chromosomes in the other. Since it seems unlikely that a different imprinting process would have evolved in these unusual cases, it follows that the sperm is not imprinted in the father but on entry at fertilization.

After imprinting, the affected chromosomes are no longer *functionally* homologous to the unaffected chromosomes, but expression of this induced alteration may not occur until a later stage in the life cycle when the imprinted chromosomes are eliminated or heterochromatized while the chromosomes which have not been imprinted are neither lost nor changed.

Differential behavior in the same nucleus of chromosomes which are *not* homologous is commonly observed; sex chromosomes often behave quite differently from autosomes during meiosis. If chromosomes are not homologous, no prior imprinting is necessary to assure that one can be regulated differently from another. It is only necessary to assume that its unique genetic endowment enables one chromosome to respond to a regulatory stimulus to which the other chromosomes are insensitive (White, 1933).

The sex chromosomes of *Drosophila* provide the most thoroughly studied examples of the differential regulation of nonhomologous chromosomes. The Y chromosome is heterochromatic and genetically inactive or very nearly so in the soma but it is necessary for the fertility of the male. In the early development of the spermatocyte, specific regions of the Y produce a series of unusual structures which are somewhat like the loops of lampbrush chromosomes and they are likewise active in RNA synthesis (Hess, 1971, review). If two Y's (e.g., one of paternal, one of maternal origin) are present in a male, both will produce the loop structures in individual spermatocytes; parental origin apparently has no influence on this function of the Y. It is possible to note only one other interesting effect of the Y chromosome, this time in the soma. The Y alters the expression of genes on other chromosomes which have become subject to ''position effect'' as the result of chromosome rearrangements. Moreover, the presence of a Y in the mother or extra Y material in the father, though not transmitted to the offspring, can alter the position effect of a rearrangement transmitted from the *other* parent. Parental origin of the rearrangement can itself also strongly influence position effect (Baker, 1968, review). Such results imply that a genetic basis for facultative behavior or gene expression may be more widespread than can usually be recognized.

The X chromosome of *Drosophila* becomes heterochromatic in spermatocytes but at present it is not clear whether the inactivation of the X is in any way directly related to the expression of the Y in the same cells (Kiefer, 1973). A regulatory site in the basal region of the X is presumably responsible for its heterochromatization, since translocations which include an appreciable part of the X but do not include the basal site result in sterility (Lifschytz and Lindsley, 1972). These authors also suggested that the constitutive heterochromatization of the X may have provided a basis from which faculative heterochromatization at other phases of the life cycle could have evolved.

The other type of differential expression known for the X chromosome of *Drosophila* is generally referred to as dosage compensation. It has been recognized for many years that the expression of genes of the X is usually related to the ratio of the number of X's to the number of sets of autosomes. Thus the genes of the single X of the male appear to be expressed twice as effectively as the genes of each of the two X's of the female, thereby compensating for the presence of only a single X in the male. These earlier conclusions have been confirmed in more recent quantitative studies on concentrations of pigments and enzymes controlled by X-linked loci and also by $^3$H-uridine labeling. In addition, quantitative studies have included the intersexes and the metasexes which result from nonnormal combinations of X chromosomes (X) and sets of autosomes (A): male, XYAA; female, XXAA; intersex, AAXXX, AAAXXXX; metamale, XAAA; and metafemale, XXXAA. Such studies have shown a general, rough parallel between the A/X ratio and the quantitative expression or transcription of the X (Lakhotia and Mukherjee, 1969; Lucchesi, 1973, review; Lucchesi *et al.*, 1974).

Studies of dosage compensation have utilized expression of autosomal genes and transcription of autosomes as a basis for determining the relative expression of the X; there has been no *a priori* reason for assuming that the expression of the autosomes is influenced by the A/X ratio, and experimental evidence for this has apparently not been found. The system of dosage compensation, therefore, is apparently based on the differential regulation of the X chromosome. Dosage compensation is not, however, under the control of a single regulatory site with pervasive effect along the X chromosome. Segments of the X inserted into autosomes also show dosage compensation. Furthermore, at several loci along the X, some alleles show dosage compensation and others, at the same loci, do not; instead the phenotype is a direct expression of the number of alleles so that expression in males is less than in females. Maroni and Plaut (1973) suggested that dosage compensation is based on competition of X chromosome loci for an autosomal product available in limited quantities. The examples of autonomy of translocated segments and of lack of dosage compensation of certain alleles indicate that receptors for the autosomal product suggested by Maroni and Plaut should occur all along the X, perhaps at each locus.

In summary, two types of chromosomal regulation have been considered, differential regulation of homologous and of nonhomologous chromosomes. The latter is far more common and chromosomes will vary in behavior or expression as a direct response to developmental processes. Since chromosomes which are not homologous are different genetically, a different response can be based on this genetic difference. The best-studied examples of different responses by genetically different chromosomes are those of the sex chromosomes, the X and Y of *Drosophila*, which, in the male, behave differently from each other and both behave differently from the autosomes. The three types of differential behavior are (1) dosage compensation of the X, (2) heterochromatization and presumed genetic inactivation of the X in the spermatocyte, and (3) loop formation by the Y in an early stage of the spermatocyte. Since fragmentation of the Y has shown that separated regions will form their specific, characteristic loops (Hess, 1971, review), the regulation of the Y is most likely on a bit-by-bit basis as is dosage compensation of the X. Only the heterochromatization of the X appears to be the result of *en bloc* regulation controlled by a region near the base of the X.

When two homologous chromosomes behave differently, there is no genetic difference enabling one to respond to a regulatory signal to which the other is insensitive. One of the two homologs must somehow become altered so that it will continue to respond to such signals differently from the other. It is this primary alteration which is indicated by the term imprinting. Once altered, the chromosome may be directed by signals at a later stage in development to behave differently from its homolog.

Three systems of facultative chromosome behavior of whole chromosomes or sets of chromosomes are now known, in *Sciara*, mammals, and certain coccids. The time and place in the life cycle at which imprinting occurs are best shown by certain parthenogenetic coccids (see Section III). In these examples, a haploid egg nucleus divides once by mitosis; the two daughter nuclei then fuse to form a diploid zygote substitute (see Fig. 6). Later one entire haploid set becomes heterochromatic in those parthenogenetic embryos destined to become males just as it does in male embryos derived from an egg and sperm. In the parthenogenetic cases, therefore, one of the two haploid sets derived from the egg nucleus must have been imprinted during this brief period of separation. Imprinting may therefore be regarded as a type of developmental signal directed toward only one of the two haploid nuclei during this brief period of separation, presumably because some differentiation of the cytoplasm of the egg permits the chromosomes of one haploid nucleus to be altered while those of the other remain unaffected. Viewed in this regard, imprinting does not seem essentially different from other types of chromosome regulation. It is, however, the extra step that is necessary when two homologs or homologous scts are to be regulated differentially. The most frequent way differential regulation of homologs has been accomplished seems to be by imprinting of one but not the other before the two are combined in the zygote nucleus.

The differential regulation of nonhomologous chromosomes may be either pervasive, as in the heterochromatization of the X chromosome in spermatocytes of *Drosophila*, or bit-by-bit, as in dosage compensation of the X of Drosophila. Similar differences occur in the differential regulation of homologous chromosomes. Regulation is apparently pervasive in the control of the X chromosome of *Sciara* (Section IV) but bit-by-bit in the heterochromatization of paternal chromosomes in males of specialized coccids (Section III). Control of heterochromatization of the mammalian X is, however, probably of a more complex, hierarchical nature.

A situation which appears to be intermediate between facultative and constitutive chromosome behavior has been observed in the behavior of the X chromosome in rare tetraploid sectors of the male germ line of a few species of Orthoptera (locusts and grasshoppers). Because of their implications in regard to mechanisms of chromosome regulation, the data will be summarized in Section V.

### C. The Three Systems of Facultative Chromosome Behavior

The three major systems requiring imprinting occur in widely divergent groups of organisms and must have been independently derived. All three systems appear to be of considerable antiquity, over 100 million years old.

There are three major groups of mammals, the egg-laying Prototheria (platypus, echidna) which formed an independent lineage about 200 million years ago, the nonplacental Metatheria (kangaroos, wombats, opossums), and the placental Eutheria (mice, whales, monkeys). The last two groups separated in the mid-Cretaceous, 120–135 million years ago. In both the marsupials and the Eutheria, the X chromosome behaves facultatively in the female; this system probably evolved before the two groups diverged. In marsupials the paternal X becomes heterochromatic in the female; in Eutheria, the X which becomes heterochromatic is either the paternal or maternal X chosen at random in cell lineages tracing back to early embryogeny. The relationship between these two systems will be considered in Section II.

*Sciara* is a genus of the family, Sciaridae, of the dipteran order of insects, and a closely related family, the Cecidomyidae (gall midges), probably has a comparable facultative system. The dipteran insects probably originated in the Permian, about 300 million years ago, and the stock ancestral to the Sciaridae and Cecidomyidae may have formed an independent lineage within the next 100 million years. Certain aspects of the *Sciara* system have been noted in Section I,A, and further details are given in Section IV.

The coccid insects are all sedentary plant parasites. The females reach sexual maturity while retaining larval morphology and are always wingless. The males undergo a series of metamorphoses to emerge as winged (usually) adults; they do not feed as adults but mate several times and die in one to a few days. Increased specialization for the sedentary, parasitic existence has often resulted in reduction of body parts and loss of motility. Facultative chromosome systems are known with certainty only in the more specialized taxonomic groups which include the great majority of species and genera. A fossil coccid more specialized morphologically than some extant forms with facultative heterochromatization was reported by Beardsley (1969) from Canadian amber of the upper Cretaceous. From an analysis of host–plant antiquity, Hoy (1962) concluded that the common ericoccid parasites of *Nothogagus*, the Antarctic beech, also dated from the upper Cretaceous. The ericoccids are well known today for the Comstockiella type of heterochromatization (see Section III). The facultative systems of the coccids are probably about 100 million years old.

In the more primitive coccids, sex is usually determined according to the well-known XX (female)–XO (male) mechanism which is found also in the aphids. In one primitive group, however, the males are haploid and, like male honeybees, come from unfertilized eggs. Unlike certain of the honeybees, these coccid species with male haploidy have evolved an additional embellishment of functional hermaphroditism (see Section IV). Certain of the primitive coccids with the XX–XO mechanism display unusual modifications of chromosome comportment during meiosis in the male; these remarkable aspects of coccid cytology are outside the scope of the present chapter and the reader is referred to the beautifully illustrated accounts of Schrader and Hughes-Schrader (see Hughes-Schrader, 1948, review).

In most of the specialized coccids, the chromosomes from the father become heterochromatized during early development of the male embryos and remain heterochromatic in most but not all tissues. The paternal chromosomes are not transmitted to the offspring but are eliminated at spermatogenesis. There are two quite different processes of meiotic elimination; in one subsystem, only one process occurs; in the other subsystem, both processes usually occur. Finally, in the most specialized of all coccids, the armored scale insects, the paternal chromosomes are not heterochromatized but simply eliminated during early embryogeny. From such evidence it seems that sex determination is based on male haploidy as seems to be the case in most Hymenoptera (White, 1973). In the specialized coccids, male haploidy would not be achieved directly from development of unfertilized eggs but indirectly, after fertilization, either by elimination of the paternal set during early development or by heterochromatization and genetic inactivation of the paternal set, thereby rendering the male functionally haploid. However, in certain relatively primitive coccids without sex chromosomes, the males are ordinary diploids and direct maternal control of sex seems to be the most likely explanation (see Section III). Regardless of the manner of sex determination in specialized coccids, it is clear that maleness and heterochromatization (or elimination) are always completely coordinated.

The function of facultative behavior will be briefly considered. It is generally accepted that the heterochromatization of one X in female mammals is a dosage compensation mechanism. Thus, one functional X chromosome is balanced against two sets of autosomes in both the male, XYAA, and the female, X(X)AA, in which (X) is the heterochromatic, genetically inactive X. (Qualification of assumption of total inactivation of the X will be considered in Section II.)

The function of facultative behavior in *Sciara* and coccids is not as immediately obvious as that of mammals, but depends on theoretical population genetics for an explanation. It will be recalled from the introduction to this section that males of *Sciara* transmit only maternal chromosomes even though the males are diploid and the paternal genes expressed during development of the male. As noted in an earlier paragraph, males of specialized coccids transmit only maternal chromosomes. Breeding systems of this sort have been referred to as parahaploid (Hartl and Brown, 1970) and their evolutionary basis is presumably the same as that of true male haploidy in which the male develops from an unfertilized egg. In the evolution of male haploidy there is a self-promotive aspect which aids the haploid males in their competition with diploid males when the haploids first appear. Given an XOAA (male)–XXAA (female) system of sex determination, the haploid males are expected to be XA, as will be their sperm. All zygotes from XA with XXAA matings would thus be female and none of the genes necessary for the production and development of haploid males would have been lost in diploid male offspring (Brown, 1964; Hartl and Brown, 1970). Finally, it may be noted that close breeding (chance mating, rather than systematic inbreeding, of closely related individuals) is preadaptive to the evolution of male haploidy since it eliminates deleterious recessives which can be sheltered in the heterozygous state in outbred diploid species but would be immediately expressed in a haploid. Once male haploidy has been established in a group of organisms it, in turn, becomes preadaptive for exploitation of ecological niches in which close breeding or inbreeding is necessary for survival. This last point seems true since the very existence of haploid males precludes reliance on heterozygosity except in special cases in which the effect of heterozygosity is limited to the female.

## II. MAMMALIAN X CHROMOSOMES

Most female mammals have two X chromosomes (XX), and males have only one (XY). This difference between the two sexes in the number of X-linked genes is nullified by an unusual regulatory mechanism which results in heterochromatization of all X chromosomes in excess of one. As a result, in the somatic cells of the XX female, the genes of only one X are expressed, those of the other are not (Lyon, 1961). Similarly, in Klinefelter males (XXY), only one X is active. In other words, of two homologous chromosomes within a cell, only one is capable of supporting transcription. As will be discussed later in this chapter, mealybugs and related coccids represent the only other group of organisms in which there is a system of differential regulation of homologous chromosomes based on heterochromatization.

## A. Sex Chromatin and the Inactive X Chromosome

The inactive X chromosome is identifiable in interphase nuclei as a heterochromatic, darkly staining body attached to the nuclear membrane; it was discovered by Barr and Bertram (1949) in neurons of female cats and has since been called the Barr body, sex chromatin or, more recently, X chromatin. Since such a body was absent from nuclei of similar cells in male cats, Barr and Bertram suggested that the densely staining body was of X-chromosome material. Similar sex chromatin was soon found in cells of females of a wide variety of mammals, including man.

It was first thought that both X's of the female might contribute to the formation of sex chromatin, but its derivation from a single X was eventually demonstrated by Ohno and Hauschka (1960) and Ohno and Makino (1961). It had become clear by then that in women as well as female mice only one X is probably active in each cell (Stewart, 1960; Beutler et al., 1962; Russell, 1961). However, it was not until Lyon (1961) recognized the significance of variegated expression of X-linked coat-color mutations in mice that an explanation was available to account for both the cytological observations and the unusual genetic data. An essential feature of the Lyon hypothesis is the random inactivation of the maternal or paternal X in each cell of the young embryo. Once inactivation is established, it

becomes part of the cell's heredity and is faithfully repeated in all descendants of the original cell. The clonal nature of perpetuation of inactivation
and an approximately equal chance of inactivation of each of the two X's
often result in a fine-grain mosaicism for X-linked alleles in adult eutherian
mammals (Lyon, 1972, review).

### 1. Sex Chromatin as Diagnostic of Sex-Chromosome Constitution

The presence or absence of sex chromatin and the number of sex-
chromatin bodies per nucleus have become useful diagnostic tools in the
identification of individuals with X-chromosome abnormalities. It was
early recognized (Polani et al., 1954) that the majority of women with the
Turner syndrome do not have sex-chromatin bodies. Several years later,
the chromosome constitution of this class of Turner patients without sex
chromatin was established as 45,XO* (Ford et al., 1959). Similarly the
discovery of sex chromatin in Klinefelter males (Bradbury et al., 1956;
Riis et al., 1956) preceded the recognition that the karyotype is 47,XXY
(Jacobs and Strong, 1959). A wide array of sex-chromosome anomalies in
humans is now recognized such as diploid females with three or four X's
instead of two; in such cases the number of sex-chromatin bodies is
generally one less than the total number of X's. The same holds for
Klinefelter males in which the 48,XXXY examples have two sex-
chromatin bodies. It is thus obvious that sexual status per se has no effect
on the mechanism of X inactivation. The Y chromosome determines
maleness (Welshons and Russell, 1959) and seemingly has no effect on the
number of sex-chromatin bodies.

Although a valuable preliminary guide, the sex-chromatin bodies do not
always yield definitive evidence of the number of inactive X chromosomes. Even in normal females, one finds significant proportions of cells
without sex chromatin and the presence or absence of sex chromatin
appears to be influenced by a variety of factors such as tissue examined
and hormonal status. In cloned cell cultures, growth rate seems to influence the proportion of cells with sex chromatin, but biochemical tests
failed to indicate more than one active X even when a fairly high proportion of cells were lacking sex chromatin (see Comings, 1972, for further
discussion and references).

---

* This notation designates *total* chromosome number and sex chromosome constitution.
For the normal human female, 46,XX means two X chromosomes and 44 autosomes; for the
male, 46, XY means two sex chromosomes, a Y and an X, and 44 autosomes. The notation
above, 45,XO, means one X chromosome, no other sex chromosome (O), and 44 autosomes.
The notation for most cases of Klinefelter's syndrome, 47,XXY, indicates three sex chromosomes and 44 autosomes.

## 2. Late Replication and Inactivation

The heterochromatic, inactive X synthesizes its DNA late in the S period after the active X and the autosomes have finished most of their synthesis (Grumbach *et al.*, 1963; Nesbitt and Gartler, 1970). The relationship between late replication and genetic inactivation has been verified in female mules ($F_1$ hybrid, horse ♀ × donkey ♂) and female hinnies ($F_1$ hybrid, donkey ♀ × horse ♂). The X's of the horse and donkey are similar in size but the centromere of the donkey X is nearly terminal and that of the horse more centrally located; thus the two X's can be distinguished in the hybrids (Mukherjee and Sinha, 1964). The locus for glucose-6-phosphate dehydrogenase (G6PD) is on the X of both the horse and the donkey. Clones of fibroblasts of female mules and hinnies showed a complete correspondence between late replication and characteristic electrophoresis of G6PD: if the donkey X was late replicating, only horse G6PD was found; if the horse X was late replicating, only donkey G6PD was present (Rattazzi and Cohen, 1972; Ray *et al.*, 1972) (Fig. 1).

## B. Regulation of the Inactive X

### 1. Time of Onset of Inactivation

The time at which inactivation first occurs in young embryos has not been precisely determined. Direct tests for expression of X-linked alleles

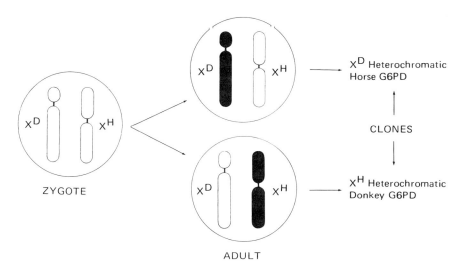

**Fig. 1.**   Heterochromatization and genetic inactivation in female mules and hinnies. The X chromosome from the donkey, $X^D$, can be clearly distinguished from that of the horse, $X^H$. In clones from adult tissues, only horse G6PD is found if the donkey X is heterochromatic, and only donkey G6PD is found if the horse X is heterochromatic.

have not been possible with very young embryos, and estimates of the time of inactivation have, therefore, been based on the cytological picture or on the late replication of the X (Lyon, 1972, 1974b, reviews). Fertilized mammalian eggs and very early cleavage embryos do not show any sex chromatin. The earliest time it has been detected is in pig blastocysts of about 50 cells (Axelson, 1968). The estimates of the latest time vary considerably but there seems to be agreement that sex chromatin is usually present no later than implantation.

It is possible that functional inactivation occurs considerably earlier than the time at which sex chromatin becomes detectable. This is suggested by the results of Issa et al. (1969) who observed late replication of X chromosomes in rabbit embryos 24 hours before sex chromatin was visible.

Takagi's (1973) work indicates that inactivation may even precede late replication. He was able to detect differential fluorescence of one of the two X chromosomes in mouse blastocysts of 40–50 cells; late replication was detectable only in embryos three days older. It is possible that the precise time at which inactivation occurs varies between species or even among different tissues of the same embryo.

Several genetic approaches have been used in attempts to obtain precise estimates of the time of inactivation. If X inactivation occurs at first cleavage, then, after destruction of one of the two blastomeres, the single surviving blastomere should give rise to a clone of cells and eventually to a full grown animal in which all somatic cells should have the same active X. Such an experiment by Hoppe and Whitten (1972) with mice showed that inactivation was random in the individuals so obtained. At least in the mouse, therefore, inactivation does not occur at first cleavage. If inactivation occurs when there are two cells, then in a population heterozygous for the $G6PD$ alleles, $Gd^A/Gd^B$, one quarter of the individuals would have only the A type enzyme (like a homozygote or hemizygote), another quarter only the B type, and the remaining half would have both A and B. This could be generalized to say that if there are $n$ cells at the time of inactivation, then, following random inactivation, $(\frac{1}{2})^n$ heterozygotes would have all $n$ cells with only the $Gd$ allele active. In other words, the frequency of heterozygotes such as $Gd^A/Gd^B$ which show only the A or B type of enzyme could be used to estimate the number of embryonic cells at the time of inactivation (Nance, 1964; Gandini et al., 1968; Nesbitt, 1971; Fialkow, 1973). Even these approaches have not yielded precise estimates because of certain limitations. For example, over 1% of women heterozygous for G6PD deficiency show one of the two homozygous phenotypes in their erythrocytes (Nance, 1964; Gandini et al. 1968; Fialkow, 1973). From these data one can say only that the number of primordial cells contributing to erythrocyte precursors is about eight, and it is not possible to say much about the time of inactivation itself (see also Section II,D).

There are several problems in extrapolating backward from such information to obtain an estimate of the time of inactivation of the X chromosome: (1) Inactivation may not occur at the same time in all sectors of the embryo. (2) The early embryo differentiates into extraembryonic and embryonic sectors and only the latter form the embryo proper. Significant differences in X inactivation between these two regions have recently been reported (Section II,D). (3) A harmful gene may be on either the active or inactive X; cells in which the harmful gene is on the active X might be expected to lose out in competition with those in which the harmful effect is suppressed by X inactivation. Such cell selection could be important, at least for some loci such as that for the enzyme, hypoxanthine phosphoribosyltransferase (HPRT) (Nyhan *et al.*, 1970).

Another interesting approach to this problem was that of Gardner and Lyon (1971) who injected single cells derived from one genetically marked mouse embryo of $3\frac{1}{2}$–$4\frac{1}{2}$ days gestation into the blastocyst of another. The donor embryos were heterozygous for different coat color genes which could later produce easily recognized patches on the uniform background of coat color developed by the recipient embryos. If one of the X's in the donor cell were inactivated prior to injection then only one color would appear on the uniform background of the recipient. If inactivation had not yet occurred at the time of injection, then the two X's would be inactivated at random in the subsequent cell lineage inside the recipient embryo. The result would be that two types of patches could occur, one with one donor color gene expressed, the other type with the other gene expressed. Most of the mice showing donor coat color showed both possible donor colors and thus indicated that inactivation had not occurred at the time of injection. This would place the time of inactivation in the mouse at a stage later than $3\frac{1}{2}$ days' gestation.

### 2. Reversibility of X Inactivation

X-Chromosome inactivation, once established, seems to be permanent in the ensuing cell lineage except in the germ line (see below). The stability of the inactive state is maintained even following long-term *in vitro* cultivation. Treatments with various substances, such as polyanions, have failed to reactivate the X; in addition reactivation has not generally been obtained by selective conditions in which a gene or genes on the inactive X are necessary for survival (Comings, 1966; Migeon, 1972). Recently, however, Kahan and DeMars (1975) have recovered reversions at the *HPRT* locus of inactive human X chromosomes at frequencies of ca. $10^{-6}$; the inactive X remained inactive at other loci tested. This information indicates that some type of mutational event occurred at or near the *HPRT* locus which released it, and only it, from the inactivation which still affected the remainder of the chromosome. There are many possibilities as to the type of mutational event which might release a single locus from inactivation.

The cytological observations on X-chromosome condensation, Epstein's (1969, 1972) results on enzyme levels, and the qualitative demonstrations by Gartler and associates (1972b, 1973) are all consistent with the view that both X chromosomes are active in oocytes. By these same methods, Gartler *et al.* (1975) have shown that only a single X is active in each oogonium in the early mitotic period of the female germ line. This is the first clear indication that, as in all somatic tissues studied so far, the X chromosome of the germ line is also subject to inactivation. Since human oocytes have two active X chromosomes, reactivation of the hitherto inactive X chromosome must occur during differentiation of oogonia into oocytes. Gartler *et al.* (1975) suggested that reactivation of the inactive X chromosome most likely occurs at about the beginning of meiosis since this is also the time of other profound changes in chromosomes.

### 3. The Problem of the Totality of X Inactivation

The first definitive biochemical evidence that only one of two homologous X-linked loci is active in any one cell of female mammals came from the studies of Davidson *et al.* (1963). Fibroblasts from women heterozygous for *G6PD* alleles, $Gd^A/Gd^B$, were cloned; each clone produced G6PD of only one type, A or B, never both. This type of analysis has since been extended to other X-linked loci analyzable at the cellular level (Lyon, 1972, review). An especially interesting study is that of Gartler *et al.* (1972a) on cultured fibroblasts from a doubly heterozygous woman, $Gd^A/GD^B$ and $Pgk^1/Pgk^2$; these loci are syntenic (on the same chromosome) but not closely linked; the *Pgk* locus codes for phosphoglycerate kinase (= PGK). If inactivation involves the whole X, alleles in a cis position should be inactivated together. Gartler *et al.* (1972a) found that each clone of fibroblasts from the double heterozygote produced either G6PDA and PGK1 or G6PDB and PGK2 and none of the other combinations. The linkage relationships in the somatic cells were thus $Gd^A$, $PGK^1$ on one X and $GD^B$, $PGK^2$ on the other. The X-linked loci in a cis configuration are coordinately repressed.

Over 170 genes are either known or suspected to be on the human X, and for over 90 of these the evidence for X linkage is convincing (McKusick, 1975). Genes coding for the following human enzymes are X linked: G6PD, HPRT, PGK, $\alpha$-galactosidase, and ornithine transcarbamylase. The expression of the loci for these five enzymes in clones derived from somatic cells is consistent with the view that only one X chromosome is capable of transcription in each cell and any remaining X or X's are inactive. The only X-linked locus apparently unaffected by inactivation is the $Xg$ blood group locus. The allele, $Xg^a$, produces an antigen [phenotype: Xg(a+)] which is not present when Xg is homo- or hemizygous [phenotype: Xg(a−)]. There is good evidence that the $Xg$ locus is not inactivated when carried on structurally normal X's (Ducos *et al.*, 1971) even though the rest of the same X is regularly inactivated for other loci tested. Lyon has recently suggested that a small segment of the human X, including $Xg$, escapes inactivation (Lyon, 1974a). Blood from heterozygous women, $Xg^a/Xg$, is not separable into the two expected antigenic types, Xg(a+) and Xg(a−), because all red blood cells react to the anti-Xg(a+) antibody and thus behave as though they were all Xg(a+). It is not possible to investigate single red cells, and Fialkow *et al.* (1970) therefore studied heterozygous women, $Xg^a/Xg$, who had developed chronic meyloid leukemia (CML). Since the evidence favors a clonal origin for CML (Fialkow *et al.* 1967), if the $Xg$ locus is subject to random inactivation, then appriximately 50% of the patients should be Xg(a−) and the others Xg(a+). Fialkow *et al.* found, however, that all eleven patients studied were Xg(a+). One interpretation of these results is that the locus escapes inactivation and only the dominant allele, $Xg^a$, is therefore detectable. Another interpretation is that the antigen produced by the $Xg^a$ allele in some cells will also coat red cells lacking it; thus red cells of $Xg$ constitution might obtain their antigen from $Xg^a$ red cells or the antigen might come from an entirely different tissue, again from those cells in which the X chromosome with $Xg^a$ is the active $X$.

On the other hand the results of Ducos *et al.* (1971) seem to support the view that the $Xg$ locus is not inactivated. They studied a twin pair who were chimeras (mixtures of genetically different tissues) for O red cells and $A_1B$ red cells. These two types of cells were separated and tested with the antibody for Xg(a+) antigen; the $A_1B$ red cells were found to be Xg(a−) and the O cells were Xg(a+). This is compelling evidence that Xg(a+) antigen is made in the ancestral red cells and that there is no intercellular transfer of antigen. These results are subject to simple genetic interpretation. The $A_1B$ red cells were homozygous for $Xg$ and made no antigen; the O cells were either homozygous $Xg^a$ or heterozygous $Xg^a/Xg$. In the latter case, present techniques would not show whether all the O cells were Xg(a+) or only the half expected if $Xg^a$ were inactivated half the time. There is no direct evidence from this experiment for the escape of the $Xg^a$ region from inactivation. But the evidence of Ducos *et al.* (1971) does show that the red cells do not acquire their Xg(a+) antigen from elsewhere. If the red cells make their own $Xg$ antigen, then the above-mentioned evidence of Fialkow *et al.* (1970) indicates that the $Xg^a$ locus is not inactivated because all the single-cell clones from hetero-zygous women tested positively for the Xg(a+) antigen; if $Xg^a$ were sub-ject to inactivation, then only half the clones should be Xg(a+) positive. The consensus, therefore, seems to be that the $Xg^a$ locus is not inac-tivated in normal chromosomes; for reasons unknown it is inactivated in structurally abnormal chromosomes (Polani *et al.,* 1970).

### C. Position Effect in X-Autosome Translocations

A number of X-autosome translocations are available in the mouse, and when a female is made heterozygous for such a translocation, she may show a variegated phenotype for the autosomal genes attached to the X (Russell and Montgomery, 1970; Cattanach, 1974). Variegation arises primarily because the autosomal segment is inactivated, at least in part, along with the X to which it is attached. There is evidence, however, that the autosomal genes may sometimes escape inactivation or, if inactivated initially, may become active at a later stage in development. Such behav-ior is best illustrated by the Cattanach translocation (Cattanach, 1961), which has been intensively studied in this regard. This translocation re-sulted from insertion of a segment of autosome 7 into an X. The insertion carries wild-type alleles of three genes affecting coat and eye color, *albino* (*c*), *pink-eye* (*p*), and *ruby-eye* (*ru-2*). Variegation is observed when the recessive alleles of any one of these genes are on both normal chromo-somes 7. Cells in which the modified X is active yield a wild-type patch: if the modified X is inactive the patch is of the mutant phenotype. Within the mutant patches, however, there may be wild-type patches which Cat-tanach has interpreted as a result of escape from inactivation of the au-tosomal segment of the translocation. Analyses of such patches indicate that this escape can apparently occur both at the time of inactivation of the X and also at later developmental stages.

### D. X-Inactivation Mosaicism as a Developmental Marker

Since the inactive-X system results in naturally occurring mosaicism, it has been used to study problems in normal as well as abnormal develop-ment. Each female contains two populations of cells in regard to the X, the one in which the paternal X is active, the other in which the maternal X is active. If inactivation precedes or occurs in a primordial group of cells, then, as Gandini and Gartler (1969) pointed out, ``If two or more cell types share common stem cells, their mosaic composition in a heterozygote should be alike; if not, these cell types should in some cases exhibit mosaic composition differences.'' Mosaic composition is tested most conveniently by determining the ratio of the A and B variants of G6PD. It is not necessary to know whether either of the alleles is of paternal or maternal origin, simply that two different tissues have the same or different proportions of A and B variants, usually expressed as percent A.

Gandini *et al.* (1968), Gandini and Gartler (1969), and Fialkow (1973) found that the blood cells (erythrocytes, granulocytes, and lymphocytes) and skin and skeletal muscle probably all came from a primordial pool of only 16 (10–25) cells (Fialkow, 1973). The ambiguities in using such estimates to determine the time of X inactivation, already mentioned in Section II,B,1 above, become especially clear in Fialkow's (1973) comment on this estimate of 16 cells, ''This number could represent progenitor cells for mesoderm or a more primordial cell pool such as one for ectoderm (from which mesoderm subsequently differentiates) or even the total number of embryoblasts* present in man at the time of X-chromosome inactivation. Thus, a minimum estimate for the number of embryoblasts present in man at the time of X inactivation is 16 (10–25).''

Of special significance for the present discussion is Fialkow's (1973) conclusions that the number of cells entering each pathway is rather large, probably over 80, when tissue-specific development commences, and that considerable cell mixing may occur during subsequent growth of skin and skeletal muscle.

In female mice heterozygous for the Cattanach translocation, Nesbitt (1971) identified cytologically the proportion of cells with the normal X inactive and with the translocation inactive. She found significant correlations in mosaic compositions among ectodermal (melanocytes) and mesodermal tissues (embryonic spleen, thymus, lung, and fascia). If X inactivation is random and occurs after formation of anlagen, no correlations in mosaic composition of different tissues within individuals would be expected. Inactivation must, therefore, occur before differentiation of mesoderm and ectoderm. The extent of variance in mosaic composition indicates that, at least in the mouse, there are relatively large precursor pools of 21–40 or more cells for all the tissues examined (Nesbitt, 1971).

Two recent reports indicate that the maternal X is active in the large majority of cells of extraembryonic tissues of rats and mice. In cells from the chorion, yolk sac, and allantois of the mouse, percentages of cells with an active maternal X ranged from 55 to 100 for averages of individual animals; for the yolk sac and chorion, averages for all animals ranged from 89 to 98% (Takagi and Sasaki, 1975). In rats, only yolk sacs were exam-

---

* Here, embryoblasts mean cells of the presumptive embryo proper.

ined; again the maternal X was active in an overall average of 91% of cells (Wake *et al.*, 1976). Since reciprocal crosses were made with both rats and mice, the data do not reflect a genetic bias. In addition, the cells of the embryos proper had an active maternal X in close to the 50% expected with random X inactivation. Among other explanations, the authors suggested that cells with an active maternal X would have a selective advantage because of closer compatibility with maternal tissues and would consequently proliferate more rapidly than those with an active paternal X. This explanation would seem reasonable since the extraembryonic membranes are formed from the same three primary tissues, ectoderm, mesoderm and endoderm, which form the embryo proper. Preliminary calculations by Muriel Nesbitt (personal communication) indicate that sufficient cell generations occur during the formation of the extraembryonic membranes to enable a relatively modest selective differential, of the order of 10%, to produce the large increases above the expected 50% of cells with an active maternal X.

The G6PD mosaic system has also been used to study the long-standing problem of unicellular versus multicellular origin of tumors (Fialkow, 1972; Gartler, 1974, reviews). For example, a tumor arising as a unique clonal event in a $Gd^A/Gd^B$ heterozygote would be expected to have only the A or B type of enzyme, not both. The first use of this system was that of Linder and Gartler (1965) with leiomyomas, common benign tumors of the uterus. Over two hundred such tumors have now been studied and all of them, except one necrotic tumor which could not be obtained free of surrounding myometrium, showed either the A or B phenotype; the normal myometrium in each case showed both the A and B types (Linder and Gartler, 1965, 1967; Linder, 1969; Townsend *et al.*, 1970). Tumors with the A type of enzyme occurred as frequently as those with the B type; thus cell selection based on G6PD type was not a significant factor in either the origin or progression of these tumors.

Burkitt's lymphoma, chronic granulocytic leukemia, and carcinoma of the cervix and colon are among a variety of neoplasms which have since been studied by these methods. The data from these neoplasms, like those from the majority of neoplasms studied to date, are consistent with clonal origin and the view that the initial neoplastic event—whether genetic or epigenetic—is of random or near-random occurrence.

In contrast to the above results, two hereditary tumors, trichoepithelioma (Gartler *et al.*, 1966) and multiple neurofibromatosis (Fialkow *et al.*, 1971), both of which are inherited as autosomal dominants, show both the A and B types of enzyme. The evidence thus favors a multiple-cell origin of such tumors. In these hereditary tumors several cells in a tissue are presumably genetically susceptible to the tumor-initiating agent or event.

## E. Paternal X Inactivation in Marsupials

That random X inactivation may not extend to marsupials was first suspected on the basis of results of autoradiography of DNA replication in X chromosomes of certain hybrid kangaroos (Sharman, 1971). Morphological differences between X chromosomes of different subspecies permitted recognition in hybrids that the paternally derived X chromosome was always late-replicating and therefore presumably inactive. This was soon confirmed by studies of X-linked enzymes, especially G6PD and a PGK variant, PGKA (Richardson *et al.*, 1971; Cooper *et al.*, 1971). Johnston *et al.* (1975) obtained similar results in a population of wallabies with naturally occurring G6PD polymorphism and thus showed that the earlier results were not an unexpected result of hybridization between subspecies. More recently, a complexity has appeared: the paternal allele may be active to varying degrees in certain tissues of the marsupial female (Table I) (Cooper *et al.* 1975a,b).

Pedigree analyses have shown that the paternal X, which is largely or completely inactive in somatic cells of the female, is transmitted to the offspring; now maternal in origin, it is fully active in the daughters. In the uterus and ovarian tissues of *Macropus giganteus*, preliminary results indicate activity of both maternal and paternal alleles of the *PGKA* locus (Cooper *et al.*, 1975a) indicating that the paternal X is probably fully active in the germ line. It either escapes inactivation in the germ line or, if inactivated, there is subsequent reversal.

**TABLE I**

**Activities of Paternally Derived and Maternally Derived X-Linked Loci (*G6PD* and *PGKA*) in Marsupial Females**[a]

| | Enzyme | |
|---|---|---|
| Tissues | *PGKA* | *G6PD* |
| Blood | Paternal allele, inactive | Paternal allele, inactive |
| Skeletal and cardiac muscle (*Macropus parryi*) and bladder muscle (*M. giganteus*) | Paternal allele, partially active | Paternal allele, inactive |
| Cultured fibroblasts | Paternal allele, partially active[b] | Paternal allele, active to some extent |
| Uterus and ovary | Both maternal and paternal alleles, active | — |

[a] From Cooper (1975a,b).
[b] Confirmed in cloned cell populations (D. W. Cooper, personal communication).

Marsupials with abnormal X chromosome constitutions are known to occur. Sharman *et al.* (1970) have described two tammar wallabies, one with an XXY constitution and a male phenotype and another with an XO constitution and a female phenotype. It is not known, however, whether these were, respectively, $X^m X^p Y$ and $X^m O$, two constitutions which might be viable in systems with paternal-X inactivation ($X^m$, $X^p$ = X's of maternal and paternal origin).

The flexibility of the marsupial system is again indicated by the complete loss, just the opposite of reactivation, of the paternal X in certain cases. In some tissues, such as bone marrow and spleen of species of the genera *Isodon* and *Parameles*, the Y chromosome is eliminated in the male and the heterochromatic X is eliminated in the female (Hayman and Martin, 1969). These observations were made prior to the demonstration that the inactive X is of paternal origin in marsupials. The only analogous example reported from Eutheria is that of the vole, *Microtus oregoni* (Ohno *et al.*, 1963, 1966). The soma of the male is XY, but the X is lost from the germ line and only O and Y sperms are formed; the female soma is XO but the germ line becomes XX through nondisjunction. Zygotes are thus XO($♀$) and XY($♂$).

## F. Chromosome Imprinting, Controlling Elements, and the Mechanism of X Inactivation

### 1. Models of the Mechanism of X Inactivation

The regularities observed in the inactivation of the mammalian X have led to several proposals of mechanisms controlling X inactivation. The fact that only one X is active among two or more X's in basically diploid individuals indicated that a single episomelike particle could be responsible for activation of a single X (Morishima *et al.*, 1962). The subsequent discovery of inactivation of the paternal X in marsupials led Cooper (1971) and Brown and Chandra (1973) to propose models of X inactivation which would account for the evolutionary change from the fixed X inactivation of marsupials to the random X inactivation of Eutheria.

In addition, data from triploids need to be considered in any model of X inactivation. In humans, both 69,XXY and 69,XXX triploids occur in two different classes: with either 1 or 2 active X chromosomes. Thus 69,XXY embryos have either one sex-chromatin body or none, and 69,XXX embryos have either one or two. Biochemical evidence has shown that both X's were indeed active in a human sex-chromatin-negative XXY triploid (Weaver *et al.*, 1975). In most cases, however, only the maximum number of sex-chromatin bodies has been determined and only rarely has confirmatory evidence been obtained from autoradiographic demonstrations of the number of late-replicating X's. As mentioned earlier in this section (A,2), formation of sex-chromatin bodies in diploids is by no means uniform. In triploids, sex-chromatin bodies seem to be equally if not more variable in their presence, and this variability leads to ambiguity in the interpretation of models of X inactivation. This point will be considered again after description of a specific model for which reliable triploid data are critical.

According to the model of Brown and Chandra (1973; Chandra and Brown, 1974, 1975), the marsupial X chromosome carries a two-part controlling element consisting of a sensitive site and a receptor site. The sensitive site of the paternal X is imprinted in the egg on or shortly after fertilization. (Evidence that imprinting occurs in the mammalian egg will be discussed below.) Imprinting results in the inactivation of the sensitive site of the paternal X. The maternal X is not imprinted; its active sensitive site influences the adjacent receptor site to maintain the activity of the X during subsequent development.

It was assumed that the sensitive site was translocated to an autosome in Eutheria during evolutionary divergence from marsupials. The sensitive site on the paternal autosome would be imprinted and made inactive; that on the maternal homolog would not be imprinted and would release an informational entity which would attach at random to the receptor site of an X; this X would become the functional X.

According to this model, in a diandric* eutherian triploid the sensitive sites of the two paternal autosomes would be imprinted; the sensitive site of the single maternal autosome would result in a single active X; thus there would be one sex-chromatin body in XXY triploids and two bodies in XXX triploids. In digynic triploids two maternal sensitive sites would not be imprinted and result in two active X's; thus there would be no sex chromatin in XXY triploids and one sex-chromatin body in XXX triploids.

The model was based in part on experiments of Bomsel-Helmreich (1971). Digynic triploid rabbit embryos were induced by colchicine treatment of the egg. As expected from the model, all XXY triploids were sex-chromatin-negative and all XXX triploids had one sex-chromatin body. Furthermore, G6PD activity in the triploid embryos was twice that of the diploid controls. Recently, Weaver et al. (1975) obtained biochemical evidence that both X's were indeed active in a human sex-chromatin-negative, XXY triploid.

On the other hand, apparent exceptions to the Brown and Chandra

---

* Diandric ("two, male") and digynic ("two, female") refer to the sex of the parent contributing two of the three sets of chromosomes of the triploid.

model have appeared among spontaneous human triploids. The diandric origin of a 69,XXY abortus was demonstrated by Niikawa and Kajii (1974) from fluorescence and C-banding of marker chromosomes. According to the model, one sex-chromatin body was expected but less than 1% of amnion cells and 5% of trophoblastic cells of the villi contained the expected sex-chromatin body which was absent from all other cells examined. Since sex chromatin is usually observed in a high proportion of amnion cells, this abortus was classified as sex-chromatin negative. In another 69,XXY abortus, HLA antigens indicated diandry and yet the cells were again sex-chromatin negative (Boué and Boué, 1974).

In conclusion, either the model is wrong or the data from human triploids do not indicate the true situation. Since there are fewer sex-chromatin bodies than expected in the human triploids just mentioned, the triploid state may have a physiological effect on their expression (Section II,A,1), an ambiguity which might be resolved by a study of late replication of the X's. Since the Bomsel-Helmreich (1971) study of digynic triploid rabbits gave consistent results, human triploids may differ from rabbit triploids in expression of sex chromatin. In addition, diandric triploids may result from either diploid sperm or double fertilization in which one of the two sperms might escape imprinting. An experimental series on diandry comparable to that of Bomsel-Helmreich (1971) on digyny would thus be of interest.

### 2. Time and Place of Imprinting

These aspects of the imprinting process have recently been discussed by Chandra and Brown (1974, 1975). Conclusive evidence that imprinting can occur in the egg comes from Nur's work on certain coccids in which fertile diploid males, with one haploid set euchromatic and the other heterochromatic, are produced parthenogenetically (Fig. 6).

Evidence from mammals is highly suggestive that imprinting here also occurs in the egg. Ovarian teratomas, which are mostly benign tumors, have a 46,XX karyotype and a single sex-chromatin body (Linder, 1969). For reasons to be explained, the teratomas were believed by Linder (1969) and Linder and Power (1970) to be parthenogenetic derivatives of oocytes in which a diploid status was achieved either by failure of the chromatids to separate at anaphase II or by fusion of the egg nucleus and polar body II (haploid). The latter explanation was favored because the fusion might be expected to stimulate development. No genetic markers were present in the teratomas which were not present in the host female; there was thus no paternal contribution. Markers present heterozygously in the female were often homozygous in the teratoma; thus origin from a somatic or a premeiotic germ line cell was ruled out.

In the case of a girl mosaic for diploid and triploid tissue, evidence suggested that the extra set in the triploid tissue was of maternal origin. According to the Brown–Chandra model, only one inactive X would be expected but two sex-chromatin bodies were present (Ellis *et al.,* 1963; Mittwoch *et al.,* 1963). A possible explanation for the extra maternal set in the triploid sector was fusion of the second polar body with a cleavage nucleus. These results, along with those from teratomas, led to the conclusion that imprinting occurs in the periphery of the egg so that both sperm on entry and the second polar body can be imprinted (Chandra and Brown, 1974, 1975). In the 1975 report, it was suggested that the entire periphery has the capacity to imprint. However, the mammalian egg is about midway in the size range of ''mosaic'' eggs, such as those of the ascidians, in which there is considerable internal differentiation. The mammalian egg might thus have sufficient peripheral cytoplasm to permit variation in capacity to imprint.*

### 3. Mechanism of X Selection in Eutheria

If homologous chromosomes are spatially separated from each other in differentiated cytoplasm, one could be imprinted, the other not. Spatial separation is clearly observed in certain parthenogenetic coccids in which imprinting occurs (see Section III,C,3). The Brown–Chandra model would require such spatial separation for imprinting the autosome presumed to carry the sensitive site. The presence of but a single activating particle (episome or information entity) which would activate one autosome but not the other has already been considered. The objection to this simple explanation is the difficulty expected when a single particle must find a single receptor site within an entire genome.

Two suggestions have been advanced to overcome the difficulty of the single-particle hypothesis. Drews *et al.* (1974) have proposed that a process such as cooperative binding of a specific protein with a specific sequence of DNA is responsible for activating a single X. In this reaction the probability of binding the first protein molecule is quite small but the second and later molecules are bound very rapidly. If, however, the system did not include primary binding of low probability, then it could be assumed that a single entity might be as successful as the second and later molecules in the cooperative binding model. A mathematical analysis of the dynamics of such a system did not show a difference between cooperative binding and attachment of a single entity on the diploid level (Mukunda *et al.* 1976). It will be recalled that there are two classes of triploids, with either one or two active X's for both the XXY and XXX

---

* Recent evidence suggests that mammalian eggs can no longer be assumed to be non-mosaic, that is without internal asymmetries. Nicosia *et al.* (1977) have shown in mouse eggs ''a marked polarity in the distribution of cortical granules and of microvilli in the egg cortex.'' Sperm penetration was found to be restricted to an area containing cortical granules and covered with short microvilli.

genotypes. The analysis of Mukunda *et al.* showed that there are "two classes of embryos differing from each other in a step-wise manner in the number of activating molecules," and that embryos with two active X's have twice the number of activating molecules as diploids and triploids with one active X. The Brown–Chandra model provides a mechanism for generating two such classes of triploid embryos. This model is consistent with the analysis of the dynamics whether each active sensitive site produced a single entity or several entities acting cooperatively.

A somewhat similar problem appears with Riggs' model based on methylation (Section V). Since activation of the first regulatory locus of the X would result in a feedback loop preventing methylation of other regulatory loci, all triploids would be expected to have a single active X. The assumption that all triploids have an average number of activating molecules which may lead to either one or two active X's seems unlikely (Mukunda *et al.*, 1976). Without a system of information transfer based on stepwise differences, the number of active X's per cell would not be constant and triploid tissues would be expected to be mosaics for different numbers of active X's in the various cell lineages. Evidence in favor of information transfer would be available if biochemical tests ruled out such mosaicism. More intensive studies of triploids may provide pertinent information.

### 4. The X-Chromosome Controlling Element in Mice

Cattanach and Isaacson (1967) reported that different alleles or states of a locus of the X of mice influence the expression of X-linked genes in heterozygous females. Several other workers have observed similar effects associated with a locus at approximately the same position on the X, and it seems likely that all have been studying the same locus, now referred to as the X-chromosome controlling element, *Xce*, by Cattanach (1975, review). Cattanach (1975) discussed two possible modes of action at the *Xce* locus. The *Xce* allele on one X may be deleterious so that cell selection results in a prepondernace of tissue with the homolog in the active state. Or differences in *Xce* alleles may lead to preferential activation of one of the X's in which case cell selection would not be a factor. Recent experiments, too complex to review here, seem to confirm the latter interpretation. It should also be noted that in some of the early experiments, *Xce* seemed to undergo changes of state similar to those of mutable loci in maize (see Chapter 2). Finally, *Xce* seems to influence the phenotype of certain X alleles in males, an unexpected effect if *Xce* were involved only in the X-activation process.

Since hypotheses on control of the X have generally included a locus on the X responsible for determining its activity, it is tempting to presume that *Xce* is indeed this locus (e.g., the receptor site in the Brown and

Chandra model). On the other hand, the locus regulating X activation could be a typical locus in regard to mutability and lack of influence on neighboring loci. In this case, *Xce* could be presumed to be a component of a mutator system imposed on the regulatory locus in mice just as such components are imposed on typical loci in maize. It will, therefore, be of considerable interest in regard to the nature of the regulatory locus if similar results are reported for other mammals.

## III. COCCID CHROMOSOME SYSTEMS

When Schrader (1921) first examined the mealybug, *Pseudococcus nipae* [*Nippaecoccus nipae* (Maskell)], he found five euchromatic and five heterochromatic chromosomes in the males. The terms, hetero- and euchromatin, were not introduced by Heitz until 1928 and 1929, but will be used here for convenience. By 1921 it was well known that sex chromosomes are often heterochromatic. Schrader proposed that the entire mealybug genome represented a compounding of the sex chromosomes to an unusual degree to give five X's and five Y's in the male and two sets of five X's in the female. During meiosis in the male, the five X's were segregated from the five Y's; transmission of these two types of sperm resulted in a 1:1 sex ratio.

That this interpretation was not correct was shown a few years later in an eriococcid, *Gossyparia spuria* (Modeer), in which the heterochromatic chromosomes were again segregated from the euchromatic chromosomes during meiosis, but here it was obvious that the heterochromatic meiotic products did not form sperm (Schrader, 1929) and could not, therefore, have been Y chromosomes. Hughes-Schrader (1935) found that the six heterochromatic chromosomes of *Phenacoccus acericola* King were segregated from the six euchromatic chromosomes during meiosis and only the euchromatic derivatives formed sperm. There is an XX(♀)–XO(♂) sex determining mechanism in the more primitive coccids. This is not the case, however, in coccids in which one chromosome set is heterochromatized in the males. The same chromosome number has been consistently found in both males and females in all the many examples of this system so far studied. Although the presence of sex chromosomes was thus ruled out, possible genetic effects on sex determination were not (see below, Section III,C, 1a). However, the absence of sex chromosomes led to consideration of other possible explanations of the heterochromatization observed in these coccids. In interpreting this system, the Schraders took into account two prior findings. In *Sciara*, Metz (1938) and his collaborators had shown that the chromosomes eliminated at meiosis in the male were of paternal origin. Males transmitted only the maternal chromosomes to their offspring, but these became paternal chromosomes in their sons, and would again be eliminated at meiosis. Somehow or other, this once maternal, now paternal chromosome set was preconditioned so that it behaved differently in sons and daughters. By analogy with *Sciara*, the Schraders proposed that a similar system operates in mealybugs and eriococcids. The paternal chromosomes become heterochromatized during early development and are later eliminated at meiosis. Second, the Schraders had helped to prove that true male haploidy (males from unfertilized eggs) occurs in a group of primitive coccids, the iceryines. They therefore concluded that the heterochromatization of the coccid chromosomes indicates genetic inactivation of a chromosome set, and that such systems could serve as intermediates between regular diploidy and true male haploidy (Schrader and Hughes-Schrader, 1931; Hughes-Schrader, 1948).

The Schraders' interpretation was later confirmed experimentally with a mealybug, *Planococcus citri* (Risso), by Brown and Nelson-Rees (1961) who showed that after fathers were X-rayed, chromosome aberrations appeared in the heterochromatic set of the sons, and in the euchromatic set after mothers were so treated. Compared to the daughters, the sons were grossly insensitive to X-ray treatment of the father. This observation could be explained on the assumption that the paternal set was euchromatic and genetically active in the daughters but heterochromatic and inactive in the sons. Further confirmation came from the use of genetic markers: male mealybugs expressed and transmitted only the genes received from the mother (Brown, 1969; Brown and Wiegmann, 1969). Finally, in the most specialized group of coccids, the armored scale insects, most species have lost the system of facultative heterochromatization and have substituted instead simple elimination of the paternal chromosome set during early embryogeny (Brown and Bennett, 1957; Bennett and Brown, 1958).

In the mealybug, analysis of tissues in sons produced after the fathers had been given high doses of ionizing radiation showed that there is a close relationship between heterochromatin and genetic inactivity. In a few tissues, such as the Malpighian tubules and the testis sheath cells, the heterochromatic set appears to revert to the euchromatic state. Nur (1967a) found that such paternal irradiation induced marked developmental disturbances in the Malpighian tubules, a result which would not have been expected unless the radiation had induced deleterious changes, such as deficiencies, in genetically *active* chromosomes. Damage to the testis sheath cells was indicated by increasing sterility in males after increasing paternal irradiation (Nelson-Rees, 1962). Tissues in which heterochromatic chromosomes did not revert to an euchromatic state showed no deleterious effects from paternal irradiation.

One apparent exception to genetic inactivity of the heterochromatic set is the occasional formation of a small nucleolus at the site of the nucleolus organizer of the heterochromatic set (S. W. Brown and S. M. Rieffel, unpublished).

## A. Chromosome Changes Ancestral to the Coccids

The coccids are at the top of an evolutionary pyramid of cytological variance and complexity. In order to understand these variations in chromosome behavior, it is necessary to review briefly their evolution.

The Hemiptera (*sensu lato*) constitute a very large order often divided into two orders. The Heteroptera (= Hemiptera *sensu stricto*) are the true bugs and include squash bugs, stink bugs, and the poisonous assassin bugs. The Homoptera include the cicadas, leaf hoppers, sharpshooters, aphids, and coccids and all are plant feeders. The coccids are the most highly specialized homopterans.

All the Hemiptera have holokinetic chromosomes (= diffuse centromeres); chromosome fragments induced by X-ray treatments do not lag behind at metaphase, as do acentric fragments of monocentric chromosomes, but show normal mitotic maneuvers (Homoptera: Coccoidea, Hughes-Schrader and Ris, 1941; Heteroptera, Hughes-Schrader, and Schrader, 1961).

Inverse meiosis is a second ancestral condition the coccids share with the closely allied aphids (Ris, 1942; Hughes-Schrader, 1944, 1948). [It also occurs in the plant family Juncaceae and probably also in the closely allied Cyperaceae (Malheiros *et al.*, 1947; Wahl, 1940).] In the inverse sequence, chromatids, not chromosomes, separate from each other and move to opposite poles at anaphase I. Two chromatids from each bivalent are present in each of the two interphase nuclei where they pair to form a dyad. Second anaphase separates the dyad chromatids one to each pole. Although the sequence is different from that of typical meiosis, the results are the same: each of the four chromatids of the meiotic bivalent reaches one of the four nuclei produced by meiosis.

In other Hemiptera, the problem of guiding holokinetic chromosomes through meiosis is solved differently. The paired chromosomes (bivalents) behave at anaphase I as though they had terminal centromeres at opposite ends which guide a typical dyad to each pole.

## B. Chromosome Systems of Primitive Coccids and *Stictococcus*

Most of the primitive coccids retain the ancestral XX ($\female$)–XO ($\male$) sex determining mechanism which also occurs in aphids; they are of interest in the context of the present chapter because of the occurrence of true male haploidy in one group, the iceryines. In the functional hermaphroditism which occurs in addition to male haploidy in a few of the iceryine species the testis sector becomes haploid by chromosome elimination which may possibly be another example of a facultative chromosome system.

The isolated genus, *Stictococcus*, is of special interest because of the possibility that sex is directly determined by the mother; there are neither sex chromosomes nor heterochromatic chromosomes.

### 1. Male Haploidy and Hermaphroditism in Iceryine Coccids

The male-haploid iceryines have four chromosomes in the female and two in the male; these numbers probably evolved on transfer of X-chromosome loci to an autosome in an ancestral stock which had six chromosomes in the female $(A_1A_1A_2A_2XX)$ and five in the male $(A_1A_1A_2A_2X)$, a constitution which is found in the closely allied, present-day llaveiines (Hughes-Schrader, 1948).

In a few iceryine species, of which the best known is *Icerya purchasi* (Mask.), functional hermaphroditism is superimposed on male haploidy. The gonad is a mosaic ovotestis and self-fertilization occurs in what otherwise is a morphological female. Males are produced by unfertilized eggs. The testis segments of the ovotestis are haploid and spermatogenesis is accomplished, as in the haploid males, by a single mitotic division; both daughter nuclei form sperm (Schrader and Hughes-Schrader, 1926; Hughes-Schrader, 1927, 1948).

The manner in which the presumptive testis segments become haploid has been difficult to analyze because the events occur during early stages of the life cycle when adjacent tissues are undergoing autolytic morphogenesis resulting in degenerating nuclei similar to those expected if chromosomes are eliminated. Nonetheless, Hughes-Schrader and Monahan (1966) have been able to provide some evidence that haploid sectors are formed by degeneration within diploid nuclei of one haploid set of chromosomes. The nuclei are normally lobulated, and one haploid set, consisting of one long and one short chromosome, may be in one lobule and the other haploid set in the other lobule. What triggers the destruction of just one haploid set is not known. The possibility that this system is also facultative and that the chromosomes destroyed are paternal is of considerable interest.

The three species of iceryines with functional hermaphroditism are widely separated geographically and appear to be unrelated taxonomically; therefore, hermaphroditism must have appeared independently on three occasions (Hughes-Schrader and Monahan, 1966). This is significant because there is also evidence for the multiple origin of facultative systems to be described later in this section.

## 2. Sex and Symbionts in Stictococcus

The genus *Stictococcus* is confined to central Africa and is so different from all other taxa that it belongs in its own major subdivision. Some years ago Buchner (1954, 1955) reported that intracellular symbionts, common in most but apparently not all coccids, are present in the females of *Stictococcus* and absent from the males. Buchner proposed that sex is determined in *Stictococcus* by the presence or absence of symbionts and extended this concept to coccids with the lecanoid chromosome system (see next section); for the lecanoid system, he believed that sex might be determined by the number of symbionts present in the oocyte rather than by their presence or absence. *Stictococcus* displays striking sexual dimorphism; males do not feed after birth; they molt several times to emerge as adults smaller than newborn larvae. Coccid symbionts may aid in the utilization of plant juices which the insect sucks in. Since the males do not feed, they presumably do not need symbionts. Failure to endow males with unneeded symbionts would be of selective advantage since the mother must expend energy to provide them. At present, this explanation seems more acceptable than the belief that presence or absence of symbionts directly determines sex. Both males and females are diploid and no sex chromosomes have been observed; meiosis in the male is reduced to a single achiasmate division in which the chromosomes pair and separate to opposite poles; two sperms are formed by each meiocyte (S. W. Brown and U. Nur, unpublished). As there are no sex chromosomes, sex may be determined physiologically in the egg and thus coordinated with presence or absence of symbionts. [Other aspects of intracellular symbiosis are discussed in Chapter 8.]

## C. Facultative Chromosome Systems of the Coccids

The facultative systems occur in the two remaining major subdivisions of the coccids. These two subdivisions are (1) the lecaniids, including mealybugs, soft scales and their several allies, and (2) the most specialized coccids, the diaspidids, *sensu lato*, which include the armored scale insects and the palm scales. The lecaniids include families almost as unspecialized as primitive coccids to those verging on the extreme specialization of the diaspidids. The armored scale insects seem to be the most highly successful in terms of numbers of genera and species, geographical distribution, and number of individuals. In most coccid lineages, specialization is accompanied by reduction and loss of parts with few if any new adaptive structures. The armored scales are a striking exception and have evolved specialized abdominal structures for making their protective coverings.

In the lecaniid assemblage, heterochromatization occurs in all but a few

## MEIOSIS

| PRE-PROPHASE | PROPHASE | ANAPHASE I | ANAPHASE II | SPERM FORMATION |
|---|---|---|---|---|

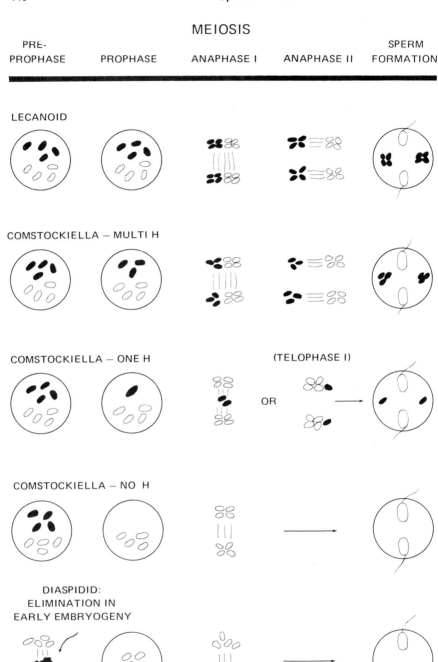

LECANOID

COMSTOCKIELLA — MULTI H

COMSTOCKIELLA — ONE H

(TELOPHASE I)

OR

COMSTOCKIELLA — NO H

DIASPIDID:
ELIMINATION IN
EARLY EMBRYOGENY

relatively primitive genera; there is some evidence that heterochromatization may have evolved independently in the "mealybug lineage" and in
the "eriococcid lineage." Heterochromatic systems occur in the diaspidids (*sensu lato*), in all known palm scales and all minor tribes of the
armored scale insects and in a few species of the two major tribes: the
Aspidiotini and the Diaspidini. In the major tribes, heterochromatization
of the paternal chromosome set has been almost completely replaced by
elimination of this set during early embryogeny of the males. Change from
heterochromatization to elimination apparently occurred several, perhaps
many, times. In one of the two major tribes, two genera are known in
which one species undergoes heterochromatization and another eliminates
the chromosomes during early embryogeny (Brown, 1965).

### 1. Systems Involving Heterochromatization

There are two different systems based on heterochromatization in the
coccids. In one of these, the lecanoid system, all the heterochromatic
chromosomes are maintained throughout the meiotic divisions; the
heterochromatic chromosomes degenerate only after they have been separated from the euchromatic chromosomes at anaphase II (Fig. 2).

The second system involving heterochromatization is the Comstockiella
system, named after the genus in which it was first discovered (Brown,
1957, 1963). Prior to meiosis, heterochromatization seems to be the same
in both the Comstockiella and lecanoid systems (Fig. 3). At preprophase
of meiosis in the Comstockiella system, one to all the heterochromatic
chromosomes are subject to intranuclear destruction. Those that are not
destroyed prior to meiosis are eliminated during or after meiosis (Fig. 2).
*Both* Comstockiella and lecanoid sequences occur in certain species, and
often in the same individual and testis. It is reasonable to assume, there-

---

**Fig. 2.** Facultative chromosome behavior in coccids. Heterochromatization of the paternal set occurs in early embryogeny in the lecanoid and Comstockiella systems (Section
III,C). *Lecanoid system*: The first meiotic division is equational for both the hetero- and
euchromatic chromosomes. During the second division the euchromatic set is segregated
from the heterochromatic set which slowly degenerates in the cytoplasm of the quadrinuc-
leate spermatids. Only the euchromatic sets form sperms. *Comstockiella system*: One to all
of the heterochromatic chromosomes are destroyed at preprophase of meiosis. If several
heterochromatic chromosomes (multi H) survive destruction, the meiotic sequence is usually
lecanoid. If one (one H) or no (no H) chromosome survives destruction, there is usually a
single equational division of the euchromatic chromosomes; a surviving heterochromatic
chromosome will be eliminated by anaphase lagging or ejection from telophase nuclei. The
Comstockiella system may vary greatly among species and within species and individuals.
*Diaspidid or elimination system*: This system is known only from the two major tribes of the
armored scale insects (diaspidids, *sensu stricto*). The paternal chromosome set is eliminated
during early embryogeny and the male develops as a haploid; spermatogenesis is a single
equational division.

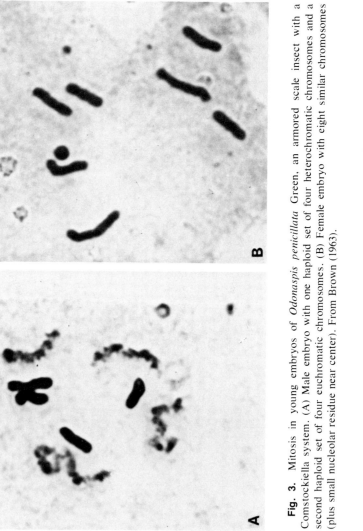

**Fig. 3.** Mitosis in young embryos of *Odonaspis penicillata* Green, an armored scale insect with a Comstockiella system. (A) Male embryo with one haploid set of four heterochromatic chromosomes and a second haploid set of four euchromatic chromosomes. (B) Female embryo with eight similar chromosomes (plus small nucleolar residue near center). From Brown (1963).

fore, that the early development is much alike in both systems, and this aspect will be considered next.

**a. Cytogenetics and Cytochemistry during Development of Species with Heterochromatic Systems.** Most studies in development, cytogenetics, and cytochemistry have been made with an example of the lecanoid system, the mealybug, *Planococcus citri* (Risso), which can be readily grown in the laboratory. But corresponding results have been obtained by Kitchin (1970) in his experiments with *Parlatoria oleae* (Colvée), an armored scale insect with a Comstockiella system.

Four different experimental approaches have been used with *P. citri*: (1) analysis of effects of irradiation, (2) determination of expression and transmission of marker genes, (3) analysis of shifting sex ratios in the progeny during oviposition, and (4) cytochemical investigations.

*i. Effects of ionizing irradiation.* When males were X-rayed prior to mating, the heterochromatic chromosomes were fragmented in their sons. If females were so treated, the euchromatic chromosome set was damaged in the sons; if both parents were treated, aberrations were found in both sets. This experiment (Brown and Nelson-Rees, 1961) showed the paternal origin of the heterochromatic set. It also demonstrated that chromosome fragments, down to the smallest of those observed after treatment, will show the expected eu- or heterochromatic characteristics. There could not be, therefore, just a single center on each chromosome which controls its heterochromatization as suggested for the heterochromatization of the X chromosome of *Drosophila* at spermatogenesis (see Section I,B). The ability to become heterochromatic may thus be dispersed or diffused along the entire chromosome much like its centric capacity. In most organisms, treatment of gametes with ionizing irradiation induces lethality in the developing zygote even though the irradiated gamete is combined with an untreated gamete from the other parent. This effect, first recognized by Muller and Settles (1927) in *Drosophila* translocations, has been called dominant lethality, and is believed to be largely due to chromosome deficiencies. Organisms with diffuse centromeres seem to be less susceptible to induction of dominant lethality because both parts of a broken chromosome may survive (Bauer, 1967; Brown, 1961; Brown and Wiegmann, 1969). Although numerous transmissible asymmetric translocations have been recovered in the mealybug, no example of a transmissible chromosome simply broken into two pieces has yet been found.

When female mealybugs were irradiated and mated to normal males, the survival of the two sexes among the progeny was similar. There was increasing dominant lethality in both sexes, i.e., both sons and daughters declined in number, with increasing dosage of ionizing irradiation. But the results were dramatically different when males were irradiated and mated to normal females. The number of daughters declined much as it did after mothers were irradiated. The number of sons, however, remained at about control percentages up to the dose of 30,000 rep; above this dose, they usually declined in number (Brown and Nelson-Rees, 1961). In general, it is believed that it is not chromosome rearrangements per se that lead to dominant lethality but that most of it is caused by loss of genes necessary in two doses for the functioning of a diploid organism. Failure to induce dominant lethality in male mealybugs after doses of ionizing radiation to the fathers sufficient to produce numerous rearrangements, including deficiencies, indicated that the heterochromatic set must be genetically inert. Above 30,000 rep, damage to the paternal chromosomes was so great that the zygote nucleus could not go through the normal mitoses of the first few cleavage divisions and its development ceased (Chandra, 1963a).

Often, after such high doses, a triploid zygote substitute was formed by fusion of polar body I (diploid) with polar body II (haploid). Normal development ensued to yield vigorous and fertile adults, all female, and these, in turn, produced further triploids, again always female. Unlike the situation in most organisms, the polar bodies do not normally degenerate in the mealybug but participate in the formation of large, polyploid cells which house the intracellular symbionts. Since the polar bodies are already active in one phase of embryogenesis, their ability to sustain normal development is not surprising (see also Chandra, 1963b).

Certain ambiguities appearing in the results of Brown and Nelson-Rees (1961) were analyzed by Nelson-Rees (1962) and by Nur (1966c, 1967a). The percentage of sterile sons increased rapidly with increasing paternal radiation dosage, from 15,000 rep upward, but the sons always showed an approximately normal amount of heterochromatin regardless of how markedly this had been rearranged (e.g., into one very large and several small chromosomes) (Nelson-Rees, 1962). Nur (1966c) found that the heterochromatic chromosomes cease to replicate in some tissues or replicate only once while the euchromatic set forms large polyploid nuclei by some type of endoreplication. In a few tissues, however, such as the gut, the Malpighian tubules, and the testis sheath cells, the heterochromatic set reverts to the euchromatic state. After high dosage paternal irradiation, the Malpighian tubules may be grossly malformed; this dominant lethal effect provides evidence that the paternal set becomes genetically active on deheterochromatization (Nur, 1967a). It also helps to account for Nelson-Rees' (1962) observation that an approximately normal amount of heterochromatin is necessary for the survival of the male after high-dosage paternal irradiation. These males would have the fewest deficiencies affecting tissues in which the paternal set was deheterochromatized. Unlike the Malpighian tubules, the testis sheath forms a simple layer so that dominant lethal-like morphological effects would not be detectable and were not observed. The testis sheath is believed to play a role either in sperm maturation or in sperm bundle formation or in both (Nur, 1962a) and it is the impairment of such functions that is presumably responsible for the sterility of the sons of irradiated fathers. Meiosis itself appears normal in such males; no matter how rearranged, the heterochromatic chromosomes follow their regular sequence of maneuvers (Nelson-Rees, 1962).

In concluding this section on the effects of ionizing radiation, it should be noted that the paternal origin of the heterochromatic set was similarly demonstrated by Kitchin (1970) for *P. oleae* in which the meiotic sequence is exclusively of the Comstockiella type (Section III,C,1,d).

*ii. Genetic markers.* Despite numerous experiments with X rays, only a few marker genes have been found in the mealybug (Brown and Wiegmann, 1969). One is an eye color mutant, salmon-eye, eye color can be observed in both males and females. Salmon-eye is recessive in the female. The males express and transmit only that allele which was received from the mother. Other genetic markers affect the wings. Since the females are wingless, these markers are expressed only in the males where they follow the same rules as those for salmon-eye: they are expressed and transmitted by the male if the mutant allele has been received from the mother regardless of whether the wild-type or mutant has been received from the father.

*iii. Sex ratios and sex determination.* Mealybug females usually produce 200–300 eggs, and progenies include, on the average, about 40% males There are no sex chromosomes, and sex ratios of progenies vary considerably among individual females (James, 1937). When aged, virgin females are mated, the sex ratio of their offspring shifts almost completely in favor of males. Environmental factors such as temperature and humidity also affect the sex ratio (James, 1938; Nelson-Rees, 1960). Such observations suggest that sex ratio is determined by the mother.

That sex determination is in large part, if not exclusively, under maternal control was shown by Nur's (1963) study of a parthenogenetic coccid in which a zygote substitute was formed by fusion of the division products of the haploid egg nucleus. Some of the embryos formed in this fashion were male. Since the mother also originated from such a zygote substitute, she herself must have been completely homozygous and there could have been no segregation of genetic factors for sex determination (see also Section III,C).

Reasons for the variability and other peculiarities of sex determination must therefore be sought in the interaction of the maternal genotype with other influences. In the armored scale insect, *Pseudaulacaspis pentagona* (Targ.), the females first lay eggs with only female embryos and then without interruption switch to eggs containing only male embryos (Dustan, 1953; Bennett and Brown, 1958).

In spite of numerous observations and experiments, the final answers to the question of sex determination in coccids with facultative chromosome systems are still missing. Since segregation of genetic factors for sex determination has been ruled out (see above), only two alternatives remain. The difficulty lies in finding evidence to discriminate between them. In both cases the determinative events are presumed to be under the control of the mother who is influenced by environmental and developmental factors. In the first alternative, the mother determines the constitution of the egg so that, on fertilization, the chromosomes in the sperm are or are not imprinted. If the paternal set is imprinted and heterochromatized, the embryo is a physiological haploid and becomes male for the same unknown reasons that unfertilized eggs develop into haploid males in Hymenoptera (bees, wasps, and ants) and several other groups of organisms including the iceryine coccids (Hartl and Brown, 1970). This may be regarded as the indirect method of sex determination. On the other hand, sex and heterochromatization could be directly but independently determined by maternal influences acting via the egg cytoplasm.

Some evidence for the direct method of sex determination comes from the triploid females which stem from fusion of polar body nuclei after high dosage paternal irradiation has prevented development from the zygote (see above). These triploid offspring may approach in frequency 40–60% of the control value for all offspring (Brown and Nelson-Rees, 1961). These data suggest that female embryos probably cannot develop in cytoplasm predetermined toward maleness thus limiting the percentage of triploid females which can be produced. Further suggestive evidence for direct determination comes from observations on other coccids.

There are three species known in which the chromosome complement is diploid in both sexes, and there is neither heterochromatization nor are there sex chromosomes. The unusual genus, *Stictococcus*, has already been considered (Section II,B). In a primitive genus, *Orthezia*, Brown (1958) found no sex chromosomes in the male at meiosis when the achiasmatic bivalents could be closely observed but in another ortheziid genus, *Newsteadia*, U. Nur (unpublished) observed what are presumably XX and XO embryos ($2n = 14$ and 13, respectively). The Australian genus, *Lachnodius*, is morphologically between the mealybug lineage and the eriococcid lineage in both of which heterochromatization occurs. In *Lachnodius*, the chromosomes appear similar in both sexes; no X chromosome was apparent at meiosis in the male which was again achiasmate (S. W. Brown, unpublished).

Direct determination of sex by morphogenetic factors in the egg cytoplasm thus seems a possibility in some coccids but its occurrence in the facultative systems is an open question.

*iv. Cytochemistry.* The molecular basis for heterochromatization and gene inactivation has been of cardinal interest in the facultative systems of mammals and coccids. The reversible, facultative heterochromatin of the mealybug is like constitutive heterochromatin in two important respects: it is late-replicating (Baer, 1965; Sabour, 1969) and it does not synthesize RNA (Berlowitz, 1965a; Sabour, 1972). There is apparently some change in the relationship between the proteins and the DNA in mealybug heterochromatin (Berlowitz, 1965b) although the histones from male and female mealybugs seem to be the same (Comings, 1967a; Pallotta *et al.*, 1970). Heterochromatic chromosomes can act as templates for *in situ* RNA synthesis after treatment with a synthetic polyanion, polystyrene sulfonate (Miller *et al.*, 1971) which combines with histones. The relationship of the RNA synthesized by heterochromatic chromosomes after polyanion treatment to the RNA synthesized by the euchromatic homologs is not known. Compared with the controls, $^3$H-uridine labeling is increased over both the eu- and heterochromatic sets by the polyanion treatment but proportionately much more so over the heterochromatic set. It is possible that some sort of nonspecific transcription is induced, at least in part, by the synthetic polyanion. Further details of this complex problem are presented in a review by Berlowitz (1974).

This question leads directly to another: To what extent is the heterochromatic appearance of a chromosome or chromosome segment simply an expression of its incapacity to serve as a template for RNA synthesis? The relationship between heterochromatization and gene action seems to be fairly constant in the coccids but not in mammals (see Section II). In both mammals and mealybugs, deheterochromatization occurs in the life cycle, but unlike the heterochromatic chromosomes of the coccids, the Barr body of man may become undetectable cytologically without restoration of genetic activity (Comings, 1966, 1967b).

**b. Meiosis in Females.** Meiosis in females of coccids with facultative chromosome systems appears to retain an unaltered inverted sequence except in those cases of parthenogenesis in which a single equational division replaces meiosis. In triploid females of *P. citri*, Chandra (1962) observed equational separation of chromatids at anaphase I and the reformation of trivalent configurations during interphase and early prophase II. Brown and Wiegmann (1969) made similar observations with several different translocations: the multivalent configurations reappeared in the second division.

**c. Meiosis in Males with the Lecanoid System.** The meiotic sequence in the lecanoid system consists of an uncomplicated series of two divisions (Hughes-Schrader, 1935; Brown, 1959). In the first division, both the euchromatic and heterochromatic chromosomes divide equationally, as in a mitotic division (Figs. 2 and 4). In both second metaphase division figures, the chromosomes arrange themselves in a double metaphase plate with

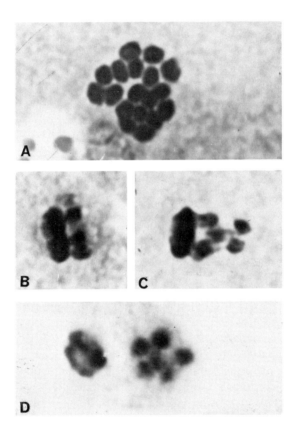

**Fig. 4.**   Selected meiotic stages of the lecanoid meiotic sequence of *Cerococcus quercus* Comstock (see also Fig. 2). (A) Metaphase I; a haploid set of nine euchromatic chromosomes at the top and a cluster of nine heterochromatic chromosomes at the bottom; the first division is equational for both types of chromosomes. (B) Metaphase II; a double plate is formed with the heterochromatic chromosomes on one side (left), the euchromatic chromosomes on the other (right). (C) Anaphase II; euchromatic chromosomes (right) are drawing away from the cluster of heterochromatic chromosomes (left). (D) Late telophase II; segregation of the heterochromatic (left) and euchromatic chromosomes (right) has been completed; only the euchromatic derivatives will form sperm (see Fig. 7 for spermiogenesis). From Brown (1959).

the heterochromatic chromosomes toward one pole and the euchromatic chromosomes toward the other; movement to the poles thus segregates the euchromatic chromosomes from the heterochromatic. Of the four nuclei resulting from the lecanoid meiotic sequence two are heterochromatic and two are euchromatic. Only the two euchromatic nuclei form sperm; the heterochromatic nuclei degenerate *in situ*.

Separation of eu- and heterochromatic chromosomes does not always occur at second anaphase if heterochromatic supernumerary (=B) chromosomes are present or a translocation has combined a eu- and a heterochromatic segment in the same chromosome. The supernumeraries undergo a sudden change of state between the first and second meiotic divisions in the male; they become less heterochromatic in appearance and usually but not always move with the euchromatic set. The second meiotic division reduces the chromosome number of the regular chromosomes by segregating a haploid euchromatic set from a haploid heterochromatic set. For the supernumeraries, the second division acts as an accumulation mechanism since most of the supernumeraries move with the euchromatic set (Nur, 1962b, 1966a,b). Supernumerary chromosomes have been reported from only a few of the many species of mealybugs examined cytologically.

If young males are irradiated and allowed to grow, testis sectors can be found in which individual chromosomes, part eu- and part heterochromatic, can be observed in meiosis. If the eu- and heterochromatic segments are approximately equal in size, then each tends to move toward the appropriate pole at anaphase II, and, in so doing, forms a bridge between the segregated groups of eu- and heterochromatic chromosomes. If one of the segments is quite small, the chromosome moves to the pole appropriate for the larger segment (Nelson-Rees, 1963; Nur, 1970). At the level of resolution of the light microscope the two types of chromatin appear to maintain their constant relative lengths. This observation is consistent with the bit-by-bit regulation of heterochromatization suggested above from the typical heterochromatization of small chromosome fragments occurring in embryos after paternal irradiation.

**d. Meiosis in Males with the Comstockiella System.** It is in the meiotic modifications of the Comstockiella system that the facultative systems of the coccids reach their greatest complexity. As a result of an experimental analysis by Kitchin (1970) much of the complex cytological variation can now be simply understood.

The Comstockiella system is characterized by the disappearance of one to all heterochromatic chromosomes during the preprophase stage immediately prior to meiosis (Fig. 2). Heterochromatic chromosomes which may remain after preprophase are eliminated during or after the meiotic sequence and, like the heterochromatic residues in the lecanoid system, they slowly degenerate during early spermiogenesis.

From a study of coccids collected in nature, Brown (1963) proposed that the heterochromatic chromosomes which disappear at preprophase of meiosis revert to the euchromatic state and pair with their homologs in the euchromatic set. In an extensive series of irradiation experiments with an armored scale insect, *P. oleae*, Kitchin (1970) induced chromosome rearrangements in the paternal chromosome set. Deheterochromatized paternal chromosomes which had paired with their maternal, euchromatic homologs would be expected to show (1) heteromorphic bivalents at meiosis, and (2) transmission of rearranged chromosomes to the offspring. Exhaustive search failed to show either; altered chromosomes appearing in meiosis were always single heterochromatic entities, never partners in heteromorphic bivalents. And no rearrangements were recovered in the progeny. This series of experiments led to only one interpretation, that the paternal chromosomes are destroyed during the preprophase period, and not deheterochromatized.

This result was of special significance for three reasons. (1) In most if not all other cases of chromosome elimination in the coccids and elsewhere, the chromosomes to be destroyed are first separated from the other chromosomes. They lag during division or are ejected from a nucleus or telophase group. (2) Preprophase disappearance of heterochromatic chromosomes in the Comstockiella system appears to be almost instantaneous, in contrast to the slow degeneration of chromosomes eliminated by lagging or ejection. (3) In most examples of the Comstockiella system, two different methods of getting rid of heterochromatic chromosomes are seen in the same cell: a part of the heterochromatic set is quickly destroyed at preprophase, the remaining chromosomes are eliminated or segregated at meiosis and degenerate slowly.

The third point just mentioned is of special interest. Evidence to be cited indicates that the heterochromatic chromosomes are individually programmed during early development of the testis to determine by which of the two possible methods each will be eliminated. But first a brief summary of meiotic sequences with the Comstockiella system is necessary and is based on studies by Brown (1957, 1963, 1965, 1967), Nur (1964, 1967b), and Kitchin (1970, 1975).

A Comstockiella system may be defined as one in which at least one chromosome undergoes preprophase destruction in at least some testis cysts of some individuals of a species. Comstockiella systems are known from only two groups of coccids, the eriococcids and the diaspidids (*sensu lato*), and species are found infrequently in which either very few or, at the other extreme, all the heterochromatic chromosomes are destroyed at preprophase. In most species, about 75% of the heterochromatic set is destroyed at preprophase and the remainder is eliminated at meiosis (Fig. 2).

When only a few chromosomes are destroyed at preprophase, meiosis is usually like that of the lecanoid system and the remaining heterochromatic chromosomes are segregated from the euchromatic chromosomes at the second meiotic division. When most or all the heterochromatic chromosomes are destroyed at preprophase, or the total chromosome number is small ($2n = 6$ to 10), the meiotic sequence usually consists of a single division. If there is only a single division, the heterochromatic chromosomes remaining after preprophase destruction are eliminated in one of two ways. They lag at anaphase and are thus excluded from either daughter nucleus, or they may divide and move successfully to the poles only to be ejected from the clumps of telophase chromosomes. If heterochromatic chromosomes escape the apparently rapid preprophase destruction, they later degenerate slowly regardless of the route of disposal, by segregation, lagging, or ejection.

In many species, especially among the eriococcids, there is considerable variation from cyst to cyst within an individual testis in the number of heterochromatic chromosomes destroyed at preprophase. Each cyst contains about 16 spermatocytes and the number of chromosomes destroyed at preprophase is usually quite constant within a cyst. That this constancy is probably determined by specific programming of individual chromosomes is indicated by observations on a palm scale, *Ancepaspis tridentatu* (Ferris), in which the haploid chromosome number is 3 and the long chromosome can be clearly recognized from the preprophase stage to the dense residues formed by the chromosomes after elimination. In this species, one chromosome regularly escapes preprophase destruction. When the large chromosome escaped preprophase destruction, its constancy within a cyst was obvious at all later stages at which residues were detectable. Measurements of residues remaining after elimination showed three size classes, each constant within a cyst, and thus indicated that any one of the three chromosomes can escape preprophase destruction (Brown, 1963). Variation in the chromosome escaping preprophase destruction was on a cyst-to-cyst basis, not on a larger sector-to-sector basis within the testis. It seems likely, therefore, that the chromosome which is later to escape destruction is chosen in the primary gonial cell of each cyst.

From the fact that species are rare in which all the heterochromatic chromosomes are destroyed at preprophase, Kitchin (1975) concluded that heterochromatic chromosomes remaining after preprophase may represent a built-in safeguard against a residual ability of the lytic enzymes to attack euchromatic chromosomes. A programming in primary gonial cells for later resistance to chromosome lysis at preprophase would provide such a safeguard.

A second type of reprogramming has been observed in an armored scale, *Nicholiella bumeliae* Ferris, in which a lecanoid system with two meiotic divisions may occur in one testis or large sector thereof, while a Comstockiella system occurs in the other testis or other sectors within the same individual. In the Comstockiella system of this species, a single chromosome escapes preprophase destruction and is ejected from the telophase nuclei after the single meiotic division (Brown, 1963). The pattern of variation, involving large sectors and individual testes, indicates that the choice of system is made very early in the development of the testis. Presumably the choice does not here involve a programming of individual chromosomes. Rather the choice should probably be regarded as alternative activation of different sets of genes in different cell lineages; activation of one set would eventually lead to a lecanoid system; activation of the other, to a Comstockiella system. Regardless of route, *N. bumeliae* successfully eliminates the heterochromatic chromosomes.

The two examples just cited from the armored scale insects indicate a chain of events following imprinting during which chromosome behavior is subject to further control. Early control appears to be developmental in nature and determines whether large sectors will have a Comstockiella or lecanoid sequence. Later control is responsible for programming individual chromosomes of each cyst for destruction either before or after meiosis.

The preprophase destruction of heterochromatic chromosomes, an intranuclear event, is highly suggestive of a process like the modification-restriction system of certain prokaryotes which functions selectively to destroy alien DNA on its entry into a cell. It has been suggested that a system similar to the modification-restriction system functions, in whole or in part, in the differential regulation of homologous chromosomes in eukaryotes (see Section V).

### 2. System Involving Elimination

In the two major tribes of armored scale insects, the majority of the species eliminate the paternal chromosomes during cleavage from both soma and germ line in male embryos. This system may be called either the diaspidid or the elimination system (Fig. 2). In most species, there is an adjunct pentaploid tissue stemming from the fusion of polar bodies and a cleavage nucleus, and the large cells so formed house the intracellular symbionts. The paternal chromosomes are not eliminated from the pentaploid tissue nor do they become heterochromatic.

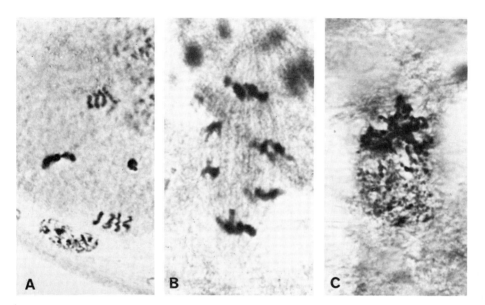

**Fig. 5.** Normal and abnormal elimination of the paternal set at early embryogeny of the armored scale insects (Section III,C,2). (A) Normal elimination; a haploid set of four is present at each pole; the eliminated chromosomes remain undivided in the equatorial zone, one at the right, three at the left. (B) Abnormal elimination; the paternal set divides and moves part way to the poles. (C) Unsuccessful elimination; rarely the paternal set succeeds in reaching the poles where it appears as heterochromatin, here at the top of the nucleus. [(A) *Aspidiotus destructor* Signoret, (B) *Aspidiotus simulans* De Lotto, (C) *Lindingaspis rossi* (Maskell)]. From Brown (1965).

Elimination is accomplished by anaphase lagging of the paternal set (Figs. 2 and 5). Paternal origin was determined by irradiating fathers or mothers and examining the chromosomes which remained after elimination (Brown and Bennett, 1957; Bennett and Brown, 1958).

Elimination occurs at a cleavage stage prior to the migration of the cleavage nuclei to the periphery of the egg and consequently at a stage earlier than that of heterochromatization. The chromosomes do not usually become heterochromatic prior to elimination. If elimination is unsuccessful and the paternal chromosomes are incorporated in telophase nuclei, they then become heterochromatic (Fig. 5). These heterochromatic chromosomes are eliminated at the next division or the nuclei degenerate; they are not observed at later stages of development.

In all the palm scales and in those armored scales which do not undergo elimination, there is a Comstockiella system, or a Comstockiella–lecanoid mosaic, except for one palm scale in which the system seems to be strictly lecanoid (Brown, 1965). It has already been mentioned that the elimination system most likely had multiple origins in each of the two major tribes of armored scale insects. It is possible that the cleavage elimination system was derived from the Comstockiella system by a shift from meiosis to cleavage in the timing of gene action responsible for chromosome elimination.

The armored scale insects in which the elimination system occurs far outnumber those with heterochromatization. In a survey of the armored and palm scales, Brown (1965) found 89 species with the elimination system and 16 with heterochromatization of any sort. (Since a special effort was made in this survey to find rare and divergent types which might show variation in chromosome system, the study was probably biased *against* examples of the elimination system.) Clearly, getting rid of heterochromatization was a distinct advantage among the armored scales.

### 3. Facultative Behavior in Parthenogenetic Development

Nur's work with parthenogenetic coccids has provided compelling evidence on the time and place of imprinting in the life cycle. As in certain other groups of organisms, various types of parthenogenesis are common in the coccids (Nur, 1971). Diploid zygote substitutes are formed from the union of two haploid, maternal chromosome sets. Typical heterochromatization may later occur in one of these two sets of maternal chromosomes (Fig. 6).

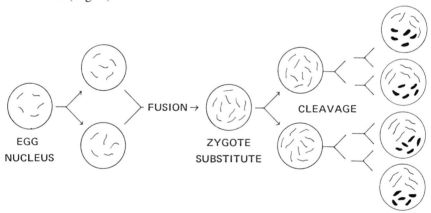

**Fig. 6.** Diagram of isoautomictic parthenogenesis. A haploid egg nucleus divides once and the daughter nuclei fuse to form a diploid zygote substitute. In some such parthenogenetic embryos of certain specialized coccids (see text), one haploid set of chromosomes becomes heterochromatic (dark chromosomes in nuclei on right) just as in sexually reproducing relatives. This sequence demonstrates that only one of the two haploid sets has been imprinted even though they were both division products of the same egg nucleus.

The species studied by Nur (1963, 1972) all belong to the soft scales, a group probably more closely allied to the eriococcids than to the mealybugs. Sperms and sperm bundles are easily recognized in the coccids. In the first species studied, *Pulvinaria hydrangeae* Steinweden ($2n = 16$), there were only females, all of which proved to be uninseminated. Examination of embryos showed that these were always female in two populations but in a third population male embryos also occurred at percentages ranging from 1 to 60 among ten of the fifteen females examined. The male embryos were identifiable because one haploid set of 8 chromosomes had become heterochromatic at an earlier embryonic stage and remained so thereafter.

Nur observed that meiosis was normal and resulted in the expected products, two polar bodies and a haploid egg nucleus. The egg nucleus divided once and the two daughter nuclei fused to form a diploid zygote substitute. Nur also established the important fact that the polar bodies could not have been involved in the development of the embryo as they are in some types of parthenogenesis. In *P. hydrangeae*, the polar bodies could be followed during the critical postmeiotic stages and were seen to divide in an irregular manner which would prevent their further participation in development. Since the polar bodies were thus removed, only the egg nucleus could have contributed to the diploid zygote substitute which was, therefore, completely homozygous.*

Of special significance for the present review is the conclusion that, in eggs destined to give male embryos, imprinting of the chromosomes of one of the daughter nuclei occurs during the brief interval between division of the egg nucleus and the subsequent fusion of the daughter nuclei to form the diploid zygote substitute. Presumably during this brief interval, one daughter nucleus moves into a cytoplasmic region where it undergoes imprinting; the other daughter nucleus does not. In eggs destined to give female embryos, it may likewise be presumed that either the egg cytoplasm is not "programmed to make male embryos" or that neither daughter nucleus moves into the requisite region. The former alternative seems at present to be the more acceptable.

A similar type of parthenogenesis was reported by Nur (1972) for two species of soft scales of the genus *Lecanium*. In their life cycles, these species differ considerably from *P. hydrangeae*. It will be recalled that in animal species with haploid males and diploid females, the females are

---

* Because of the genetic identity of the two haploid nuclei which fuse to form the diploid zygote substitute, this type of parthenogenesis may be referred to as *isoautomixis* (new term). It is known with certainty only from the coccids (White, 1973) where it occurs in one primitive species with only females as well as in the specialized coccids mentioned here.

produced bisexually, from sperms and eggs, and the males are produced unisexually by parthenogcnetic development of unfertilized eggs. A similar sort of life cycle occurs in the two *Lecanium* species. All females are produced bisexually, by fusion of eggs and sperms. All males are produced unisexually by parthenogenesis of the same type as that described for *P. hydrangeae*. But in the two *Lecanium* species, imprinting of one of the daughter nuclei always occurs with parthenogenesis because only diploid males, with one haploid chromosome set heterochromatic, are produced.

The two *Lecanium* species represent a reversal of the situation found in the common facultative systems of completely bisexual coccids. In the bisexual coccids, the sperm is imprinted if the embryo is to be a male; the maternal contribution of chromosomes is never imprinted. In the two *Lecanium* species, the chromosomes of the sperm are never imprinted while the maternal chromosomes for the parthenogenetically derived diploid sons are *always* imprinted. The net genetic result is the same: parthenogenetically derived diploid males transmit only maternal chromosomes as do sexually derived diploid males (Fig. 7). The difference is the companion set of inactive, heterochromatized chromosomes which is not transmitted is of paternal origin in sexually derived males and of maternal origin in those derived parthenogenetically.

### 4. A Duplex Facultative System in Certain Armored Scales

The system to be described here occurs in three allied genera: *Aulacaspis*, *Phenacaspis*, and *Pinnaspis* in the Diaspidini, one of the two major tribes of armored scales. The sexual forms in these genera all show the elimination system. In most but not all species of these genera, heterochromatization occurs in the pentaploid nuclei destined to form the adjunct tissue housing the intracellular symbionts, and occurs in both sexes (Brown, 1965). In the armored scale insects, the adjunct tissue is derived from a single pentaploid nucleus formed from union of the two polar bodies and a cleavage nucleus.

The sequence of events could be followed clearly in *Phenacaspis pinifolieae* (Fitch) ($2n = 8$). After completion of several mitotic divisions, the pentaploid nuclei begin endoreplication resulting in large nuclei and cells. At the start of endometaphase of the first round of endoreduplication, eight chromosomes are considerably more condensed than the remaining twelve. The eight chromosomes then decondense and disappear from view. The twelve chromosomes next condense and then clump together to form a chromocenter (Fig. 8). Because the chromocenter becomes quite large and is intensely Feulgen-positive during subsequent enlargement of the nucleus and cell, it seems likely that it is also undergo-

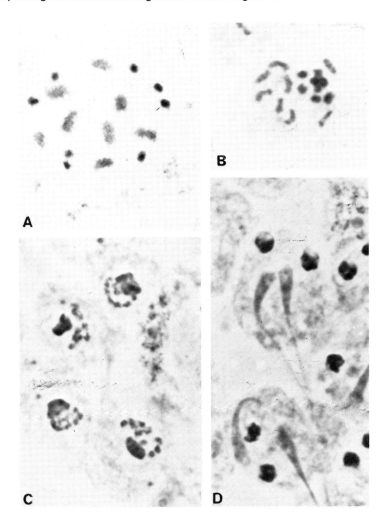

**Fig. 7.** Cytology of males derived from isoautomictic parthenogenesis in the soft scale, *Lecanium putnami* Phillips (Section III,C,3 and Fig. 6). (A) Mitotic prophase from the balstula stage with haploid sets of eight hetero- and 8 euchromatic chromosomes. (B) Prophase I of meiosis with eight euchromatic chromosomes surrounding the eight condensed and clustered heterochromatic chromosomes. (C) Interphase of meiosis with the euchromatic chromosomes to one side of the dense groups of the heterochromatic chromosomes. (D) Spermiogenesis; two complete spermatids (left) each with two dense heterochromatic derivatives and two partially formed sperms. From Nur (1972).

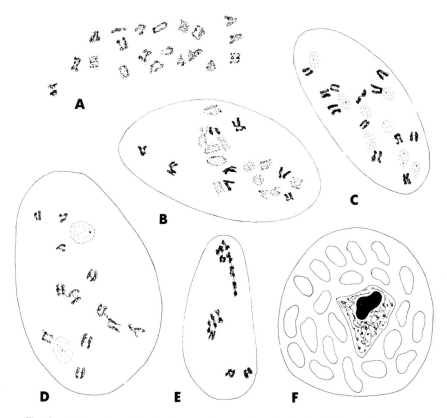

**Fig. 8.** Heterochromatization in mycetocytes in a duplex facultative system in an armored scale insect, *Phenacaspis pinifoliae* (Fitch) (Section III,C,4). The mycetocyte contains five haploid sets of four chromosomes each: two sets come from a cleavage nucleus, and three from polar bodies I and II. (A) The endomitotic division which signals the onset of heterochromatization. (B) Two sets become condensed; three sets are not condensed. (C) Three sets become condensed while the two previously condensed sets become diffuse. (D) Three sets remain condensed while two sets continue to decondense; here only two of the diffuse chromosomes were still visible. (E) The three sets of heterochromatic chromosomes alone are visible and are beginning to cluster. (F) Mycetocyte from an adult female at smaller magnification; the cytoplasm is filled with intracellular symbionts (shown in outline); the polygonal nucleus contains a large chromocenter. From Brown (1965).

ing endoreplication. In the common unisexual race of *P. pinifoliae*, mitosis presumably replaces meiosis to give a single polar body with eight chromosomes and an egg nucleus with eight chromosomes; the latter serves as a zygote substitute. Sixteen chromosomes instead of twenty are found in the polyploid sector. Chromocenters are again present in the polyploid sector but the mode of their formation is not clear. Evidence from the bisexual race indicates that the polar body derivatives are hetero-chromatized and that from the unisexual race shows that a paternal complement is not necessary for heterochromatization.

When these results are considered in view of the conclusions on parthenogenesis, the following picture emerges for the bisexual species with the duplex system. (1) The eggs imprint the sperm chromosomes if the embryo is destined to become a male, but not if it is destined to become a female. If the sperm is imprinted, the paternal chromosomes are eliminated at cleavage. (2) The eggs always imprint both polar bodies so that the chromosomes derived from them become heterochromatic during the first round of endoreplication in the adjunct polyploid tissue in both male and female embryos. The second type of imprinting can occur only in unisexual strains.

## D. Concluding Remarks

The coccid chromosome systems reveal a wide range of types within a small and relatively homogeneous group of organisms. All coccids are plant parasites and all are relatively sedentary. It may be presumed that variation in their chromosome systems represents attempts at more precise adjustment to a mode of life favoring colonial distribution and inbreeding when colonies are small. These are believed to be the circumstances conducive to the evolution of male haploidy (Hartl and Brown, 1970). Yet true male haploidy is represented but once, in the iceryines, one of the primitive tribes of coccids. There is no true male haploidy among the specialized coccids, but their facultative systems represent various types of parahaploidy (see Section I).

It is of interest to note that the inverted meiotic sequence seems to be preadapted for both haploidy and parahaploidy. Since the first meiotic division is equational, suppression of the second division in a haploid male would result in two haploid nuclei capable of forming genetically complete sperm. Heterochromatization of the paternal complement would again lead to no change in the first meiotic division. During the second meiotic division, the marked tendency of the heterochromatic chromosomes to clump together would block interphase pairing of homologous chromatids and favor differential behavior at meta- and anaphase II. Parahaploidy rather than haploidy may have evolved because of chance occurrence of mutations favoring a facultative system rather than for development of unfertilized eggs.

Paraphaploidy also occurs in *Sciara* in which the paternal set remains genetically active but is eliminated at meiosis. Presumably the differential behavior is controlled in *Sciara* via the centromere. Such control is not possible in the coccids because the centromeres arc diffuse.

In all cases in the coccids in which imprinting occurs, the homologous chromosomes are separated at or about the time of fertilization. Entities which, on the basis of their subsequent behavior, appear to have been imprinted are (1) sperm, (2) one of two identical nuclei in isoautomictic parthenogenesis, and (3) polar bodies in a restricted group of armored scale insects. In regular sexual reproduction some sperms are imprinted, some are not. If sex is not determined by the physiological haploidy resulting from heterochromatization, then sex and heterochromatization are at least predetermined together. Only eggs destined to produce males will imprint sperm.

Once imprinted and heterochromatized, coccid chromosomes are subject to what is essentially heterologous regulation during development. The active euchromatic set is presumably responsible for the tissue differentiation in which the heterochromatic set divides in step with the euchromatic set or, in endoreplication sequences, either fails to divide, or reverts to the euchromatic state. In the Comstockiella system, gene action on the part of the euchromatic set is again presumably responsible for a series of variable and flexible controls, Comstockiella versus lecanoid in those species in which both systems occur, and the later cyst-by-cyst programming of preprophase destruction. Evidence from the parthenogenetic coccids indicates that this complex, differential regulation of homologous chromosomes is the end result of events believed to occur in the egg during the brief period when two sets of chromosomes, egg and sperm or daughter nuclei, are spatially separated.

## IV. *SCIARA* AND THE CECIDOMYIDAE

Both the Sciaridae and the Cecidomyidae belong to the division Nemocera of the Diptera, which also includes the mosquitoes and the chironimids, but is only distantly related to the division which includes *Drosophila*. As dipterans the life cycles and morphology of the two families will be a more familiar to biologists than are those of the coccids. Both families have very clear-cut patterning in the giant chromosomes of the salivary glands in which the X chromosomes and the autosomes can be seen but not the chromosomes limited to the germ line (see below). Precise cytogenetic studies of *Sciara* have been pursued for several decades but were limited to a few species of the genus. Nearly all the work with cecidomyids has been cytological, both descriptive and experimental, but numerous genera have been studied. Because of the lack of genetic evidence, interpretation of the cecidomyid chromosome systems must be based on a presumed analogy with those of *Sciara*.

## A. Chromosome Systems of *Sciara*

About twenty of the several hundred species of *Sciara* have been examined cytologically, but most of the cytogenetic work has been carried out with *S. coprophila*.

The basic cystem is simple in outline (Metz, 1938). Each zygote contains two sets of autosomes and three X chromosomes: $A^mA^pX^mX^pX^p$ (m, maternal; p, paternal in origin). During development of the female soma and germ line, one paternal X is eliminated leaving $X^mX^pA^mA^p$ in both. Meiosis is essentially orthodox in the female and crossing-over has been demonstrated by genetic tests. During development of the male, one paternal X is eliminated from the germ line which thus has the same constitution as the germ line and soma of the female. Both paternal X's are lost from the male soma which has the constitution $X^mA^mA^p$. X chromosomes are eliminated from the soma by lagging at anaphase in cleavage divisions; they are eliminated from the primordial germ cells after these have reached the gonadal site and by the unusual process of extrusion through the nuclear membrane (Berry, 1939, 1941). Sex is thus determined by an $XX(♀)-XO(♂)$ system operative in the surrounding somatic tissue and determining differentiation of the gonads into ovaries or testes which themselves are alike chromosomally (Table II).

Meiosis in the males is unusual (Fig. 9). A monopolar spindle is present at the first division. The maternal chromosomes move toward the single pole; the paternal chromosomes, which are also oriented toward the pole, appear to move backward and are eliminated. A bipolar spindle is present during the second division and the maternal autosomes divide equationally. The maternal X moves precociously to one pole where its two chromatids remain. Only the meiotic product containing the two X chromatids forms a sperm which has the constitution, $X^mX^mA^m$ in regard to the origin of the chromosomes, but will be the paternal contribution to the zygote: $X^mX^pX^pA^mA^p$.

Some species of *Sciara* produce bisexual broods; in others, such as *S. coprophila*, the broods are unisexual. There are two types of females in *S. coprophila*: XX' females which usually produce only daughters and XX which usually produce only sons. Mating of any male with an XX' female results in a 1 : 1 ratio of the two types of females, XX and XX', and, with equal fertility of these two types, a 1 : 1 sex ratio will result. The dominant marker, Wavy, can be used to follow the X' chromosome in crosses

**TABLE II**

Chromosome Eliminations during Embryogeny of *Sciara*[a,b]

| Chromosome number and type in zygote | Female[c] | | Male[c] | |
|---|---|---|---|---|
| | Germ line | Soma | Germ line | Soma |
| Three L's | 2 L's (L) | 0 L's (3 L's) | 2 L's (L) | 0 L's (3 L's) |
| Three X's: $X^p X^p X^m$ | $X^p X^m (X^p)$ | $X^p X^m (X^p)$ | $X^p X^m (X^p)$ | $X^m (X^p X^p)$ |

[a] Chromosomes retained are cited first, those eliminated are in parentheses.

[b] Genetic constitution of the mother governs elimination of X's; females come from XX' mothers; males, from XX mothers.

[c] The L chromosomes are limited to the germ line of both sexes. They are absent from some species and may vary in number in those species in which they are present. The L's do not seem to be regulated on the basis of parental origin. One L is eliminated in the germ line of both sexes. At meiosis in the male, both L's enter the single sperm formed by each spermatocyte. Apparently the L's pair and disjoin regularly during meiosis in the female since only one is usually transmitted via the egg. The X's are eliminated later in development than L's and are much more regular in behavior. See text for further details and references for both the L's and X's.

(Crouse, 1943); the X' chromosome also includes a long paracentric inversion which may help to account for the maintenance of differences between X and X' (Crouse, 1960b).

### 1. Sex Determination or Predetermination

A series of experiments with translocations showed that the XX'–XX system determines the pattern of sex chromosome elimination in the offspring: the XX' constitution leads to somatic elimination of one X; the XX constitution directs two X's to be eliminated (Table II). The resulting number of X's in the soma of the embryo determines its sex. Crouse (1960b) utilized a translocation (PT1) between chromosomes II and X in studying this problem. (The notation, $II^X$, indicates that the centromere is in the II segment, not in the X segment, regardless of the size of the segments. It should also be noted that in $X^{II}II^X$, the two X segments add to equal one entire X, and the two II segments add to equal one entire II, barring small deletions occurring at the sites of the original breakages and reunions responsible for the translocation.) Because of irregular disjunction in the translocation heterozygote, some eggs receive either one or three chromosomes instead of two. Eggs from $X'X^{II}II\ II^X$ mothers sometimes contained only chromosome II from the translocation and were thus

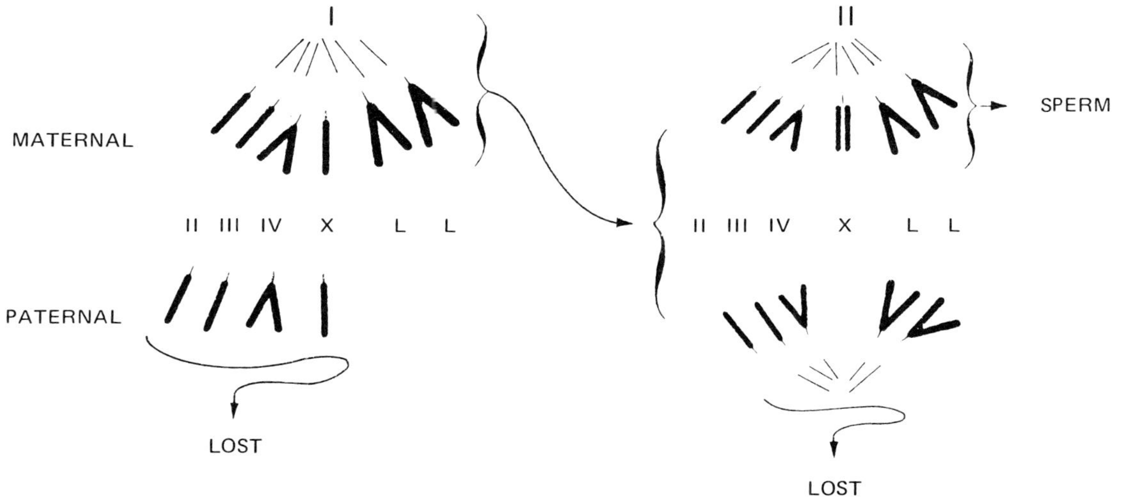

**Fig. 9.** Meiosis in males of *Sciara*. At anaphase of the first division (I) there appears to be a single pole toward which one set of chromosomes moves. The other set appears to move backward, away from this pole, and is eventually pinched off in a small bud of cytoplasm. Genetic studies showed that the lost chromosome set was of paternal origin. At anaphase of the second division (II), the chromosomes divide regularly except for the X which moves precociously to one pole where both of its chromatids remain. Most of the cytoplasm is associated with the meiotic product containing the two X chromatids, and it alone forms a sperm: the chromosomes going to the opposite pole are cast off in a rudimentary process. The single sperm thus contains a maternal set of autosomes and two maternal X's; these then become the paternal chromosomes of the next generation. Chromosomes X, II, and III have quite small short arms which are not shown in the diagram. In species in which limited (L) chromosomes occur, usually two in number, both go with the maternal chromosomes at I and divide with the autosomes at II. Meiosis in males serves as an accumulation device for the L's as these are not reduced in number, as chromosomes usually are, by the meiotic process. (Diagram adapted from Metz, 1938.)

$A^m$, with no X or X segment present.* On fertilization with XXA sperm, an $X^pX^pA^mA^p$ zygote is produced. If $X'$ acts to eliminate *one* $X^p$ from an $X^mX^pX^pA^mA^p$ zygote and thus produces XX females, the same elimination process in the case of the exceptional zygote, $X^pX^pA^mA^p$, should produce $X^pA^mA^p$ embryos. These XO males were indeed produced from XX′ mothers with a frequency which ranged from 2.5 to 17.3%, and an average of 9.1%; the XO males were infertile but otherwise normal (see below).

Exceptional females were obtained from XX females (normally male producing) when the egg received three chromosomes, $XX^{II}II^X$, from the four included in the translocation heterozygote, $XX^{II}II^XII$; on fertilization by an XX sperm, the zygote would contain four X chromosomes, two from the mother, $XXII^X$ ($= X X II$), and two from the sperm. If XX acts to direct the elimination of two X chromosomes from the somatic cells of the progeny, then in this exceptional case the elimination of two X's will leave two X's remaining, and the individuals will be females, not males. An average of 4.9% of such exceptional females was thus obtained from XX mothers heterozygous for the translocation. Of these, about a third were wild-type, two-thirds showed the X-linked gene, swollen (sw), for which the mother had been homozygous. From among the four X chromosomes, the two paternal chromosomes had thus been eliminated in two-thirds of the cases; in the other third, one paternal (carrying the wild-type allele) and one maternal chromosome had been retained, and one paternal and one maternal chromosome had been eliminated.

Since only one X is normally eliminated from the germ line, the exceptional females would in this case be expected to be trisomic in their germ lines, with two X's from the mother $XX^{II}II^X$, sw/sw, and one from the father, X,+. In breeding tests, the exceptional females proved to be trisomic as expected; they produced exceptional daughters (XX instead of X eggs) and the + allele appeared in both sons and daughters indicating retention of a paternal X chromosome.

This series of experiments thus conclusively showed that sex is determined by the chromosomes remaining in the soma after elimination.

The infertility of the exceptional males is perhaps attributable to the atypical XO state of the germ line, but a persistent effect of the "wrong" maternal cytoplasm cannot be ruled out. In a later study, Crouse (1966) found that two among 380 exceptional males were fertile and these produced a total of 5 offspring. Thus, functional sperm can be formed but only rarely. *Crouse concluded that the change in state (imprinting) of the paternal X, destined to be eliminated at the first meiotic division, prevented it from*

* See Fig. 10 for an example of such irregular disjunction, but note that Fig. 10 is about quite a different translocation between II and X (OR T-1) than the one considered here (PT1).

*functioning properly in the germ line.* Thus the exceptional males are infertile, not because they are XO instead of XX in the germ line, but because they are $X^pO$ instead of $X^mX^p$. Further support for this interpretation came from "XO" testis sectors in which the "X" was present in two segments as the result of a translocation. The bulk of the X was in $X^{IVm}$ and a much smaller segment in $IV^{Xp}$. In these "XO" sectors in which the bulk of the "X" was maternal in origin, meiosis appeared to be normal (Crouse, 1965).

## 2. *The Controlling Element of the X Chromosome*

Crouse (1943) showed with translocations between autosomes and X chromosomes that the rearranged chromosome which contained the centromere of the X moved precociously at anaphase II, and both its chromatids were thus included in the functional sperm. She was later able to separate the controlling element of the X from the nearby X centromere and move the element to autosome II (Crouse, 1960a). The controlling element proved to be in the short arm of the X and adjacent to the centromere (Fig. 10). The X, II, and III are all acrocentric, with quite small short arms; IV is metacentric. In the translocation in question, OR T-1, a break near the centromere of the X and another, cutting off the distal two-thirds of II, resulted in a grossly asymmetric rearrangement. The translocated segment of the short arm of the X was now placed some distance from the centromere of II, a distance equal to one-third of the original length of the long arm of II. Nonetheless, typical $X^m$ behavior was induced in $II^X$ but not in $X^{II}$. $II^X$ moved precociously in most (but not all) anaphase II divisions; two division products of $II^X$ were included in the sperm, and $II^{Xp}$ was subject, like $X^p$, to elimination in the developing embryo (Tables IIIA and IIIB).

Several questions remain, at least in part. Are there similar controlling elements adjacent to the centromeres of the normal autosomes which govern the behavior of the chromosomes at the first meiotic division when the entire paternal set appears to move backward? Does such an autosomal element act by inhibiting or otherwise influencing the adjacent centromere? Are these controlling elements the sites which permit imprinting, later expressed as differential behavior at anaphase I but without effect on the expression of paternal genes in the soma? When the X controlling element is placed some distance from the centromere, as in $II^X$ (OR T-1), does its regulatory activity somehow bridge the distance because something moves along the chromosome to the centromere? Is it the occasional loss of this something from the $II^X$ chromosome which leads to occasional equational division of $II^X$ at second anaphase?

One paternal X is lost from all tissues, germ line and soma of both sexes. It will be recalled that the two paternal X's present in the sperm were

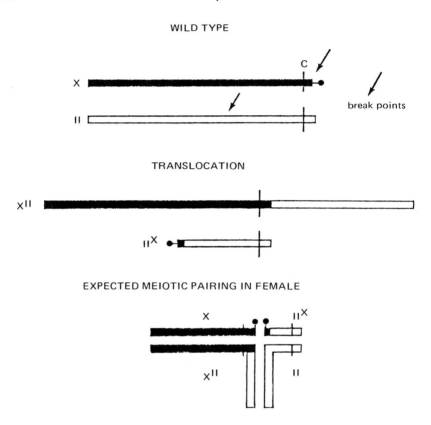

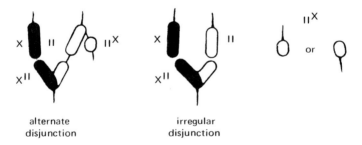

**Fig. 10.** Translocation in *Sciara* (based on Crouse, 1960a, on translocation OR T-1). (Centromeres are indicated by short, vertical lines; see C in uppermost chromosome.) In the ORT-1 rearrangement, a small distal segment of the short arm of the X was moved to the end of the proximal third of II. II$^X$ then behaved like an X and X$^{II}$ did not, thus indicating that the locus dictating the differential behavior of the X is normally located in the distal portion of the short arm of the X (see Tables IIIA and IIIB). In the giant salivary gland chromosomes, the

**TABLE IIIA**

**Chromosome Constitution in *Sciara* with the Translocation OR T-1[a,b]**

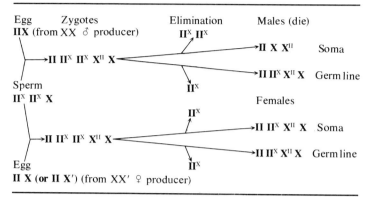

[a] Adapted from Crouse, 1960a.

[b] This translocation shifted the X-controlling element to an autosome (chromosome II). The scheme outlined here shows the results when the male parent carries translocated chromosomes ($II^X X^{II}$) derived from its mother. See Table IIIB for origin of such males, and sperm production by them. Only the chromosomes involved in the translocation are shown; autosomes III and IV and L (limited) chromosomes are not shown (see Fig. 9). The males in the above diagram die because of genetic imbalance. Note that $II^X X^{II}$ is genetically complete and equivalent to II X. The $II X X^{II}$ constitution of the soma is deficient for part of II on comparison with the normal II II constitution of the diploid soma, and is X plus an X segment so that the soma is neither XO (male) nor XX (female). The same imbalance would occur if the males had received the translocated chromosomes, $II^X X^{II}$, from the mother as well as the father. The zygote would be $II^{Xm} II^{Xp} II^{Xp} X^{IIm} X^{IIp}$. After elimination of the two $II^{Xp}$ chromosomes, the constitution of the soma would be $II^X X^{II} X^{II}$, genetically equivalent as noted above to II X $X^{II}$, and suffering from the same imbalance.

distal portion of the short arm of the X shows up as a small heterochromatic bleb (see Fig. 5 in Crouse, 1960a); this bleb is indicated in the present diagrams as a small knob. In heterozygotes for a reciprocal translocation, pairing of homologous segments leads to formation of a cross-shaped configuration with four limbs. In the present translocation, the limb represented by the distal segment of the short arm of the X (upper limb in pairing diagram) would be exceedingly small. Chiasma formation in the other three limbs would give a chain of 4; at anaphase, alternate disjunction of the chromosomes in the chain would result in genetically balanced meiotic products, X II and $X^{II} II^X$. Chiasma formation may be reduced in relatively short limbs such as that on the right formed by pairing of homologous segments of II and $II^X$. If a chiasma is not formed in this limb, $II^X$ will move independently of the other three chromosomes; if $II^X$ moves to the upper pole, then $X^{II}$ will be the only chromosome at the lower pole from this group of four. Such irregular disjunction of chromosomes in a translocation heterozygote is responsible for production of gametes with aberrant chromosome numbers, such as one or three of the chromosomes involved instead of two.

**TABLE IIIB**

**Origin of the Sperm in Crosses Outlined in Table IIIA**[a]

Female (soma and germ line)[b] (right, Table IIIA)
$$\text{II II}^X \text{ X}^{II} \text{ X}$$
↓          Wild-type male (see Fig. 9)

Eggs[c] $\text{II}^X \text{ X}^{II}$ + Sperms II X X
↓
Zygotes $\text{II}^X \text{ X}^{II}$ II X X
↓
Elimination of paternal X's in embryo yields:
Soma $\text{II}^X \text{ X}^{II}$ II (=II II X O[d])
Germ line $\text{II}^X \text{ X}^{II}$ II X (=II II X X[d])
↓
Spermatogenesis (see Fig. 9)

I. Maternal chromosomes $\text{II}^X \text{ X}^{II}$
II. Precocious behavior (centromere of II under control of translocated X-controlling element) $\text{II}^X \text{ II}^X$ X = Sperm (left, Table IIIA)

---

[a] Based on Crouse (1960a).

[b] To be male producers, females must come from an X egg in order to have the constitution XX; females from X' eggs are XX' and produce only daughters.

[c] $\text{II}^X \text{ X}^{II}$ is one possible meiotic product (see Fig. 10).

[d] Equivalent genetic constitution without translocation.

---

sister chromatids at second meiotic metaphase and moved together precociously to one pole. Is one of them marked or altered in some way at this time so that it is always the eliminated paternal X and its unaltered sister always the retained paternal X? In other words: Why the complication of the two X's in the sperm? An $X^P X^m A^P A^m$ zygote could eliminate $X^P$ from the male soma with apparently the same results for all tissues. An answer may emerge from an understanding of the probable evolutionary events which led to the establishment of this remarkable chromosome system.

### 3. The L Chromosomes, Heterochromatization, and Chromosome Elimination in the Germ Line

These three topics may be considered together since critical events involving all three occur at about the same stage in the germ line (Rieffel and Crouse, 1966) (see also Table II).

The L (= limited) chromosomes are large, apparently heterochromatic chromosomes and are so called because they are limited to the germ line. They are eliminated from the soma by lagging on the metaphase plate during a cleavage mitosis. L chromosomes occur only in certain species of *Sciara*. The number of L chromosomes in the germ line of *S. coprophila* may vary from zero to four. All of the L chromosomes present at meiosis in the male move with the maternal chromosomes to the single pole at anaphase I, and divide equationally during the second division (Fig. 9). The transmission of an *unreduced* number of L chromosomes would result in continued accumulation during successive generations unless compensation occurred by (1) inviability or infertility of individuals with extra L chromosomes, or (2) a compensatory elimination of L chromosomes elsewhere in the life cycle. Compensation is apparently achieved in the germ line with elimination of L's by the unusual mechanism of ejection through the nuclear membrane after about 30 primordial nuclei have migrated from the polar plasm to the definitive site of the future gonad. This elimination occurs along with two other events of considerable interest: (1) elimination of one paternal X chromosome by the same mechanism, and (2) decondensation of the paternal chromosome set (identifiable by use of translocations) so that the maternal chromosomes appear heterochromatic in contrast. The L's are also decondensed for a brief period, but condense again while all but two L chromosomes are eliminated at this stage in the germ line. This number is usually maintained in the female germ line, and the two L's form a regular bivalent at meiosis, and one L reaches the egg. If the two L's remained constant in the male germ line each zygote would have three L's and a regular system would occur in which the accumulation (= failure of reduction) at meiosis in the male would be precisely balanced by elimination in the early germ line. But irregularities occur in the further development of the male germ line so that cells entering meiosis may have 0, 1, 3, or 4 L's instead of the most common number, 2 L's, which is present in 78% of the spermatocytes. Since the deviations in opposite directions are about the same (0 L's, 0.3% or 4 L's, 0.9%; and 1 L, 11.3% or 3 L's, 9.5%) (Rieffel and Crouse, 1966; Crouse *et al.*, 1971), they are most likely due to nondisjunction in which both daughter gonial cells have about equal chance of survival (11.6% losing L's versus 10.4% gaining L's).

In another species, *S. impatiens*, Crouse *et al.* (1971) made reciprocal crosses between strains with and without L chromosomes. The L's showed no evidence of prior imprinting; whether maternally or paternally derived, they were regularly found in the single meiotic product destined to form a sperm. And progeny tests proved that these paternal L's were indeed transmitted by the fathers to their offspring.

Failure of the heterochromatic L's to undergo imprinting may possibly provide a clue to the differential diffuse state of the paternal chromosome set in the primordial germ cells. Since heterochromatization is elsewhere associated with inactivation and elimination of the affected chromosomes, it is puzzling that the paternal set, soon to be eliminated, has a "euchromatic" appearance and the maternal set, which will be retained, appears "heterochromatic." A possible explanation is that the paternal chromosomes of the germ line require a second set of regulatory instructions not applicable to those of the soma.

If this view is valid, it would then seem likely that such regulatory instructions are recognizable by chromosomes in a euchromatic state but are not recognized by chromosomes in a condensed, heterochromatic state. The L's, which are also decondensed during this brief period in the development of the germ line (but more briefly so than the paternal set), may also be programmed at this time for their subsequent behavior in meiosis. Since the X's, the autosomes, and the L's are all different genetically, they can respond to different regulatory signals or differentially to the same regulatory signal (see Section I). As noted in a preceding paragraph, the L's, unlike the X's and autosomes, are not subject to the basic imprint affecting paternal chromosomes. According to the ideas presented here, the L's could escape from the basic imprint either (1) directly because their genetic constitution would not permit imprinting, or (2) indirectly because their genetic constitution would enable them to remain condensed during the period of basic imprinting and therefore not susceptible.

Evidence bearing on the supernumerary (= B or accessory) status of L chromosomes has been discussed by Nur (1962b), Crouse et al. (1971), and White (1973). The occurrence of an accumulation mechanism, limitation of the L's to the germ line, mitotic irregularities of the L's, variation in number and size of L's among individuals within a population, and presence in some strains and species and absence from others are all indicative of the supernumerary nature of these chromosomes. One or more L's are present in all individuals of some species; these L's may have either acquired an essential role in the life cycle, or they have become successful, though regulated "parasites" (Östergren, 1945).

Crouse et al. (1971) noted a possible, puzzling relationship between L chromosomes and sex determination in S. impatiens, which had previously been reported to have an XX' mechanism, to produce unisexual progeny, and to contain L's. In one laboratory culture, S. impatiens changed from unisexual to bisexual progeny at the same time that the L chromosomes were undergoing fragmentation and loss. The L-free culture then reverted to unisexuality and was lost because all the offspring were male. (A new, maintainable L-free culture was obtained from an L-fragment line.) Two other, apparently related species, S. ocellaris and S. reynoldsi, produce both unisexual or bisexual broods and have no L chromosomes.

It will be recalled that there is an accumulation mechanism for L's in males (transmission of L's at an unreduced number; see Fig. 9) but not in females (transmission of L's at the reduced number expected after an orthodox meiosis). One way, then, in which the presence of L's (or too many L's) could be countered would be to make the sex of the embryo dependent on the presence of L's, a change which could be accomplished by sensitizing the process of X elimination to the presence of L's. If L's (or too many L's) were present, only one X would be eliminated from the soma; the individual would be a female and not permit accumulation of L's. If L's were absent (or few in number), two X's would be eliminated from the soma as the individual could safely become a male. The occurrence of all-male progeny on loss of L's in the laboratory culture of *S. impatiens* indicates that such a process of sensitization would have to evolve gradually in nature to assure production of both sexes. The relationship would thus not be that the L's directly determined sex, but that there was a selective advantage in eliminating the L's and this in turn resulted in a modification of the control of X elimination and thereby of sex determination.

## B. Chromosome Systems of the Cecidomyidae

The chromosome systems of the gall midges or cecidomyids seem to parallel those of *Sciara* but are more complex and have not been studied cytogenetically like those of *Sciara*. They will be only briefly noted and the reader is referred to White (1973) for further details and references. Information given in the following synopsis will apply to one or more cecidomyids but not necessarily to all.

Instead of the L chromosomes of *Sciara*, the germ line of both sexes contains numerous E chromosomes, so designated because they are eliminated from the soma at the fourth cleavage. Geyer-Duszyńska (1959, 1961) and Nicklas (1959) showed that all embryonic nuclei eliminate E chromosomes if prevented from reaching that part of the egg in which the germ line usually develops. Embryos thus manipulated develop normally, but meiosis in both sexes is abnormal and degeneration follows. During the normal meiosis of control embryos, the regular complement becomes heterochromatic prior to meiosis, while the E chromosomes appear more diffuse. One interpretation of the experimental results and the cytological observations is that the E chromosomes are essential for normal meiosis.

As in *Sciara*, the constitution of the germ line is the same in both sexes for the X's and the autosomes. Together the X's and the autosomes have been called the S (somatic) chromosomes because they are present without the E's in the somatic tissues. Sex determination in the more specialized cecidomyids is essentially equivalent to the XX ($\female$)–XO($\male$) of *Sciara* but two nonhomologous X chromosomes, instead of two homologous X's, are eliminated at the seventh cleavage during development of the male soma. The somatic chromosome constitutions of the two sexes are thus $X_1X_1X_2X_2AA$ ($♀$) and $X_1X_2AA$ ($♂$). At meiosis in the male one haploid set, presumably $X_1X_2A$, is separated from the other haploid set (also $X_1X_2A$), which remains with the E chromosomes. The nucleus with only a set of S chromosomes divides again to form two sperms; the nucleus with a set of S chromosomes and the E's neither divides nor forms sperms. By analogy with *Sciara*, $X_1X_2$ chromosomes eliminated during development of the male, and the $X_1X_2A$ set remaining with the E chromosomes at meiosis in the male would be of paternal origin.

In the female, the two sets of S chromosomes undergo normal meiosis. The behavior of the E's is not fully understood, but it seems likely that all of the E chromosomes are of maternal origin, except in a few species in which a few E's manage to move along with the haploid set of S chromosomes that will form sperm.

In the more primitive cecidomyids, there is an important variation in the life cycle and another in chromosome elimination in male embryos. In the primitive gall midges one or more cycles of pedogenesis* alternate with a generation of sexual adults. The pedogenetic cycles cease on deterioration of the immediate environment and winged adults emerge which are capable of moving to a new site. In development of male embryos, an entire haploid set is eliminated from the soma so that there are twice as many S chromosomes in the female as in the male, instead of only two more (the X's) as in the more specialized cecidomyids. Such somatic haploidy can apparently also occur in at least some females among certain of the pedogenetic species. This somatic haploidy probably occurs for strictly developmental reasons, since the cecidomyid males, by analogy with *Sciara*, presumably transmit only the maternal chromosomes and thus always breed as haploids whether the soma is haploid or diploid.

### C. Imprinting in *Sciara* and the Cecidomyidae

There is no direct evidence from either of these two groups of insects on the time and place of imprinting such as that available for mammals (Section II,F,2) and for coccids (Section III,C,3). The existence in *Sciara* of two types of females (XX, which produce sons; and XX', which produce daughters) indicates at least maternal control over the expression of imprinting even if imprinting occurs in the sperm prior to copulation (Chandra and Brown, 1975). The problem therefore remains unresolved as

* Pedogenesis (= paedogenesis) is parthenogenetic reproduction by a larva or pupa rather than an adult; paedo-, from Greek, = child; here, origin (genesis) from an immature stage.

to whether *Sciara* has a two-step system with imprinting in the male and its expression regulated by the female or a simple system with the female determining directly whether the paternal chromosomes are imprinted on entry into the egg.

It will be recalled that both *Sciara* and the coccids with systems of heterochromatization or elimination are examples of parahaploidy since both transmit only maternal chromosomes (Section I,C). In *Sciara*, the paternal chromosomes are not heterochromatized and remain genetically active. This difference in accomplishing the same net genetic effect—failure of transmission of paternal chromosomes—may depend on the difference in chromosome structure. *Sciara* chromosomes are monocentric, but coccid chromosomes are holokinetic; there is no single site on the coccid chromosomes by which their movements may be controlled. Furthermore, in translocations between eu- and heterochromatic chromosomes of coccids, both types of chromatin are maintained in single chromosomes (Section III,C,1). Thus coccid chromosomes appear to be susceptible to differential regulation only if affected in their entirety. The evidence from the experiments of Crouse on the controlling element of the X of *Sciara* (Section IV,A,2) implicates the centromere as the pivotal point in the *Sciara* system. At present, imprinting of *Sciara* chromosomes may be viewed as confined to controlling elements of the X and the autosomes. The controlling elements of the autosomes may be similar to those of the X in that they determine the differential behavior of the centromeres.

In regard to imprinting, the cecidomyids are of special interest because of pedogenesis in which one or more generations stem from parthenogenetic reproduction of larvae or pupae. All the individuals reproducing pedogenetically are female and no reductional divisions occur in the germ line. Eventually the pedogenetic cycles cease and male and female sexual adults appear. It will be recalled from the previous section that sex is determined in the cecidomyids by whether or not X's are eliminated in early development. If the cecidomyid system is analogous to that of *Sciara*, the eliminated X's are of paternal origin. Thus the original imprint would be maintained through the generations of pedogenetic females, and recognized only in the last pedogenetic generation, the one which produces the male and female embryos which mature into the sexual adults. The pedogenetic cecidomyids would thus differ in an important respect from *Sciara* in which paternal chromosomes seem never to be differentially regulated in the female. Other explanations of chromosome regulation in cecidomyids are certainly possible and study of the genetic systems of pedogenetic species will be of considerable interest.

## V. MOLECULAR MODELS OF IMPRINTING

Little is known with certainty about the molecular basis of the regulation of genetic expression in eukaryotes. An analysis of the type of control systems which would be necessary to guide multicellular organisms through their life cycles indicates that these are probably of a hierarchical nature (Davidson and Britten, 1973). Sensor sites detect changes in the environment and monitor the progress of differentiation and development. Somehow these sensor sites then instruct regulatory loci, which in turn are responsible for controlling the expression of batteries of structural genes.

The process of imprinting probably represents some modification of a regulatory mechanism that is widespread in living organisms. As noted in Section I, the potentiality for facultative systems seems to exist in a wide variety of living organisms, but such systems have been selectively advantageous and have therefore evolved in only a few groups of organisms. The cytogenetic data on imprinting do, however, put certain constraints on models of this process, and it is these constraints which may also provide clues to the molecular basis of imprinting.

Molecular models will be discussed later, but first certain suggestions on the nature of imprinting, its biological basis, and the evolution of facultative behavior will be considered.

A first point of concern is that imprinting probably results in a change in a regulatory locus, or loci, which is then perpetuated in subsequent development. Later differential behavior is a consequence of this change at the regulatory locus and will influence processes under the control of this locus. The eventual differential behavior itself is thus not apt to be indicative of the molecular nature of the imprinting process.

The *Sciara* system may be taken as an example of the difference between imprinting and subsequent facultative behavior. The paternal set of chromosomes is imprinted in the egg at or about the time of fertilization (or, according to other interpretations, is already imprinted in the sperm). After elimination of one or two paternal X chromosomes in early embryogeny, the paternal set thereafter divides in step with the maternal set during development. The effects of imprinting are suddenly expressed at anaphase I of meiosis: only the maternal set is capable of moving to the single pole; the paternal set appears to move backward and is soon eliminated (Section IV). Thus, a capacity for anaphase movement, maintained during earlier development, has suddenly been lost. It seems unlikely, therefore, that the original imprint could have directly affected this capacity. A more likely explanation may be outlined as follows. At meiosis, sister centromere regions do not separate at anaphase I, as they do at mitotic anaphase, but are responsible for holding together pairs of chromatids (the dyads) from late anaphase I to prophase II. It seems likely that it is this change from mitotic to meiotic regulation of the centric region which has been exploited in the *Sciara* system to prevent movement of the paternal chromosomes toward the single pole of anaphase I (Fig. 9). Somehow or other, the paternal chromosomes are made incompetent, and are segregated from the maternal chromosomes and eventually eliminated.

Two different approaches toward an understanding of the evolution of the heterochromatization of the mammalian X have been made by Lyon (1974a,b) and by Lifschytz and Lindsley (1972), and their ideas have relevance for molecular models.

Lyon (1974a,b) suggested that the mammalian X evolved after the primitive X was duplicated by translocation of a homologous segment of the primitive Y. After inactivation in the female of one of the two new chromosomes, each consisting of two homologous segments, the other active X would be equivalent to two X chromosomes. In support of the idea that the present-day active X is functionally equivalent to two X's, Lyon cited examples of apparently duplicate loci on the X which have similar effects such as those for color blindness and hemophilia. This idea provides for preadaptation for the later appearance of inactivation; it requires, however, that the female be able to tolerate the equivalent of four active X's in the meanwhile. In view of the general inability of animals to tolerate large duplications, this original "four-X" female would probably not have had much of an evolutionary future.

An alternative explanation is that X loci are found duplicated on the X itself because expression of these loci had been adapted to the unusual requirements of the X chromosome: (1) Prior to the heterochromatization of one X in the female, the X, or its less differentiated ancestral chromosome, would have been present in one dose in the male and two in the female. Presumably only certain sorts of loci could function appropriately under these circumstances. (2) The X eventually became heterochromatized and presumably inactivated genetically in the spermatocyte. Loci would be adapted to the X only if the organism could tolerate their heterochromatization at this stage. (3) If duplication occurred after the evolution of heterochromatization of the X, reason (2) would still hold. There would be the additional reason that the genetic system had already been adjusted to heterochromatization of one X in the female. Under this circumstance, there would be one effective "dose" of each X-chromosome locus per diploid cell, male or female. Duplication on the X would give two doses in both sexes. Duplication on an autosome (of which both homologs are expressed) would give three doses per cell. Duplication on the X itself would thus be more readily tolerated and therefore more apt to remain available for mutation to a new, selectively advantageous form.

Lifschytz and Lindsley (1972) suggested that the heterochromatization of an X in female mammals may somehow have been derived from the heterochromatization of the X which occurs in the spermatocyte (Section I,C). A direct derivation for the eutherian system is difficult to envisage, since all but one X will become inactive in otherwise diploid XXX and XXXX females regardless of the parental origin of the extra X's. It would seem more reasonable to presume an indirect derivation such as the duplication and modification of a key regulatory gene so that the gene action occurring in the spermatocyte could also occur in the embryogeny of the female. According to this view, all X's in the female would be automatically heterochromatized unless otherwise protected; such protection is normally accorded a single X in the diploid female.

Of special interest in regard to the evolution of facultative systems is the evidence from X chromosome behavior in tetraploid cells or sectors in the male germ line of grasshoppers and locusts (White, 1933, 1970; U. Nur, unpublished). If a mitosis is aborted late in development so that only one or a few tetraploid nuclei are formed, the two X's in these nuclei are both heterochromatic just as they would have been if normally separated into different diploid nuclei. If doubling of the chromosomes occurs much earlier in the germ line so that all the nuclei of one or two cysts are tetraploid, *only one of the two X's* is heterochromatic; the other X closely resembles the autosomes. Since tetraploid cells have little or no reproductive potential, the differential behavior of the two X's after an early doubling must reflect the regulatory processes occurring in diploid nuclei at the same stage. The fact that only one X is heterochromatized in either diploid or tetraploid nuclei of an early origin indicates stringent limitations on the quantity of regulatory entities. It may be that these limitations are necessary because of possible harmful effects elsewhere in the genome. In the diploid germ line there is no question of a facultative, differential regulation of homologous chromosomes because only one X is present, but this type of regulation is somewhat imitated by the two X's present in the abnormal circumstance of early occurring tetraploidy. A similar mechanism, but preventing heterochromatization, would account for one active X among two or more in diploid eutherian females; only one active X would also be expected in triploid embryos and this is not the case (Section III). Although regulation of X chromosomes in male grasshoppers and locusts is different from that of the X chromosomes in eutherian females, the two cases seem to show a common feature in the apparently limited quantity of regulatory entities. The evidence from the grasshoppers and locusts thus helps in visualizing a possible evolutionary origin of facultative systems from constitutive antecedents.

The four molecular models to be briefly outlined here involve (1) regulatory feedback loops and histones (Lyon, 1972), (2) feedback loops and methylation of DNA (Riggs, 1975), (3) methylation of DNA and nuclease action without feedback (Sager and Kitchin, 1975), and (4) perseverance of a state of coiling of DNA (Cook, 1973, 1974). The feedback models were primarily concerned with the regulation of the eutherian X; that of Sager and Kitchin attempted to explain the functioning of a wide variety of genetic systems including those of mammals, coccids, and *Sciara*.

The feedback models of regulation of the eutherian X depended on certain assumptions in regard to the activity of the X's during early embryogeny. Both X's are active in the oocyte. However, there seems to be little or no gene action during early embryogeny but such quiescence could occur for reasons other than the transient inactivation of the chromosomes. A fairly high proportion of embryos of XO mice die at the two-cell stage and were presumed to be OY embryos (Morris, 1968). Recently, Luthardt (1976) concluded from a detailed cytogenetic study of one and two-celled embryos from XO mothers, "that, in order for development to continue, at least one functional X chromosome is necessary as early as the two-cell stage." Models based on the assumption that both X's are active in early embryogeny (Lyon, 1972; see below) are unaffected by these results. However, models such as that of Riggs (1975) require both X's to be inactive (unmethylated) at the start (see below); such models thus need to be modified to limit the inactivity to key sites directly involved with heterochromatization.

In the model of Lyon (1972) the two active X's present in early embryogeny cooperate in inducing an autosomal locus or loci to produce repressor substances, such as histones. In the diploid, these repressor substances combine with one or, abnormally, more X's until there remains only a single active X. The remaining X stays active because it is incapable by itself of inducing the autosomal locus to produce more repressors. A necessary adjunct to the model is that the repressor substance, once added to an X, continues to accumulate on that chromosome until it is saturated. Otherwise there would be piecemeal inactivation scattered at random along the X's. Or the scheme could be pictured as involving just a regulatory locus or loci which in turn would determine whether the chromosome as a whole would be subject to heterochromatization.

Holliday and Pugh (1975) considered both controlled base-pair changes and methylation of DNA as possible regulatory devices in eukaryotes and briefly discussed the applicability of a methylation model to the differential regulation of homologous chromosomes. Chromosome regulation was examined in much greater detail by both Riggs (1975) and Sager and Kitchin (1975) who looked to the modification-restriction system of bacteria as a likely model. The DNA of the host bacteria is methylated at critical regulatory sites. If invading DNA, such as that of a phage, is not methylated at identical sites, it is attacked by endonucleases highly specific for the DNA sequences involved. Following this attack, the invading DNA can be successfully broken down by other nucleases. The invading DNA can thus be differentiated from the host DNA and destroyed without endangering the host DNA. There is thus a clear parallel to the differential regulation of homologous chromosomes.

Riggs (1975) pointed out another important aspect of methylation which seems to make it an ideal candidate for a feedback model. Unmethylated DNA is methylated only after seven to ten hours. However, if one strand is methylated, the other will be methylated within a minute or so. The sequences of the sites at which methylation occurs are reverse repeats of the general form:

$$T \ \ C^* \ \ A \ \ T \ \ A \ \ T \ \ G \ \ A$$
$$A \ \ G \ \ T \ \ A \ \ T \ \ A \ \ C' \ \ T$$

If neither $C^*$ or $C'$ is methylated, the slow reaction occurs; if either $C^*$ or $C'$ is methylated, the fast reaction quickly methylates the other. The methylase thus seems to be allosteric: it can effectively methylate cytosine only if another part of the enzyme somehow recognizes an already methylated cytosine in the proper position. The great difference in the time required for the slow and fast reactions led Riggs (1975) to propose that the first X to become active by methylation initiated the production of a methylase inhibitor which would prevent methylation of other X's by the slow reaction but would not be sufficiently potent to prevent methylation by the fast reaction of DNA replicated from the methylated X. Inactive and active X's thus could coexist in the same nuclei during subsequent development. Riggs' model would not be dependent on assumptions on the activity of the X during early embryogeny if the methylation were confined to a regulatory locus on the X. Such a locus could be assumed to have no direct relationship to the activity of the remainder of the X but to be solely responsible for preventing heterochromatization that would occur automatically if the locus were not methylated. The reactivation of an X which occurs in the female germ line (Section II,B,2) could presumably result from a cessation in production of methylase inhibitor. However, Riggs' model would still require demethylation of the maternal X (or its regulatory site) in the zygote.

Sager and Kitchin (1975) assumed that sperm DNA is unmethylated. In *Sciara* and coccids, if the embryo is to be female, the sperm DNA is methylated; if the embryo is to be male, the DNA is not methylated and is attacked by an endonuclease which makes the changes necessary for later differential behavior. The authors assumed that changes made by endonucleases are responsible for imprinting, but the changes themselves were not specified. At the present time, it seems unlikely that essential DNA, such as that of the mammalian X, would be subjected to gross remodeling by an endonuclease. However, if the action of the enzyme were restricted to a regulatory site, then some of the peculiarities observed at the X-chromosome controlling element of mice (Section II,F,4) might be explained.

As described earlier, diploid males are produced by an unusual type of parthenogenesis in certain coccids. A normal meiotic sequence results in a haploid egg which divides once; the two daughter nuclei fuse to form a zygote substitute (see Fig. 6). During this brief interval of separation, one of the daughter nuclei may or may not be preconditioned so that one haploid set becomes heterochromatic in some or all of the embryos. In some species, fertile adult males originate in this manner. According to the Sager and Kitchin (1975) model, imprinting is the result of the action of endonucleases on unmethylated DNA. In the present example from coccids, one of the two daughter nuclei may be imprinted. Either the DNA of one daughter nucleus was demethylated between division and fusion or an endonuclease had attacked methylated DNA in one but not the other daughter nucleus.

In summary, a system of control based on methylation would serve as an acceptable molecular model if demethylation could be included. Although the action of endonucleases cannot be ruled out, it does not seem necessary to assume that it occurs since methylation alone should enable differential recognition by regulatory molecules. Although methylation has long been regarded as a possibility in the differential regulation of homologous chromosomes, there is no direct evidence of its involvement in gene or chromosome regulation in eukaryotic nuclei. An attempt to find methyl cytosine or other unusual bases in a mealybug (a coccid) (Loewus *et al.*, 1964) was unsuccessful but technical improvements since then warrant further search. There seems to be some evidence that modification–restriction systems can occur in eukaryotic cells, but such evidence has so far been limited to the prokaryonlike DNA of chloroplasts (Sager and Kitchin, 1975).

Cook (1973, 1974) proposed that the state of coiling of DNA, once initiated, somehow perseveres during replication of chromosomes. In coccids and other organisms with heterochromatic chromosomes, such chromosomes are almost never seen in a diffuse state in interphase nuclei of dividing tissues; they seem, therefore, to replicate while in a condensed state. However, sex chromatin is visible in mammalian cells under certain circumstances but not others and the inactive X does not become active when the sex chromatin is not visible (Section II,A). In certain coccids, the same developmental process, such as a shift from mitosis to endomitosis, may be accompanied by quite different changes in the state of the chromosomes: the onset of heterochromatization (Section III,C,4), the reversal of heterochromatization, or the failure of heterochromatin to replicate (Section III,C,1). Such facts do not rule out that perseverance of coiling does not contribute to the maintenance of facultative heterochromatization, but they do indicate that other factors, closely controlled by development, are significantly involved.

In conclusion, much information has become available about imprinting and subsequent chromosome behavior, and it might appear that such information would help identify possible molecular mechanisms. At present, the reverse has occurred. Poorly understood phases in the life cycle (e.g., gene action in early mammalian embryogeny) or unusual events in certain organisms (e.g., parthenogenesis in soft scale insects) have either left models biologically untestable or made them unacceptable in their present forms. The models are, however, of considerable intrinsic interest, and are susceptible to refinement. It is to be hoped that the molecular events which begin with imprinting and are eventually expressed as differential behavior will soon be understood. Differential behavior leads to the elimination of chromosomes from the germ line in coccids and *Sciara*. The mammalian X is an exception; it is not eliminated but is restored, after deheterochromatization, to equal participation with the X which was not inactivated. The DNA of the mammalian X has been carefully conserved during the course of evolution (Ohno, 1967). Whatever changes are invoked in it by imprinting and heterochromatization, it seems likely that they would be the molecular equivalent of "a very light touch," and, for that reason, difficult to identify.

## ACKNOWLEDGMENTS

The authors are indebted to Dr. B. R. Seshachar, Dr. Marta S. Walters, Dr. Sally M. Rieffel, Miss Jerilyn Hirshberg, and Miss Yvonne Lee for reading all or parts of the manuscript. Thanks are due Dr. William E. Berg, Mr. Glenn Hall, Dr. David Hayman, Dr. J. L. Kermickle, Mr. Andrei Laszlo, Dr. Dan L. Lindsley, Dr. Barbara McClintock, Dr. Muriel Nesbitt, Dr. Uzi Nur, Dr. Evert I. Schlinger, and Dr. M. J. D. White for consultation on specific items which were included in the present review or proved beyond its scope. Miss Judy Aizuss and Miss Betsy Llosa typed the manuscript and Mr. E. Jay Ryon prepared the diagrams. Part of the work on the coccids reported here was aided by grants from the National Science Foundation to SWB. Preparation of the manuscript was facilitated by a grant from the Indian Council of Medical Research and by a travel grant from the U.S.–India Exchange of Scientists program, both to HSC. Special thanks are due the Institute of Biochemistry and Biophysics, University of Tehran, for enabling the authors to work together on the final revision of the manuscript.

## REFERENCES

Axelson, M. (1968). Sex chromatin in early pig embryos. *Hereditas* **60**, 347–354.

Baer, D. (1965). Asynchronous replication of DNA in a heterochromatic set of chromosomes in *Pseudococcus obscurus*. *Genetics* **52**, 275–285.

Baker, W. K. (1968). Position-effect variegation. *Adv. Genet.* **14**, 133–169.

Barr, M. L., and Bertram, E. G. (1949). A morphological distinction between neurones of the male and female, and the behavior of the nucleolar satellite during accelerated nucleoprotein synthesis. *Nature (London)* **163**, 676–677.

Bauer, H. (1967). Die kinetische Organisation der Lepidopterenchromosomen. *Chromosoma* **22**, 101–125.

Beardsley, J. W. (1969). A new fossil scale insect (Homoptera: Coccoïdea) from Canadian amber. *Psyche* **76**, 270–279.

Bennett, F. D., and Brown, S. W. (1958). Life history and sex determination in the diaspine scale, *Pseudaulacaspis pentagona* (Targ.) (Coccoidea). *Can. Entomol.* **90**, 317–325.

Berlowitz, L. (1965a). Correlation of genetic activity, heterochromatization, and RNA metabolism. *Proc. Natl. Acad. Sci. U.S.A.* **53**, 68–73.

Berlowitz, L. (1965b). Analysis of histone *in situ* in developmentally inactive chromatin. *Proc. Natl. Acad. Sci. U.S.A.* **54**, 476–480.

Berlowitz, L. (1974). Chromosomal inactivation and reactivation in mealybugs. *Genetics* **78**, 311–322.

Berry, R. O. (1939). Observations on chromosome elimination in the germ cells of *Sciara ocellaris*. *Proc. Natl. Acad. Sci. U.S.A.* **25**, 125–127.

Berry, R. O. (1941). Chromosome behavior in the germ cells and development of the gonads in *Sciara ocellaris*. *J. Morphol.* **68**, 547–583.

Beutler, E., Yeh, M., and Fairbanks, V. F. (1962). The normal human female as a mosaic of X-chromosome activity: Studies using the gene for G-6-Pd deficiency as a marker. *Proc. Natl. Acad. Sci. U.S.A.* **48**, 9–16.

Bomsel-Helmreich, O. (1971). The fate of heteroploid embryos. *Adv. Biosci.* **6**, 381–403.

Boué, A., and Boué, J. (1974). Chromosome abnormalities and abortion. *In* "Physiology and Genetics of Reproduction" (M. Coutinho and F. Fuchs, eds.), Part B, pp. ix–454. Plenum, New York.

Bradbury, J. T., Bunge, R. G., and Boccabella, R. A. (1956). Chromatin test in Klinefelter's syndrome. *J. Clin. Endocrinol. Metab.* **16**, 689.

Bridges, C. B. (1916). Non-disjunction as proof of the chromosome theory of heredity. *Genetics* **1**, 1–52, 107–163.

Brown, S. W. (1957). Chromosome behavior in *Comstockiella sabalis* (Comstk.) (Coccoidea-Diaspididae). *Genetics* **42**, 362–363.

Brown, S. W. (1958). The chromosomes of an *Orthezia* species (Coccoidea-Homoptera). *Cytologia* **23**, 429–434.

Brown, S. W. (1959). Lecanoid chromosome behavior in three more families of the Coccoidea (Homoptera). *Chromosoma* **10**, 278–300.

Brown, S. W. (1961). Fracture and fusion of coccid chromosomes. *Nature (London)* **191**, 1419–1420.

Brown, S. W. (1963). The Comstockiella system of chromosome behavior in the armored scale insects (*Coccoidea: Diaspididae*). *Chromosoma* **14**, 360–406.

Brown, S. W. (1964). Automatic frequency response in the evolution of male haploidy and other coccid chromosome systems. *Genetics* **49**, 797–817.

Brown, S. W. (1965). Chromosomal survey of the armored and palm scale insects (Coccoidea: Diaspididae and Phoenicococcidae). *Hilgardia* **36**, 189–294.

Brown, S. W. (1966). Heterochromatin. *Science* **151**, 417–425.

Brown, S. W. (1967). Chromosome systems of the *Eriococcidae* (*Coccoidea-Homoptera*). *Chromosoma* **22**, 126–150.

Brown, S. W. (1969). Developmental control of heterochromatization in coccids. *Genetics* **61**, No. 1, Part 2, Suppl., 191–198.

Brown, S. W., and Bennett, F. D. (1957). On sex determination in the diaspine scale *Pseudaulacaspis pentagona* (Targ.) (Coccoidea). *Genetics* **42**, 510–523.

Brown, S. W., and Chandra, H. S. (1973). Inactivation system of the mammalian X chromosome. *Proc. Natl. Acad. Sci. U.S.A.* **70**, 195–199.

Brown, S. W., and Nelson-Rees, W. A. (1961). Radiation analysis of a lecanoid genetic system. *Genetics* **46**, 983–1007.

Brown, S. W., and Wiegmann, L. I. (1969). Cytogenetics of the mealybug *Planococcus citri* (Risso) (Homoptera: Coccoidea): Genetic markers, lethals, and chromosome rearrangements. *Chromosoma* **28**, 255–279.

Buchner, P. (1954). Endosymbiosestudien an Schildläusen. I. *Stictococcus sjoestedti*. *Z. Morphol. Oekol. Tiere* **43**, 262–312.

Buchner, P. (1955). Endosymbiosestudien an Schildlüsen. II. *Stictococcus diversiseta*. *Z. Morphol. Oekol. Tiere* **43**, 397–424.

Cattanach, B. M. (1961). A chemically-induced variegated type position effect in the mouse. *Z. Indukt. Abstamm.- Vererbungsl.* **92**, 165–182.

Cattanach, B. M. (1974). Position effect variegation in the mouse. *Genet. Res.* **23**, 291–306.

Cattanach, B. M. (1975). Control of chromosome inactivation. *Annu. Rev. Genet.* **9**, 1–17.

Cattanach, B. M., and Isaacson, J. H. (1967). Controlling elements in the mouse X chromosome. *Genetics* **57**, 331–346.

Chandra, H. S. (1962). Inverse meiosis in triploid females of the mealybug, *Planococcus citri*. *Genetics* **47**, 1441–1454.

Chandra, H. S. (1963a). Cytogenetic studies following high dosage paternal irradiation in the mealybug, *Planococcus citri*. I. Cytology of $X_1$ embryos. *Chromosoma* **14**, 310–329.

Chandra, H. S. (1963b). Cytogenetic studies following high dosage paternal irradiation in the mealybug, *Planococcus citri*. II. Cytology of $X_1$ females and the problem of lecanoid sex determination. *Chromosoma* **14**, 330–346.

Chandra, H. S., and Brown, S. W. (1974). Regulation of X-chromosome inactivation in mammals. *Genetics* **78**, 343–349.

Chandra, H. S., and Brown, S. W. (1975). Chromosome imprinting and the mammalian X chromosome. *Nature* (*London*) **253**, 165–168.

Comings, D. E. (1966). The inactive X chromosome. *Lancet* **ii**, 1137–1138.

Comings, D. E. (1967a). Histones of genetically active and inactive chromatin. *J. Cell Biol.* **35**, 699–708.

Comings, D. E. (1967b). Sex chromatin, nuclear size and the cell cycle. *Cytogenetics* **6**, 120–144.

Comings, D. E. (1972). The structure and function of chromatin. *Adv. Hum. Genet.* **3**, 237–431.

Cook, P. R. (1973). Hypothesis on differentiation and the inheritance of gene superstructure. *Nature* (*London*) **245**, 23–25.

Cook, P. R. (1974). On the inheritance of differentiated traits. *Biol. Rev. Cambridge Philos. Soc.* **49**, 51–84.

Cooper, D. W. (1971). A directed genetic change model for X-chromosome inactivation in eutherian mammals. *Nature* (*London*) **230**, 292–294.

Cooper, D. W., VandeBerg, J. L., Sharman, G. B., and Poole, W. E. (1971). Phosphoglycerate kinase polymorphism provides further evidence for paternal X inactivation. *Nature* (*London*), *New Biol.* **230**, 155–157.

Cooper, D. W., Johnston, P. G., Murtagh, C. E., Sharman, G. B., VandeBerg, J. L., and Poole, W. E. (1975a). Sex-linked isozymes and sex chromosome evolution and inactivation in mammals. *In* "Proceedings of the Third International Isozyme Conference" (C. L. Markert, ed.), pp. 559–573. Yale Univ. Press, New Haven, Connecticut.

Cooper, D. W., Johnston, P. G., Murtagh, C. E., and VandeBerg, J. L. (1975b). Sex chromosome evolution and activity in mammals, particularly kangaroos. *In* "The Eukaryotic Chromosome" (W. J. Peacock and R. D. Brock, eds.), pp. 381–393. Aust. Natl. Univ. Press, Canberra.

Crouse, H. V. (1943). Translocations in *Sciara;* their bearing on chromosome behavior and sex determination. *Mo., Agric. Exp. Stn., Res. Bull.* **379,** 1–75.

Crouse, H. V. (1960a). The controlling element in sex chromosome behavior in *Sciara*. *Genetics* **45,** 1429–1443.

Crouse, H. V. (1960b). The nature of the influence of X-translocations on sex of progeny in *Sciara coprophila*. *Chromosoma* **11,** 146–166.

Crouse, H. V. (1965). Experimental alterations in the chromosome constitution of *Sciara*. *Chromosoma* **16,** 391–410.

Crouse, H. V. (1966). An inducible change in state on the chromosomes of *Sciara*: Its effects on the genetic components of the X. *Chromosoma* **18,** 230–253.

Crouse, H. V., Brown, A., and Mumford, B. C. (1971). L-Chromosome inheritance and the problem of "imprinting" in *Sciara* (*Sciaridae, Diptera*). *Chromosoma* **34,** 324–339.

Davidson, E. H., and Britten, R. J. (1973). Organization, transcription, and regulation in the animal genome. *Q. Rev. Biol.* **48,** 565–613.

Davidson, R. G., Nitowsky, H. M., and Childs, B. (1963). Demonstration of two populations of cells in the human female heterozygous for glucose-6-phosphate dehydrogenase variants. *Proc. Natl. Acad. Sci. U.S.A.* **50,** 481–485.

Drews, U., Blecher, S. R., Owen, D. A., and Ohno, S. (1974). Genetically directed preferential X-activation seen in mice. *Cell* **1,** 3–8.

Ducos, J., Marty, Y., Sanger, R., and Race, R. R. (1971). $X_g$ and X chromosome inactivation. *Lancet* **ii,** 219–220.

Dustan, A. J. (1953). A method of rearing the oleander scale, *Pseudaulacaspis pentagona* (Targ.) on potato tubers. *Dep. Agric. Bermuda Bull.* **27,** 1–7.

Ellis, J. R., Marshall, R., Normand, I. C. S., and Penrose, L. S. (1963). A girl with triploid cells. *Nature (London)* **198,** 411.

Epstein, C. J. (1969). Mammalian oocytes: X chromosome activity. *Science* **163,** 1078–1079.

Epstein, C. J. (1972). Expression of the mammalian X-chromosome before and after fertilization. *Science* **175,** 1467–1468.

Fialkow, P. J. (1972). Use of genetic markers to study cellular origin and development of tumors in human females. *Adv. Cancer Res.* **15,** 191–226.

Fialkow, P. J. (1973). Primordial cell pool size and lineage relationships of five human cell types. *Ann. Hum. Genet.* **37,** 39–48.

Fialkow, P. J., Gartler, S. M., and Yoshida, A. (1967). Clonal origin of chronic myelocytic leukemia in man. *Proc. Natl. Acad. Sci. U.S.A.* **58,** 1468–1471.

Fialkow, P. J., Lisker, R., Giblett, E. R., and Zavala, C. (1970). $X_g$ locus: Failure to detect inactivation in females with chronic myelocytic leukemia. *Nature (London)* **226,** 367–368.

Fialkow, P. J., Sagebiel, R. W., Gartler, S. M., and Rimoin, D. L. (1971). Multiple origin of hereditary neurofibromas. *N. Engl. J. Med.* **284,** 298–300.

Ford, C. E., Jones, K. W., Polani, P. E., de Almeida, J. C., and Briggs, J. H. (1959). A sex chromosomal anomaly in a case of gonadal dysgenesis (Turner's syndrome). *Lancet* **i,** 711–713.

Gandini, E., and Gartler, S. M. (1969). Glucose-6-phosphate dehydrogenase mosaicism for studying the development of blood cell precursors. *Nature (London)* **224**, 599–600.

Gandini, E., Gartler, S. M., Angioni, G., Argiolas, N., and Dell'Aqua, G. (1968). Developmental implications of multiple tissue studies in glucose-6-phosphate dehydrogenase deficient heterozygotes. *Proc. Natl. Acad. Sci. U.S.A.* **61**, 945–948.

Gardner, R. L., and Lyon, M. F. (1971). X chromosome inactivation studied by injection of a single cell into the mouse blastocyst. *Nature (London)* **231**, 385–386.

Gartler, S. M. (1974). Utilization of mosaic systems in the study of the origin and progression of tumors. *In* "Chromosomes and Cancer" (J. German, ed.), pp. 313–334. Wiley, New York.

Gartler, S. M., Ziprkowski, L., Krakowski, A., Ezra, R., Szienberg, A. and Adam, A. (1966). Glucose-6-phosphate dehydrogenase mosaicism as a tracer in the study of hereditary multiple trichoepithelioma. *Am. J. Human Genet.* **18**, 282–287.

Gartler, S. M., Chen, S. H., Fialkow, P. J., and Giblett, E. R. (1972a). X chromosome inactivation in cells from an individual heterozygous for two X-linked genes. *Nature (London), New Biol.* **236**, 149–150.

Gartler, S. M., Liskay, R. M., Campbell, B. K., Sparkes, R. and Gant, N. (1972b). Evidence for two functional X chromosomes in human oocytes. *Cell Differ.* **1**, 215–218.

Gartler, S. M., Liskay, R. M., and Gant, N. (1973). Two functional X chromosomes in human foetal oocytes. *Exp. Cell Res.* **82**, 464–466.

Gartler, S. M., Andina, R., and Gant, N. (1975). Ontogeny of X-chromosome inactivation in the female germ line. *Exp. Cell Res.* **91**, 454–457.

Geyer-Duszyńska, I. (1959). Experimental research on chromosome elimination in Cecidomyiidae (Diptera). *J. Exp. Zool.* **141**, 391–448.

Geyer-Duszyńska, I. (1961). Spindle disappearance and chromosome behavior after partial-embryo irradiation in Cecidomyiidae (Diptera). *Chromosoma* **12**, 233–247.

Grumbach, M. M., Morishima, A., and Taylor, J. H. (1963). Human sex chromosome abnormalities in relation to DNA replication and heterochromatinization. *Proc. Natl. Acad. Sci. U.S.A.* **49**, 581–589.

Hartl, D. L., and Brown, S. W. (1970). The origin of male haploid genetic systems and their expected sex ratio. *Theor. Popul. Biol.* **1**, 165–190.

Hayman, D. L., and Martin, P. G. (1969). Cytogenetics of marsupials. *In* "Comparative Mammalian Cytogenetics" (K. Benirschke, ed.), pp. 191–217. Springer-Verlag, Berlin and New York.

Hess, O. (1971). Lampenbürstenchromosomen. *Handb. Allg. Pathol.* **II,** Part 2, 215–281.

Holliday, R., and Pugh, J. E. (1975). DNA modification mechanisms and gene activity during development. *Science* **187**, 226–232.

Hoppe, P. C., and Whitten, P. K. (1972). Does X chromosome inactivation occur during mitosis of first cleavage? *Nature (London)* **239**, 520.

Hoy, J. M. (1962). Eriococcidae (Homoptera: Coccoidea) of New Zealand. *N. Z. Dep. Sci. Ind. Res., Bull.* **146**, 1–219.

Hughes-Schrader, S. (1927). Origin and differentiation of the male and female germ cells in the hermaphrodite of *Icerya purchasi* (Coccidae). *Z. Zellforsch. Mikrosk. Anat.* **6**, 509–540.

Hughes-Schrader, S. (1935). The chromosome cycle of Phenacoccus (Coccidae). *Biol. Bull. (Woods Hole, Mass.)* **69**, 462–468.

Hughes-Schrader, S. (1944). A primitive coccid chromosome cycle in *Puto* sp. *Biol. Bull. (Woods Hole, Mass.)* **87**, 167–176.

Hughes-Schrader, S. (1948). Cytology of coccids (Coccoidea-Homoptera). *Adv. Genet.* **2**, 127–203.

Hughes-Schrader, S., and Monahan, D. F. (1966). Hermaphroditism in *Icerya zeteki* Cockerell, and the mechanism of gonial reduction in iceryine coccids (Coccoidea: Margarodidae Morrison). *Chromosoma* **20**, 15–31.

Hughes-Schrader, S., and Ris, H. (1941). The diffuse spindle attachment of coccids, verified by the mitotic behavior of induced chromosome fragments. *J. Exp. Zool.* **87**, 429–456.

Hughes-Schrader, S., and Schrader, F. (1961). The kinetochore of the *Hemiptera*. *Chromosoma* **12**, 327–350.

Issa, M., Blank, C. E., and Atherton, G. W. (1969). The temporal appearance of sex chromatin and of the late-replicating X chromosome in blastocysts of the domestic rabbit. *Cytogenetics* **8**, 219–237.

Jacobs, P. A., and Strong, J. A. (1959). A case of human intersexuality having a possible XXY sex-determining mechanism. *Nature (London)* **183**, 302–303.

James, H. C. (1937). Sex ratios and the status of the male of Pseudococcinae (Hem. Coccidae). *Bull. Entomol. Res.* **28**, 429–461.

James, H. C. (1938). The effect of humidity of the environment on sex ratios from over-aged ova of *Pseudococcus citri* Risso (Hemipt. Coccidae). *Proc. R. Entomol. Soc. London, Ser. A* **13**, 73–79.

Johnston, P. G., VandeBerg, J. L., and Sharman, G. B. (1975). Inheritance of glucose 6-phosphate dehydrogenase in the red-necked wallaby, *Macropus rufogriseus* (Desmarest) consistent with paternal X inactivation. *Biochem. Genet.* **13**, 235–242.

Kahan, B., and DeMars, R. (1975). Localized derepression on the human inactive X chromosome in mouse–human cell hybrids. *Proc. Natl. Acad. Sci. U.S.A.* **72**, 1510–1514.

Kiefer, B. I. (1973). Genetics of sperm development in *Drosophila*. *In* "Genetic Mechanisms of Development" (F. H. Ruddle, ed.), pp. 47–102. Academic Press, New York.

Kitchin, R. M. (1970). A radiation analysis of a Comstockiella chromosome system: Destruction of heterochromatic chromosomes during spermatogenesis in *Parlatoria oleae* (Coccoidea: Diaspididae). *Chromosoma* **31**, 165–197.

Kitchin, R. M. (1975). Intranuclear destruction of heterochromatin in two species of armored scale insects. *Genetica (The Hague)* **45**, 227–235.

Lakhotia, S. C., and Mukherjee, A. S. (1969). Chromosomal basis of dosage compensation in *Drosophila*. I. Cellular autonomy of hyperactivity of the X-chromosome in salivary glands and sex differentiation. *Genet. Res.* **14**, 137–150.

Lifschytz, E., and Lindsley, D. L. (1972). The role of X-chromosome inactivation during spermatogenesis. *Proc. Natl. Acad. Sci. U.S.A.* **69**, 182–186.

Linder, D. (1969). Gene loss in human teratomas. *Proc. Natl. Acad. Sci. U.S.A.* **63**, 699–704.

Linder, D., and Gartler, S. M. (1965). Glucose-6-phosphate dehydrogenase mosaicism: Utilization as a cell marker in the study of leiomyomas. *Science* **150**, 67–69.

Linder, D., and Gartler, S. M. (1967). Problem of single cell vs. multiple cell origin of a tumor. *In* "Proceedings of the Fifth Berkeley Symposium on Mathematical Statistics and Probability," pp. 625–633.

Linder, D., and Power, J. (1970). Further evidence of postmeiotic origin of teratomas in the human female. *Ann. Hum. Genet.* **34**, 21–31.

Loewus, M. W., Brown, S. W., and McLaren, A. D. (1964). Base ratios in DNA in male and female *Pseudococcus citri*. *Nature (London)* **203**, 104.

Lucchesi, J. C. (1973). Dosage compensation in *Drosophila*. *Annu. Rev. Genet.* **7**, 225–237.

Lucchesi, J. C., Rawls, J. M., Jr., and Maroni, G. (1974). Gene dosage compensation in metafemales (3X; 2A) of *Drosophila*. *Nature (London)* **248**, 564–567.

Luthardt, F. W. (1976). Cytogenetic analysis of oocytes and early preimplantation embryos from XO mice. *Dev. Biol.* **54**, 73–81.

Lyon, M. F. (1961). Gene action in the X-chromosome of the mouse. *Nature (London)* **190**, 372–373.

Lyon, M. F. (1972). X-chromosome inactivation and developmental patterns in mammals. *Biol. Rev. Cambridge Philos. Soc.* **47**, 1–35.

Lyon, M. F. (1974a). Evolution of X-chromosome inactivation in mammals. *Nature (London)* **250**, 651–653.

Lyon, M. F. (1974b). Mechanisms and evolutionary origins of variable X-chromosome activity in mammals. *Proc. R. Soc. London, Ser. B* **187**, 243–268.

McKusick, V. A. (1975). "Mendelian Inheritance in Man: Catalogs of Autosomal Dominant, Autosomal Recessive, and X-linked Phenotypes," 4th ed. Johns Hopkins Univ. Press, Baltimore, Maryland.

Malheiros, N., deCastro, D., and Câmara, A. (1947). Cromosomas sem centrómero localizado. O caso da *Luzula purpurea* Link. *Agron. Lusit.* **9**, 51–71.

Maroni, G., and Plaut, W. (1973). Dosage compensation in *Drosophila melanogaster* triploids. II. Glucose-6-phosphate dehydrogenase activity. *Genetics* **74**, 331–342.

Metz, C. W. (1938). Chromosome behavior, inheritance and sex determination in *Sciara*. *Am. Nat.* **72**, 485–520.

Migeon, B. R. (1972). Stability of X chromosomal inactivation in human somatic cells. *Nature (London)* **239**, 87–89.

Miller, G., Berlowitz, L., and Regelson, W. (1971). Chromatin and histones in mealybug cell explants: Activation and decondensation of facultative heterochromatin by a synthetic polyanion. *Chromosoma* **32**, 251–261.

Mittwoch, U., Atkin, N. B., and Ellis, J. R. (1963). Barr bodies in triploid cells. *Cytogenetics* **2**, 323–330.

Morishima, A., Grumbach, M. M., and Taylor, J. H. (1962). Asynchronous duplication of human chromosomes and the origin of sex chromatin. *Proc. Natl. Acad. Sci. U.S.A.* **48**, 756–763.

Morris, T. (1968). The XO and OY chromosome constitutions in the mouse. *Genet. Res.* **12**, 125–137.

Mukherjee, B. B., and Sinha, A. K. (1964). Single-active-X hypothesis: Cytological evidence for random inactivation of X-chromosomes in a female mule complement. *Proc. Natl. Acad. Sci. U.S.A.* **51**, 252–259.

Mukunda, N., Chandra, H. S., Gadgil, M., Rajagopal, A. K., and Sudarshan, E. C. G. (1976). On the dynamics of activation of mammalian X chromosomes. *Genet. Res.* **28**: 147–162.

Muller, H. J., and Settles, F. (1927). The nonfunctioning of genes in spermatozoa. *Z. Indukt. Abstamm.- Vererbungsl.* **43**, 285–312.

Nance, W. E. (1964). Genetic tests with a sex-linked marker: Glucose-6–phosphate dehydrogenase. *Cold Spring Harbor Symp. Quant. Biol.* **29**, 415–425.

Nelson-Rees, W. A. (1960). A study of sex predetermination in the mealybug *Planococcus citri* (Risso). *J. Exp. Zool.* **144**, 111–137.

Nelson-Rees, W. A. (1962). The effects of radiation damaged heterochromatic chromosomes on male fertility in the mealybug, *Planococcus citri* (Risso). *Genetics* **47**, 661–683.

Nelson-Rees, W. A. (1963). New observations on lecanoid spermatogenesis in the mealybug, *Planococcus citri*. *Chromosoma* **14**, 1–17.

Nesbitt, M. N. (1971). X chromosome inactivation in the mouse. *Dev. Biol.* **26**, 252–263.

Nesbitt, M. N., and Gartler, S. M. (1970). Replication of the mouse sex chromosomes early in the S period. *Cytogenetics* **9**, 212–221.

Nicklas, R. B. (1959). An experimental and descriptive study of chromosome elimination in *Miastor* spec. (Cecidomyiidae, Diptera). *Chromosoma* **10**, 301–336.

Nicosia, S. A., Wolf, D. P., and Inoue, M. (1977). Cortical granule distribution and cell surface characteristics in mouse eggs. *Dev. Biol.* **57**, 56–74.

Niikawa, N., and Kajii, T. (1974). A triploid human abortus due to dispermy. *Humangenetick* **24**, 261–264.

Nur, U. (1962a). Sperm, sperm bundles and fertilization in a mealybug *Pseudococcus obscurus* Essig (Homoptera: Coccoidea). *J. Morphol.* **111**, 173–199.

Nur, U. (1962b). A supernumerary chromosome with an accumulation mechanism in the lecanoid genetic system. *Chromosoma* **13**, 249–271.

Nur, U. (1963). Meiotic parthenogenesis and heterochromatization in a soft scale, *Pulvinaria hydrangeae* (Coccoidea: Homoptera). *Chromosoma* **14**, 123–139.

Nur, U. (1964). A modified Comstockiella chromosome system in the olive scale insect, *Parlatoria oleae* (Coccoidea: Diaspididae). *Chromosoma* **17**, 104–120.

Nur, U. (1966a). Harmful supernumerary chromosomes in a mealybug population. *Genetics* **54**, 1225–1238.

Nur, U. (1966b). The effect of supernumerary chromosomes on the development of mealybugs. *Genetics* **54**, 1239–1249.

Nur, U. (1966c). Nonreplication of heterochromatic chromosomes in a mealybug, *Planococcus citri* (Coccoidea: Homoptera). *Chromosoma* **19**, 439–448.

Nur, U. (1967a). Reversal of heterochromatization and the activity of the paternal chromosome set in the male mealybug. *Genetics* **56**, 375–389.

Nur, U. (1967b). Chromosome systems in the Eriococcidae (Coccoidea Homoptera). II. *Gossyparia spuria* and *Eriococcus araucariae*. *Chromosoma* **22**, 151–163.

Nur, U. (1970). Translocations between eu- and heterochromatic chromosomes and spermatocytes lacking a heterochromatic set in male mealybugs. *Chromosoma* **29**, 42–61.

Nur, U. (1971). Parthenogenesis in coccids (Homoptera). *Am. Zool.* **11**, 301–308.

Nur, U. (1972). Diploid arrhenotoky and automictic thelytoky in soft scale insects (Lecaniidae: Coccoidea: Homoptera). *Chromosoma* **39**, 381–401.

Nyhan, W. L., Bakay, B., Connor, J. D., Marks, J. F., and Kelle, D. K. (1970). Hemizygous expression of glucose-6-phosphate dehydrogenase in erythrocytes of heterozygotes for Lesch-Nyhan syndrome. *Proc. Natl. Acad. Sci. U.S.A.* **65**, 214–218.

Ohno, S. (1967). "Sex Chromosomes and Sex-Linked Genes." Springer-Verlag, Berlin.

Ohno, S., and Hauschka, T. S. (1960). Allocycly of the X-chromosome in tumors and normal tissues. *Cancer Res.* **20**, 541–545.

Ohno, S., and Makino, S. (1961). The single-X nature of sex chromatin in man. *Lancet* **i**, 78–79.

Ohno, S., Jainchill, J., and Stenius, C. (1963). The creeping vole (*Microtus oregoni*) as a gonosomic mosaic. The OY/XY constitution of the male. *Cytogenetics* **2**, 232–239.

Ohno, S., Stenius, C., and Christian, L. (1966). The XO as the normal female of the creeping vole (*Microtus oregoni*). *Chromosomes Today* **1**, 182–187.

Östergren, G. (1945). Parasitic nature of extra chromosome fragments. *Bot. Not.* pp. 157–163.

Pallotta, D., Berlowitz, L., and Rodriguez, L. (1970). Histones of genetically active and inactive chromatin in mealybugs. *Exp. Cell Res.* **60**, 474–477.

Polani, P. E., Hunter, W. F., and Lennox, B. (1954). Chromosomal sex in Turner's syndrome with coarctation of the aorta. *Lancet* **ii**, 120–121.

Polani, P. E., Angell, R., Giannelli, F., de la Chapelle, A., Race, R. R., and Sanger, R. (1970). Evidence that the $X_g$ locus is inactivated in structurally abnormal X chromosomes. *Nature (London)* **227**, 613–616.

Rattazzi, M. C., and Cohen, M. M. (1972). Further proof of the genetic inactivation of the X-chromosome in the female mule. *Nature (London)* **237**, 393–395.

Ray, M., Gee, P. A., Richardson, B. J., and Hamerton, J. L. (1972). G6PD expression and X chromosome late replication in fibroblast clones from a female mule. *Nature (London)* **237**, 396–397.

Richardson, B. J., Czuppon, A. B., and Sharman, G. B. (1971). Inheritance of glucose-6-dehydrogenase variation in kangaroos. *Nature (London), New Biol.* **230**, 154–155.

Rieffel, S. M., and Crouse, H. V. (1966). The elimination and differentiation of chromosomes in the germ line of *Sciara*. *Chromosoma* **19**, 231–276.

Riggs, A. D. (1975). X inactivation, differentiation, and DNA methylation. *Cytogenet. Cell Genet.* **14**, 9–25.

Riis, P., Johnsen, S. G., and Mosbech, J. (1956). Nuclear sex in Klinefelter's syndrome. *Lancet* **i**, 962–963.

Ris, H. (1942). A cytological and experimental analysis of the meiotic behavior of the univalent X-chromosome in the bearberry aphid *Tamalia* (= *Phyllaphis*) *coweni* (Ckll.) *J. Exp. Zool.* **90**, 267–326.

Russell, L. B. (1961). Genetics of mammalian sex chromosomes. *Science* **133**, 1795–1803.

Russell, L. B., and Montgomery, C. L. (1970). Comparative studies on X-autosome translocations in the mouse. II. Inactivation of autosomal loci, segregation, and mapping of autosomal breakpoints in five T (X; 1)'s. *Genetics* **64**, 281–312.

Sabour, M. (1969). Nucleic acid and protein synthesis during early embryogenesis of a mealybug, *Pseudococcus obscurus* Essig (Homoptera: Coccoidea). Doctoral Thesis, University of California, Berkeley.

Sabour, M. (1972). RNA synthesis and heterochromatization in early development of a mealybug. *Genetics* **70**, 291–298.

Sager, R., and Kitchin, R. (1975). Selective silencing of eukaryotic DNA. *Science* **189**, 426–433.

Schrader, F. (1921). The chromosomes of *Pseudococcus nipae*. *Biol. Bull. (Woods Hole, Mass.)* **40**, 259–270.

Schrader, F. (1929). Experimental and cytological investigations of the life cycle of *Gossyparia spuria* (Coccidae) and their bearing on the problem of haploidy in males. *Z. Wiss. Zool.* **134**, 149–179.

Schrader, F., and Hughes-Schrader, S. (1926). Haploidy in *Icerya purchasi*. *Z. Wiss. Zool.* **128**, 182–200.

Schrader, F., and Hughes-Schrader, S. (1931). Haploidy in metazoa. *Q. Rev. Biol.* **6**, 411–438.

Sharman, G. B. (1971). Late DNA replication in the paternally derived X chromosome of female kangaroos. *Nature (London), New Biol.* **230**, 231–232.

Sharman, G. B., Robinson, E. S., Walton, S. M., and Berger, P. J. (1970). Sex chromosomes and reproductive anatomy of some intersexual marsupials. *J. Reprod. Fertil.* **21**, 57–68.

Stewart, J. S. S. (1960). Genetic mechanisms in human intersexes. *Lancet* **i**, 825–826.

Sturtevant, A. H. (1923). Inheritance of direction of coiling in Limnaea. *Science* **58**, 269–270.

Takagi, N. (1973). Differentiation of X chromosomes in the female mouse. *Genetics* **74**, No. 2, Part 2, 269.

Takagi, N., and Sasaki, M. (1975). Preferential inactivation of the paternally derived X chromosome in the extraembryonic membranes of the mouse. *Nature (London)* **256**, 641–642.

Townsend, D. E., Sparkes, R. S., Baluda, M. C., and McClelland, G. (1970). Unicellular histogenesis of uterine leiomyomas as determined by electrophoresis of glucose-6-phosphate dehydrogenase. *Am. J. Obstet. Gynecol.* **107**, 1168–1173.

Wahl, H. A. (1940). Chromosome numbers and meiosis in the genus *Carex*. *Am. J. Bot.* **27**, 458–470.

Wake, N., Takagi, N., and Sasaki, M. (1976). Nonrandom inactivation of X chromosome in the rat yolk sac. *Nature (London)* **262,** 580–581.

Weaver, D. D., Gartler, S. M., Boué, A., and Boué, J. G. (1975). Evidence for two active X chromosomes in a human XXY triploid. *Humangenetik* **28,** 39–42.

Welshons, W. J., and Russell, L. B. (1959). The Y-chromosome as the bearer of male determining factors in the mouse. *Proc. Natl. Acad. Sci. U.S.A.* **45,** 560–566.

White, M. J. D. (1933). Tetraploid spermatocytes in a locust, *Schistocerca gregaria. Cytologia* **5,** 135–139.

White, M. J. D. (1970). Asymmetry in heteropyknosis in tetraploid cells of a grasshopper. *Chromosoma* **29,** 51–61.

White, M. J. D. (1973). "Animal Cytology and Evolution," 3rd ed. Cambridge Univ. Press, London and New York.

# 5

# Use of Mutant, Hybrid, and Reconstructed Cells in Somatic Cell Genetics

*Nils R. Ringertz and Thorfinn Ege*

## I. INTRODUCTION

Classic genetics is based on the analysis of individuals that arise from sexual crosses. Although this is still a central area of genetics, recent technical developments in the analysis of genes at the molecular, chromosomal, and cellular levels have established new subfields of great importance. Among these, *somatic cell genetics* is a recent one, and it owes its origin to new techniques which make it possible to generate somatic cells with altered genomes such as *mutant* cells with specific genetic defects, *hybrid cells* containing genetic material from two different cell types, and *reconstructed cells* where the nucleus may be derived from one cell type and the cytoplasm from another.

The aim of this chapter is to review briefly some of the techniques which have made it possible to generate these cell types and to indicate how they can be used to study genetic mechanisms in somatic cells. Various sections and illustrations in this chapter have been taken from the monograph "Cell Hybrids" by N. R. Ringertz and R. E. Savage (1976). We thank Bob Savage and Academic Press for permission to let us do so.

Within the field covered by this chapter there have been several excellent reviews which discuss various aspects of somatic cell genetics such as the generation and selection of mutants (M. Harris, 1964; Thompson and Baker, 1973; Basilico and Meiss, 1974; Kao and Puck, 1974; Naha, 1974), the use of cell hybridization in the analysis of gene expression (Ephrussi, 1972; H. Harris, 1970), cell differentiation (Davidson, 1974), and gene mapping (Ruddle and Creagan, 1975).

## II. MUTANT CELLS

Mutants of somatic mammalian cells represent an interesting material for biochemical and cell biological studies since they provide information about metabolic pathways and regulatory interactions in the cell. Mutants are also useful in cell hybridization experiments where they make it possible to dissect metabolic pathways and complex cell functions by means of gene complementation tests. Unfortunately there are still many difficulties in isolating and defining mutant eukaryotic cells, and these difficulties also complicate the interpretation of data obtained by cell hybridization. This section therefore briefly discusses the methods used in generating and isolating mutant cells.

### A. Definition of Mutant Cells

The term *mutant* has often been used to describe cell lines showing stable and heritable alterations in phenotype. In many cases these cells may well have been true mutants in the sense that changes in the composition of their nuclear DNA had occurred. It is, however, often difficult to distinguish mutants from variant cells with stable phenotypic changes ("epigenetic changes").

Cell differentiation is the result of stable phenotypic changes, but these changes generally are not associated with detectable modifications of DNA. Rather they are believed to involve changes in nuclear protein components and alterations in regulatory circuits. In the following sections, cells will be described as mutants if the altered phenotype (a) occurs randomly and with a frequency that is increased by mutagens; (b) is inherited by daughter cells and persists after mutagens and selective pressures have been removed; (c) is due to changes in specific gene products, e.g., proteins; (d) can be assigned to a certain chromosome or be mapped to a specific chromosome region; (e) can be reversed and the reversion frequency is increased by the addition of mutagens.

Most mutants used in cell hybrid work satisfy criteria (a) and (b) and some also meet criterion (c). The number of mutations which can be assigned to specific chromosomes is still small. Some cell lines described as mutants fail to show revertants. In these cases it is probable that the altered phenotype is due to the loss of a specific chromosome region thus making the change irreversible (deletion mutants).

## B. Mutant Frequency and Mutation Rate

The frequency of mutant cells (*mutant frequency*) in a cell population is the fraction of cells showing a variant phenotype. The mutant frequency depends on the *mutation rate*, the selection pressure for the mutant phenotype, and time. Mutants (with the exception of deletion mutants) also undergo back mutations (*reversion*) to wild-type at a low rate. The mutation rate (reversion rate) for a particular property is defined as the average number of mutations occurring per cell per generation. Under normal conditions the spontaneous mutation frequency in tissue culture ranges from $10^{-6}$ to $10^{-7}$. The rate of mutation can be determined in several different ways. A commonly used method is the classic fluctuation test developed by Luria and Delbrück in studies of mutations in microorganisms.

The chances that a mutant will arise and manifest itself in cell culture depends on the number of genes involved in the control of the property examined. In this context it is important to remember that cultured vertebrate cells, with the notable exceptions of some haploid frog cell lines, are either diploid, tetraploid, octoploid, or heteroploid. In diploid cells, where all autosomes occur in pairs, there are homozygous (*AA* or *aa*) and heterozygous (*Aa* or *aA*) states for any given gene depending on whether the two homologous chromosomes involved carry the same or different alleles.

In addition to factors affecting gene dosage relationships the frequency observed for a given type of mutant will depend on whether the mutant gene shows a dominant or recessive expression. For dominantly expressed mutations one would expect an increase in mutation rate with increasing levels of ploidy, since each copy of a gene should represent an independent target where a mutation could occur. For recessive mutations the increase in polyploidy should decrease the observed mutation rate since a larger number of genes would have to mutate before the recessive mutant phenotype expresses itself. Morgan Harris (1971, 1974), however, found the surprising result that the mutation rate for two independent recessive markers (temperature sensitivity and azaguanine resistance) was essentially constant for cells at different ploidy levels. This and other discrepancies between expected and observed mutation rates discussed by M. Harris (1974) indicate that at least some of the variations established in cell lines may not be due to "honest mutations" but have another and as yet unknown basis. Thus it is likely that some phenomena which are interpreted as gene mutations are in fact stable phenotypic changes which arise as a consequence of changes in complex regulatory mechanisms. The most important reason why the expected relationship between mutation rate and ploidy is not observed, however, is the heteroploid nature of established cell lines. Although the cells may show a near-diploid or near-tetraploid karyotype, individual chromosomes may be present in greatly varying numbers. Even when the number of copies of a specific chromosome can be ascertained and the mutation studied is one that normally maps on this chromosome, the comparison of mutation rate and chromosome number can be misleading since the chromosomes may have undergone submicroscopic deletions, translocations or other forms of aberrations.

The incidence of back mutations in many other cases is lower than that expected from the forward mutation rate. The frequency with which this phenomenon is encountered may be understood if the forward mutations in question arise from losses of parts or whole chromosomes, thereby rendering the change essentially irreversible.

## C. Isolation of Mutants

### 1. Mutagenesis and Expression of Mutations

Although spontaneous mutants arise in cell cultures, X-irradiation or chemical mutagens are usually applied in order to increase the mutation rate. Among the commonly used chemical mutagens are ethyl methanesulfonate (EMS), methyl methanesulfonate (MMS) and $N$-methyl $N'$-nitro-$N$-nitrosoguanidine (MNNG). Treatment with these agents increases the frequency of drug-resistant, auxotrophic, and temperature-sensitive mutants and can also be used to induce reverse mutations.

The yield of viable mutants depends on a number of factors, e.g., dose and time of exposure to the mutagen, cell density, and the treatment of the cells after exposure to mutagen. Usually the cells are maintained under nonselective conditions for a few days after the mutagen has been removed in order to allow the mutations to be expressed. This period is followed by selection for the mutant cells on special media in order to kill the normal cells and allow a special class of mutants to accumulate by cell proliferation.

### 2. Drug-Resistant Mutants

Drug-resistant mutant cells are obtained by including the drug in the normal tissue culture medium. A wide range of drugs affecting a variety of metabolic pathways and cell functions have been used in the isolation of mutant mammalian cells (Table I).

The cells used in Littlefield's selective system (Fig. 5, p. 210) are examples of mutants obtained by drug selection. Cells deficient in hypoxanthine-guanine phosphoribosyltransferase (HGPRT cells) are isolated by exposing cell populations to azaguanine (AG) or thioguanine (TG). These drugs kill normal (HGPRT$^+$) cells. Only HGPRT$^-$ cells, which fail to metabolize the drug, survive. Azaguanine-resistant mutants ($AG^R$) survive

**TABLE I**

**Examples of Drug-Resistant Mutant Cells**[a]

| Drug | Alteration in resistant cell line |
|------|-----------------------------------|
| Azaguanine (AG), thioguanine (TG) | Hypoxanthine-guanine phosphoribosyltransferase deficiency (HGPRT$^-$) |
| Fluoroadenine | Adenine phosphoribosyltransferase deficiency (APRT$^-$) |
| 5-Bromodeoxyuridine (BUdR) | Thymidine kinase deficiency (TK$^-$) |
| Cytosine arabinoside (ara C) | Deoxycytidine kinase deficiency (dCK$^-$) |
| 5-Bromodeoxycytidine (BCdR) | Deoxycytidine deaminase deficiency (dCD$^-$) |
| $\alpha$-Amanitin | Resistant RNA polymerase II (nucleoplasmic) |
| Aminopterin | Overproduction of tetrahydrofolate reductase |
| Steroid resistance | Receptor protein deficiency |
| Ouabain | Resistant Na$^+$K$^+$ activated ATPase |
| Colchicine | Permeability barrier |

[a] Modified from Ringertz and Savage (1976).

10- to 100-fold higher doses of AG than do normal wild-type cells and remain resistant under nonselective conditions, i.e., when cultured on normal media in the absence of the drug. Spontaneous back mutations to azaguanine sensitivity do, however, occur at a low frequency.

Several lines of evidence indicate that AG-resistant mutants ($AG^R$) form a heterogeneous group. Thus, some of the enzyme deficient mutants appear to be due to mutations in structural genes, while others are due to mutations in regulatory genes. One class of $AG^R$ mutants appears to be due to altered permeability to the drug rather than to enzyme deficiency.

Much of the complexity which is associated with $AG^R$, HGPRT⁻ cells applies also to the thymidine kinase deficient cells which are used as the other type of parental cells in the selective system of Littlefield. These cells are obtained by treating cell populations with the thymidine analog bromodeoxyuridine (BUdR). In normal cells this drug will first be phosphorylated by thymidine kinase (TK) and then incorporated into DNA. This usually results in the death of the cell. The exact mechanism of killing is not known but there is evidence that chromosome breaks and mutations are induced and that cell growth and differentiation are affected. Mutant cells deficient in thymidine kinase (TK⁻) fail to phosphorylate and incorporate BUdR into DNA and therefore are drug resistant. The level of BUdR resistance and the reduction in TK activity may vary from one mutant cell line to another. Usually the highly resistant lines are obtained by stepwise selection at increasing concentrations of the drug. Even among those mutant lines which show a drastic reduction in TK activity, it is possible to demonstrate some residual TK activity. Kit and co-workers (1973) have found that this residual activity differs from the main cytoplasmic TK activity in its electrophoretic mobility and appears to be due to a mitochondrial TK activity.

As with $AG^R$, HGPRT⁻ cells there are indications that the TK⁻ phenotype may have a complex genetic background. Some mutants may be due to base changes or deletions in the chromosome segment carrying the *TK* genes.

### 3. Auxotrophic Mutants

Auxotrophic mutants can be isolated by several different methods. Kao and Puck (1967) used a technique where the cells were allowed to multiply rapidly on a medium from which one amino acid had been excluded (minimal medium). Under these conditions mutant cells requiring this amino acid do not proliferate. The cultures were then supplied with BUdR which was incorporated into DNA by the growing, DNA-synthesizing cells. DNA which has incorporated BUdR is more sensitive to light-induced chromosome breakage than is normal DNA. When the cultures were irradiated with UV light, the rapidly proliferating, normal cells were therefore selectively killed, whereas the mutant cells survived. When the cells were then shifted to an enriched medium, the mutant cells proliferated. An alternative approach to using BUdR and light is to kill all DNA synthesizing cells with highly radioactive thymidine.

### 4. Temperature-Sensitive Mutants

The methods used in the isolation of *ts* mutants are similar to those employed in the isolation of auxotrophic mutants. Cells are first cultured and exposed to a chemical mutagen at the permissive temperature, usually 33°–36°C (Fig. 1). After removal of the mutagen some time is allowed for the recovery of the cells and fixation of the mutations. At the end of this period the cultures are shifted to the nonpermissive temperature, which is usually at or slightly above 38°C. Normal cells, but not the *ts* mutants, will grow at this temperature. After allowing some time for the mutant phenotypes to express themselves, a selective agent is added to kill all cells capable of growing at the nonpermissive temperature. Strongly radioactive thymidine and BUdR have been used as selective agents to kill dividing cells as discussed above. Nongrowing *ts* mutants which fail to grow at 38°C are not killed. The time of treatment with the selective agent must be long enough to kill normal (''wild-type'') cells but not so long that the nonmultiplying *ts* mutants are irreversibly damaged. Commonly one uses repeated 48 hours treatment cycles or a single cycle of 24–72 hours. At the end of a treatment cycle the cells are returned to the permissive temperature (e.g., 34°C) where the *ts* mutants will grow. The cells are usually tested by plating different numbers of cells at the permissive and nonpermissive temperatures, followed by analysis of plating efficiencies, growth characteristics, and reversion frequencies.

The biochemical basis of *ts* mutants in eukaryotic cells is not known. Presumably the *ts* mutations are very heterogeneous and involve changes

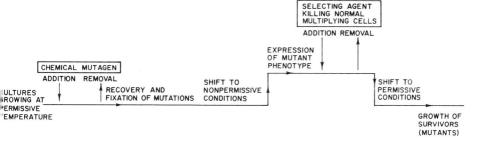

**Fig. 1.** Schematic diagram illustrating the isolation of temperature-sensitive mutant cells (from Ringertz and Savage, 1976).

in many different proteins which cause them to denature or become inactive when the temperature is increased from the permissive to the nonpermissive level. The observation that *ts* mutants frequently show gene complementation in cell fusion experiments indicates that they constitute a very heterogeneous group and suggests that cell hybridization with *ts* mutants may be a useful genetic tool in the analysis of many different metabolic pathways.

### 5. Mutant Cells Obtained from Humans

A large number of inborn errors of metabolism are known in humans and many of these conditions can be detected *in vitro* in fibroblasts growing out from skin biopsies. One advantage of using mutant cells from patients is that their genetics is often well documented on the basis of pedigree analysis. There are many syndromes which may be of interest in future cell fusion work and which involve defects in carbohydrate, lipid, amino acid, and nucleotide metabolism. For five cell types (galactosemia, orotic aciduria, citrullinemia, arginosuccinyluria, and Lesch Nyhan syndrome) there are selective media which prevent the proliferation of normal cells but permit growth of the homozygous mutant cells. For example, patients with the Lesch Nyhan syndrome lack the enzyme HGPRT, and their cells can therefore grow in 8-azaguanine or thioguanine containing media.

### 6. Mitochondrial Mutants

So far mitochondria are the only cytoplasmic organelles in animal cells which are known to carry a detectable amount of DNA. The mitochondrial genome has the form of a small circular double helix with a length of 4–5 $\mu$m and a molecular weight of about $1 \times 10^7$ daltons (Sager, 1972). This genome codes for mitochondrial transfer RNA and RNA components of mitochondrial ribosomes. Most of the proteins of mitochondria appear to be specified by nuclear genes but there are indications that mitochondrial DNA specifies some mitochondrial proteins (for a review, see Sager, 1972). In protozoa and in yeast there is evidence that resistance to certain antibiotics (chloramphenicol, erythromycin, mikamycin) is determined by mitochondrial genes or is cytoplasmically inherited. The molecular mechanisms causing the resistance have not yet been fully explored but in some cases there is a decreased permeability of the mitochondria to the drug while in others mitochondrial ribosomes appear to be altered.

In mammalian cells it is known that chloramphenicol and ethidium bromide impair mitochondrial protein and nucleic acid syntheses and cause characteristic changes in the ultrastructure of mitochondria. These observations have been exploited in attempts to isolate mitochondrial mutants of human cells. Using ethidium bromide as a mutagen and chloramphenicol as a selective agent, Spolsky and Eisenstadt (1972) and Mitchell *et al.* (1975) have isolated mutants which are resistant to chloramphenicol. The resistant cells continue to multiply at concentrations of the drug at which the sensitive parental cell is inhibited, and they also show less alterations of mitochondrial ultrastructure in the presence of the drug than do the sensitive cells. The resistance appears to be stable for many cell generations in the absence of the drug.

## 7. Use of Mutant Cells in Somatic Cell Genetics

The most important use of mutants in somatic cell genetics is to test the direct involvement of a specific gene product in the control of a given phenotype. Drug-resistant, enzyme deficient cells have been useful in testing the relative importance of different pathways of purine and pyrimidine synthesis and in studies of metabolic cooperation and cross-feeding between cells. *In vitro* studies of cells sensitive or resistant to cytostatic drugs have provided information about the mechanism of action of these drugs. This type of information is important for cancer chemotherapy. Some types of drug-resistant mutants show decreased permeability, not only to the drug used for selection, but also to other compounds. Further analysis of such phenomena will no doubt contribute information about the role of membrane proteins and membrane architecture in barrier and transport functions. In a similar way auxotrophic mutants of different types can be used for a genetic analysis of nutritional requirements of cells. Temperature-sensitive mutants represent a particularly interesting class of mutants since they can affect a variety of different cell functions. Most *ts* mutants are believed to be associated with single base changes which cause single amino acid substitutions in proteins, thereby causing them to become nonfunctional when the temperature is raised from the permissive to the nonpermissive temperature. The protein affected may be one which is important for DNA, RNA, or protein synthesis in general or one which affects an enzyme of a specific pathway. Other *ts* mutants affect complex cell functions, for instance, progress through the cell cycle, mitosis, susceptibility to viral infection, or ability to undergo transformation. To identify the specific change in a given *ts* mutant may require a lot of work and detective talent but can be quite rewarding.

In addition to the mutants which have been mentioned above there are others which can be used to analyze transformation, malignancy, cell differentiation, and specific virological questions. Some of these applications will be touched upon in the next section, which discusses cell fusion with mutant cells.

## III. SOMATIC CELL HYBRIDS

The relatively new technique of somatic cell hybridization has already proved itself to be an extremely powerful method with applications in cell biology, genetics, developmental biology, tumor biology, and virology. Basically the technique involves a spontaneous or induced fusion of two different types of cells (A and B) into hybrid cells, called *heterokaryons*, which contain both A and B type nuclei within a common cytoplasm (Fig. 2). Usually fusion between cells of the same type also occurs. Multinucleate cells which contain only one type of nuclei (e.g., only A nuclei) are referred to as *homokaryons*.

Most heterokaryons die within a few days after fusion, but among the

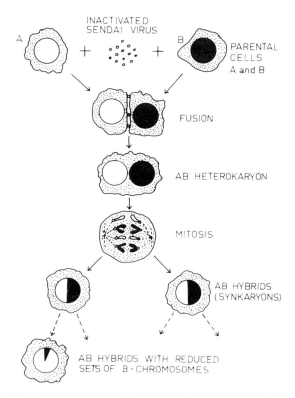

**Fig. 2.**   Schematic representation of Sendai virus-induced fusion of two mononucleate cells (A and B) from different species into a binucleate heterokaryon which then divides and gives rise to two mononucleate hybrid cells (synkaryons). These AB hybrids then divide repeatedly. Some cells eliminate chromosomes originating from parental cell B and gradually overgrow other cells. Intraspecific hybrids often lose chromosomes from both parental genomes (from Ringertz and Savage, 1976).

smaller ones and in particular among those which contain only one A nucleus and one B nucleus, many survive and complete a normal cell division. During mitosis the two chromatids from each chromosome migrate to the opposite poles of a cell, which then divides. The binucleate heterokaryon thereby gives rise to two *synkaryons*, which are mononucleate hybrid cells that often contain a complete set of both A and B chromosomes. Synkaryons are now commonly referred to as *hybrids*. In many cases, these cells have been found to show great vitality and to be capable of multiplication in tissue culture for many years.

A wide variety of animal, human, and even plant cell types have been used as parental cells in these fusions. When cells of different organisms are fused, *interspecific* hybrid cells are produced. In these cases, the parental cells obviously differ with respect to genotype, if not phenotype as well. *Intraspecific hybrids* are obtained by fusing two different cell types from one species. In these instances, the parental cells share a common genotype but differ in morphological, biochemical, immunological, or functional properties (i.e., they differ in phenotype).

## A. Cell Fusion

As part of the program of myogenic differentiation, mononucleate myoblasts of a specific stage fuse to form multinucleate myotubes in which the synthesis of contractile proteins occurs. Macrophages fuse to produce giant cells as part of mechanisms which degrade large foreign particles. Spontaneous cell fusion, with these exceptions, is however, a very uncommon event both *in vivo* and *in vitro*, and only rarely does it give rise to cell hybrids.

The first cell hybrids were discovered in 1960 by Barski and his collaborators in Paris. These authors discovered a third cell type in a mixed culture of two different mouse cell lines. Analysis of the chromosomal content of these cells revealed that the new cells were hybrids which contained chromosomes characteristic of the two original cell types. The hybrid cells, which had formed by spontaneous cell fusion, multiplied rapidly *in vitro* and spontaneously overgrew the two parental cell types. This situation, however, is unusual, and it was only after methods of inducing fusion (Okada *et al.,* 1957; Harris and Watkins, 1965; Okada and Murayama, 1965), and methods for selecting hybrid cells (Littlefield, 1964) were introduced that cell hybridization became widely used.

The most common method of inducing cell fusion is to add inactivated Sendai virus to a suspension or monolayer of cells, but chemical methods of fusing cells are now available. The use of inactivated Sendai virus as a fusing agent is based on the observation that this virus, when added to a suspension of mononucleate cells *in vitro*, will cause the formation of large multinucleate cells (*polykaryons*). The fusion process can be divided into two main steps: agglutination and coalescence. Agglutination requires that the cells carry receptors for Sendai virus and that ionic conditions are appropriate. The concentration of calcium ions appears to be particularly important. Coalescence of the cells starts as a breakdown of cell membranes at points of contact. The exact molecular mechanism involved is not known but it is believed that this process results from a change in the molecular organization of the membrane lipids. As a result, narrow cytoplasmic bridges form between the cells, and widening of these bridges ultimately brings about coalescence. While agglutination can occur at low temperatures, coalescence requires physiological temperatures and a normal energy metabolism. Other factors also play a role in the reorganization of intracellular structures during coalescence. At the end of the fusion process, nuclei tend to accumulate in the center of the polykaryon. This process can be blocked by colchicine, suggesting that microtubular structures are actively engaged in the reorganization of the internal cell structure. There is also a rapid reorganization of cell surface components. Thus within a few minutes after the fusion of two cells carrying different surface antigens, there is a rapid intermixing of the antigens over the entire cell surface.

Several chemical methods of inducing cell fusion have been suggested. These involve the use of lysolecithin, polyethylene glycol, and artifically produced liposomes. So far these methods have not been tested widely but there is hope that chemical methods of fusing cells will replace the use of inactivated Sendai virus.

## B. Polykaryons

### 1. DNA Synthesis

Cells at all stages of the cell growth cycle may fuse with each other when exposed to appropriate viruses. In experiments with asynchronous cultures, the polykaryons will be formed, therefore, from different combinations of $G_1$, S, $G_2$, and mitotic cells. Heterokaryons or homokaryons containing nuclei of more than one stage are said to be *heterophasic* whereas those which by chance or by design, contain nuclei from only one stage of the cell growth cycle are classified as *homophasic*.

As a rule homokaryons and heterokaryons examined some time after cell fusion show synchronization of DNA synthesis (for references, see Johnson and Rao, 1971; Rao and Johnson, 1974). In homophasic cells containing $G_1$ nuclei all nuclei usually start DNA synthesis at the same time. In heterophasic $G_1$/S homokaryons DNA synthesis is induced in more than 50% of the $G_1$ nuclei within 2 hours after fusion. The greater the proportion of S nuclei, the faster DNA synthesis is induced in the $G_1$ nuclei. This dose-effect relationship suggests that the triggering of DNA synthesis depends on the concentration of inducing factors present in the cytoplasms of the S cells. In homophasic homokaryons the nuclei retain the cell cycle times characteristic of the parental cells. In homophasic $G_1$ heterokaryons, however, the time point at which DNA synthesis is initiated is determined by the cell with the shorter $G_1$ phase. On the other hand, each nucleus in a heterokaryon retains the S phase duration typical of its mononucleate parental cell. Nuclei with a short S phase will therefore have completed their DNA replication before those with a longer S phase even though they all reside in a common cytoplasm.

## 2. Mitosis

Mitotic synchrony is induced in most polykaryons. The best synchrony is found in homophasic and heterophasic homokaryons while heterophasic heterokaryons show a low degree of synchrony.

Studies of the mechanisms by which synchrony of mitosis is established (for reviews, see Johnson and Rao, 1971; Rao and Johnson, 1974), suggest that more than one factor is involved: (a) Late $G_2$ and mitotic cells seem to contain factors which induce accelerated entry of $G_1$ or S nuclei into mitosis. These factors will be discussed in the following section where we consider fusions between mitotic and interphase cells. (b) S phase cells contain factors which prevent or delay the entry of $G_2$ nuclei into mitosis. (c) Mitosis itself may be prolonged so that lagging nuclei catch up with advanced nuclei during an extended metaphase.

## 3. Premature Chromosome Condensation (PCC)

Fusion of a mitotic cell with a cell in interphase results in a precocious attempt of the interphase nucleus to enter mitosis. Chromosome-like structures, sometimes with a fragmented appearance, form in the interphase nucleus, and the nuclear membrane disappears. This phenomenon has been called *premature chromosome condensation* (PCC).

There is considerable variation from cell to cell in the morphology and extent of PCC. The nature of the different forms of PCC and their relationship to the stage of the cell cycle was clarified by Johnson and Rao (1971) in experiments where one population of synchronized interphase cells was fused with another in mitosis. $G_1$ nuclei produced thin, extended filaments whereas $G_2$ nuclei produced thicker filaments which were similar to normal mitotic chromosomes. Furthermore, chromosome filaments derived from $G_1$ nuclei were single-stranded while filaments of $G_2$ nuclei consisted of two chromatids (Fig. 3). S phase nuclei condensed only partially and

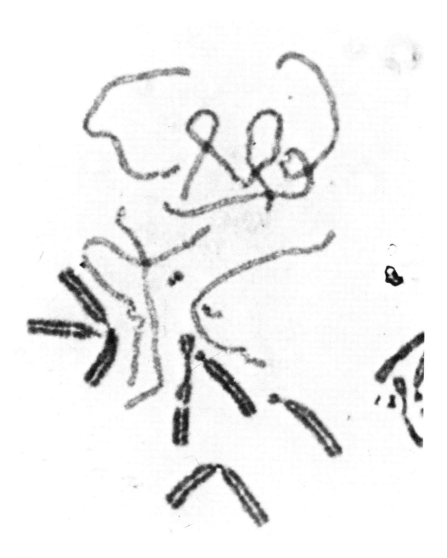

**Fig. 3.** Premature chromosome condensation after fusion of a mitotic Indian muntjac cell (condensed chromosomes) with a $G_2$ interphase cell, also from Indian muntjac. The prematurely condensed $G_2$ chromosomes are much longer than the normal mitotic chromosomes of the "inducer" cell and consist of two chromatids (courtesy Dr. Dan Röhme).

gave rise to irregular, fragmented chromatin masses. Nuclei blocked at the $G_1$–S transition produced patterns intermediate between those obtained with $G_1$ and S nuclei.

Further studies of S phase PCC by Röhme (1974) showed that these chromosomes may not be as fragmented as they first appeared. He labeled S phase cells with [3]H-thymidine prior to fusion with metaphase cells and then performed autoradiography. In such preparations the chromosome fragments appear to be linked to each other by radioactively labeled DNA filaments which are too thin to be visible in the light microscope. The fragments seen in light microscopic preparations probably represent chromatin which has already replicated or which has not yet started to replicate while the [3]H-thymidine-labeled "invisible" regions between these "fragments" are greatly extended chromosome fibers which are in the process of DNA synthesis.

The induction of PCC promises to be useful for a number of purposes not directly related to the study of mitosis, DNA replication, or the cell cycle (for a more detailed discussion, see Sperling and Rao, 1974). Among these applications are the following: (1) to visualize chromosomes in nondividing cells; (2) to study the molecular architecture of interphase chromatin during $G_1$, S, and $G_2$; (3) for detailed analysis of banding patterns and chromosome structure; (4) to detect chromosome damage arising from alkylating agents or radiation, and to monitor the repair of such chromosome damage.

### 4. RNA and DNA Synthesis in Heterokaryons

Soon after the introduction of Sendai virus as an agent to produce multinucleated hybrid cells, it was noted that inactive cell nuclei, among them chick erythrocyte nuclei and mouse macrophage nuclei, undergo a marked reactivation and show an accelerated synthesis of RNA (Harris *et al.*, 1966; Harris, 1970) when fused with a transcriptively active cell. This reactivation in heterokaryons has provided a valuable system for the analysis of nucleocytoplasmic interaction and its role in the control of nuclear activity.

One special feature of erythrocyte fusions is that these cells often undergo virus-induced lysis and lose their cytoplasmic contents before the fusion takes place. The erythrocyte *ghosts* which are formed as a result of this lysis consist of a small condensed nucleus surrounded by a plasma membrane. In the presence of Sendai virus the erythrocyte ghosts will agglutinate to each other and to other cells present.

Once it has been incorporated into the cytoplasm of an actively growing cell (Fig. 4) the erythrocyte nucleus responds to its new environment by undergoing a reactivation process which includes the resumption of tran-

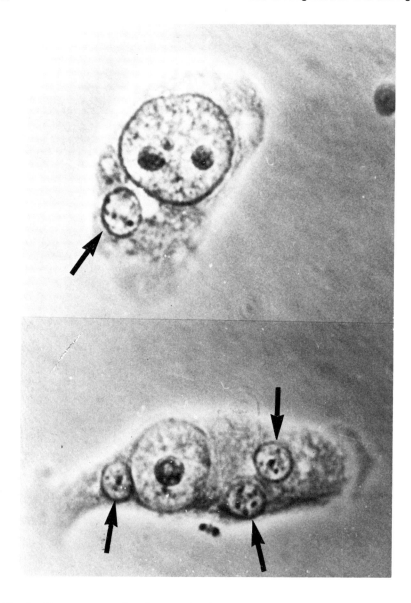

**Fig. 4.** Phase contrast microphotographs of two heterokaryons formed by fusing chick erythrocytes with human HeLa cells. The chick nuclei (arrows) have been partially reactivated and are beginning to form nucleoli (from Ringertz and Bolund, 1974).

scription and replication and, with some lag, the synthesis of new chick proteins. The reactivation process involves a series of morphological and chemical changes for references (see Appels and Ringertz, 1975). The main events that occur during reactivation of chick erythrocyte nuclei in heterokaryons made with human tumor cells (HeLa) (or other actively growing cells) can be summarized as follows: (1) early changes in the physicochemical properties of deoxyribonucleoprotein; (2) increase in nuclear volume and dry mass and a dispersion of the condensed chromatin; (3) migration of *human* nucleospecific proteins into the chick nucleus; (4) initiation of HnRNA synthesis and an increase in nucleoplasmic RNA polymerase activity; (5) formation of a nucleolus, an increase in nucleolar RNA polymerase activity, and initiation of ribosomal RNA (rRNA) synthesis; (6) initiation of chick specific protein synthesis including chick nucleolar antigens, surface receptors and antigens, enzymes; (7) DNA synthesis; (8) mitosis and formation of mononucleate synkaryons.

A reactivation phenomenon similar to that observed for chick erythrocyte nuclei occurs also in other types of heterokaryons where an inactive nucleus is confronted with the cytoplasm of an active cell. One example of this is macrophage heterokaryons. Macrophages do not normally synthesize DNA but appear to be arrested in the $G_1$ phase in the same way as many other highly differentiated cells. Their kidney-shaped cell nuclei do, however, synthesize RNA and have small nucleoli. The nucleus is smaller than that of more active nuclei from other tissues. When macrophages are fused with more active cells such as HeLa cells (Harris *et al.*, 1966) or melanoma cells (Gordon and Cohn, 1970), the macrophage nucleus becomes swollen, the nucleoli become more prominent, and RNA synthesis accelerates. The macrophage nuclei are also triggered to resume DNA synthesis, and some heterokaryons enter mitosis. Results obtained by Gordon and Cohn indicate that reactivation of macrophage DNA synthesis is independent of macrophage RNA and protein synthesis, but requires RNA and protein synthesis specified by the "active" nucleus. This indicates that the "active" cell provides the macrophage nucleus with products which trigger DNA synthesis.

### 5. *Phenotypic Expression in Heterokaryons*

Most hybrid cells express properties characteristic of both parental cells (*coexpression* or *codominance*). Thus, Watkins and Grace (1967) found surface antigens characteristic of both parental cells in heterokaryons made of human HeLa cells and mouse Ehrlich cells. Coexpression was also observed by Carlsson *et al.* (1974) in studies of chick–rat hybrid myotubes, which synthesize both chick and rat myosin. In other types of heterokaryons it has been observed that properties characteristic of one parental cell are *extinguished* and cannot be detected in the heterokaryons. Extinction of differentiated properties has been observed by Gordon and Cohn (1971) in macrophage × melanoma heterokaryons, where the macrophage marker properties disappear. In heterokaryons of rat hepatoma cells and rat epithelial cells, Thompson and Gelehrter (1971) observed that the inducibility of tyrosine aminotransferase (TAT), a property characteristic of liver differentiation, disappeared. These examples illustrate the two most important patterns of phenotypic expression, *coexpression* and *extinction* of properties characteristic of the parental cells.

### 6. Gene Complementation Analysis with Heterokaryons

Studies of heterokaryons made with fibroblasts from patients suffering from inborn errors of metabolism clearly show the value of cell fusion as a general system for gene complementation analysis of genetically defective cells. Fusion experiments with cells from cases of a rare skin disease, xeroderma pigmentosum (XP), illustrate this point. In this disease the skin is extremely sensitive to sunlight because of genetic defects which affect the repair of base changes in DNA induced by ultraviolet light. Biochemical evidence suggests that the repair process involves excision of UV-induced thymine dimers by combined action of endo- and exonucleases, synthesis of new polynucleotide segments by a special DNA polymerase, and linking of the newly synthesized DNA to the preexisting DNA by special ligases. In principle the DNA repair can be impaired if any one or a combination of these functions is impaired, e.g., lack of, or presence of, a defective enzyme.

DeWeerd-Kastelein *et al.* (1973) found that in homokaryons containing two fibroblast nuclei from a single xeroderma pigmentosum patient, both nuclei were unable to repair their DNA. However, in heterokaryons containing nuclei from patients with different forms of the disease or nuclei from normal and XP cells, both types of nuclei showed DNA repair synthesis after UV-irradiation. The most likely interpretation of this result is that different genes are involved in the basic defect of the two types of XP cells and that the nuclei, therefore, complement each other. At the moment the xeroderma syndrome appears to fall into 4–6 different complementation groups, but further analysis of more cases may reveal additional groups.

Heterokaryons clearly offer interesting possibilities for the analysis of genetic defects in human cells, and they have the advantage over synkaryons that chromosome elimination is very unlikely. In principle the analysis of heterokaryons should be applicable to a number of other syndromes, e.g., glycogen storage diseases and disorders related to amino acid, carbohydrate, lipid, and purine and pyrimidine metabolism. The main requirements for an analysis of these conditions in heterokaryons are that the genetic defects manifest themselves in cultured cells and that methods suitable for single cell analysis are available. The latter requirement, however, is not an absolute one. Biochemical methods can be used as long as the efficiency of fusion is high and the culture contains a large number of heterokaryons.

## C. Isolation of Mononucleated Hybrid Cells (Synkaryons)

Multinucleated cells, both homokaryons and heterokaryons, have a limited life span *in vitro*. Most die as they try to divide, perhaps due to the formation of several spindles from many centrioles present, and a resulting aberrant mitosis. In a population of multinucleated cells, there are, however, usually some cells in which only one spindle forms. In such cases, mononucleated cells, which contain two or more sets of chromosomes in the same nucleus, arise. In this way, a homokaryon gives rise to two mononucleated polyploid cells, while a heterokaryon gives rise to two mononucleated hybrids (synkaryons), in which the daughter nuclei contain chromosomes from two different animal species or from cells with different forms of differentiation.

The first hybrid cells were discovered and isolated because they proliferated more rapidly than either of the two parental cell types and became the dominant cells in the culture. This phenomenon, however, is unusual, and in order to study hybrid cells they have to be isolated from the fusion mixture by a different technique. If the number of hybrid cells is great relative to the number of parental cells, the hybrid cells may be isolated by single cell cloning, but only seldom can such nonselective isolation methods be utilized. Usually the experimenter is faced with the problem of isolating a small number of slow-growing hybrid cells from a large number of rapidly proliferating parental cells. By far the easiest solution to this problem is to use a selective medium which favors the growth of the hybrid cells, while killing or inhibiting the parental cells. On the other hand, differentiated cells (e.g., nerve cells, lymphocytes, spermatozoa), which grow poorly or not at all *in vitro*, form proliferating hybrids when fused with more active cells (such as tumor cells or cells of established lines). Such hybrids can be isolated by semiselective media in which the selection is directed only against the rapidly proliferating parent.

In the selection system developed by Littlefield (1964), hybrid cells are isolated on a selective medium after fusion of thioguanine resistant (TG$^R$) cells, which lack the enzyme hypoxanthine-guanine phosphoribosyltransferase (HGPRT$^-$), with bromodeoxyuridine resistant (BUdR$^R$) cells, which are deficient in the enzyme thymidine kinase (TK$^-$). The genetic defects are of little consequence during growth on normal tissue culture media, since the relevant enzymes are only involved in salvage (reserve) pathways for nucleotide synthesis. When the main biosynthetic pathways for purine and pyrimidine nucleotides are blocked by the folic acid analog aminopterin (Fig. 5), normal cells (HGPRT⁺, TK⁺) can survive if supplied with exogenous hypoxanthine and thymidine whereas the mutant cells die because of their inability to synthesize nucleotides from hypoxanthine (HGPRT⁻ cells) or from thymidine (TK⁻ cells).

Littlefield demonstrated that hybrid cells formed from fusion of HGPRT⁻ with TK⁻ cells could be isolated by selection in a medium containing hypoxanthine, aminopterin, and thymidine (HAT medium). Hybrid cells obtaining one chromosome set which was HGPRT⁻ but TK⁺ and one that was HGPRT⁺ and TK⁻, were able to produce HGPRT and TK and to utilize exogenous hypoxanthine and thymidine for nucleotide synthesis. Thus when combined in one cell the two parental genomes complement each other and make it possible for the cell to survive on HAT medium.

The development of the HAT medium represented an introduction into eukaryotic cell biology of techniques already in use in microbial genetics, and it has had a great impact on the development of somatic cell genetics. The technique has already been utilized in the isolation of many different types of somatic cell hybrids and is now extensively used in connection with chromosome mapping (see below). In addition to the HAT selective system there are now other selective systems based on the fusion of drug-resistant cells and selection on special media.

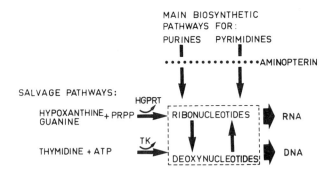

**Fig. 5.** Selection of hybrid cells according to Littlefield (1964). On the selective HAT medium the main biosynthetic pathways for purines and pyrimidines are blocked by the folic acid analog aminopterin. Normal cells survive by utilizing hypoxanthine and thymidine in the medium. Mutant cells lacking HGPRT or TK enzymes die on the selective medium. Hybrid cells made by fusing HGPRT⁻, TK⁺ cells with HGPRT⁺, TK⁻ cells produce some HGPRT and TK and therefore survive because of gene complementation (from Ringertz and Savage, 1976).

Auxotrophic mutants (see Section II) have been very useful in the isolation of hybrid cells. If, for instance, a cell line which requires the addition of glycine to the medium for continued cell proliferation is fused with a line auxotrophic for proline, the hybrid cells, by means of gene complementation, will be prototrophic (nonrequiring) and grow on a minimal medium deficient in both glycine and proline. At the same time both the parental cell types will die because of their defects. Sometimes two different cell lines require the same substance but differ with respect to their gene mutations. For instance, the mutations may affect two different enzymes in the same metabolic pathway or two different polypeptide chains of the same enzyme. If this is the case, fusion of two auxotrophic cell lines requiring the same substance will give hybrids which are prototrophic. It is possible, therefore, to use cell hybridization with mutant cells to dissect metabolic pathways and cell functions.

Another approach to the isolation of hybrids is to use temperature-sensitive mutants as parental cells. Cultured mammalian cells are capable of multiplying in a temperature range from about 32° to about 40°C, the optimum temperature being about 37°C. Using the selection procedures discussed in a previous section, it is possible to isolate temperature-sensitive (ts) mutants which differ from normal (wild-type) cells in that they are unable to grow at 38°–39°C (nonpermissive temperature) but grow like normal cells at 33°–36°C (permissive temperature). These properties have been used to select hybrids from crosses of different ts mutants and from crosses of ts mutants with drug-resistant and auxotrophic mutants. In crosses where both parental cells are ts mutants the hybrid cells frequently show gene complementation and grow at the nonpermissive temperature.

Temperature can also be used as a selection pressure for the isolation of interspecific hybrids from crosses of nonmutant cells. In these cases the parental cells have been derived from species differing in body temperature.

### D. Chromosome Segregation in Hybrid Cells

Intraspecific hybrids isolated and analyzed relatively soon after fusion often show chromosome numbers which approximate or are a little below the expected sum if one assumes that 1 + 1 fusion has occurred. The early workers in somatic cell hybridization noted that intraspecific hybrid lines which had been growing for some time *in vitro* often had lost a few chromosomes but it was not until interspecific hybrids were studied that it became clear that hybrid cells may show extensive chromosome elimination. This phenomenon, also referred to as *chromosome segregation* frequently involves the preferential elimination of chromosomes of one species. Thus human chromosomes are selectively eliminated in man × rodent hybrids. Weiss and Green (1967) were the first to isolate hybrids which retained practically all the mouse chromosomes but had lost almost all of the human chromosomes. Since it was already known that many homologous isozymes of man and mouse could be distinguished by differences in their electrophoretic mobility, it then became apparent that these methods, if applied to somatic cell hybrids undergoing chromosome segregation, would make it possible to assign genes to specific human chromosomes.

Other interspecific hybrids also have been found, in general, to eliminate preferentially the chromosomes of one of the two parental genomes. In hybrids between rodents and primates the primate chromosomes are usually eliminated. For other crosses, such as those between different rodents, the direction of segregation varies. In rat × mouse hybrids the rat chromosomes are selectively eliminated whereas in hamster × mouse crosses both parental sets are reduced but with a preferential loss of mouse chromosomes.

In most interspecific crosses which had been examined for chromosome segregation patterns before 1971, the parental cells had been heteroploid cells of established lines. As hybrids involving diploid cells began to be studied it became apparent that the direction of chromosome loss depended not only on the species but also on the type of parental cells used. Reverse chromosome segregation was observed by Minna and co-workers (Minna and Coon, 1974) in a large number of independent hybrid clones produced by fusing mouse or rat primary cells with human aneuploid cells from some human cell lines. In these studies the rodent chromosomes were preferentially lost, which is the opposite of what is observed when established cell lines from the two species are used.

Attempts have also been made to influence the direction of chromosome elimination by damaging the chromosomes of one of the parental cells before fusion. Pontecorvo (1971) succeeded in directing the chromosome losses by pretreating one of the partner cell types of a fusion with bromodeoxyuridine (BUdR) or X-rays and then fusing with untreated cells from another species. Chromosomes were preferentially lost from the pretreated species, but there were severe drawbacks with the method. Mild pretreatment caused only a moderate chromosome loss, while more extensive treatment resulted in chromosome aberrations.

Back selection with drugs has in some cases been successful in forcing the elimination of the complementing chromosomes of hybrids from $HGPRT^- + TK^-$ crosses. In these cases the hybrids first have been isolated in HAT-medium and then been transferred to normal medium containing either thioguanine or BUdR. In the presence of thioguanine only hybrids which eliminate the chromosome carrying the *HGPRT* gene will survive whereas in the presence of BUdR the elimination of the thymidine kinase gene is forced. Selection with TG or BUdR may not only cause the loss of the two chromosomes carrying the *HGPRT* or *TK* genes but may also provoke extensive losses of seemingly unrelated chromosomes. Wiblin and Macpherson (1973) in a study of hybrids between SV40-transformed Syrian hamster cells (BHK) and mouse 3T3 cells, combined BUdR selection against the hamster chromosomes with immunological selection against surface antigens of hamster cells by including cytotoxic antihamster antibodies in the medium. In this way elimination of hamster chromosomes was obtained in an interspecies combination which, in other cases, have shown elimination of mouse chromosomes.

From these examples it appears that chromosome segregation in populations of interspecific hybrids can to some extent be directed either by choosing the appropriate parental cells, by damaging the undesired chromosome complement before fusion, or by applying selective pressures against variants expressing the genes on the chromosome one is seeking to eliminate.

The rate at which chromosome losses occur may to some extent reflect the phylogenetic distance between the two parental cells in an interspecific fusion. With closely related species, such as two different rodents, (e.g., mouse × Chinese hamster) the rate is usually slow, whereas with cells from different classes (man × chick) chromosome segregation is very rapid. Within the life history of a hybrid cell line, chromosome elimination may be rapid during an early phase followed by a period during which the chromosome number is stabilized and chromosome segregation is slow. It is possible that man × Chinese hamster hybrids are the most favorable material for human gene mapping, since hybrids containing a hamster complement plus only one or two human chromosomes can be obtained within 1–2 weeks. This rapid chromosome segregation is advantageous since it minimizes the risk of chromosomal rearrangements. The fact that the Chinese hamster cell has only 22 chromosomes, which can easily be distinguished from those of man, is also an advantage.

Chromosome segregation in interspecific hybrids is clearly a nonrandom process since chromosomes of one of the parental genomes are preferentially lost. It is also of interest and importance to know whether individual human chromosomes in a man × mouse hybrid are randomly eliminated. There have been several reports describing selective elimination and retention of individual chromosomes, but the phenomenon has not been studied extensively enough to allow the establishment of any rules by which one could make predictions about the segregation pattern likely to be found in a given fusion.

### E. Phenotypic Expression in Hybrid Cells

One of the main applications of the cell hybridization technique in cell biology has been in the analysis of gene expression and cell differentiation. This field has recently been reviewed by Davidson (1974) and Ringertz and Savage (1976), and therefore, in this chapter we restrict ourselves to some of the main principles and a few examples which illustrate these principles. For more complete lists of references these reviews should be consulted, and for another viewpoint see Chapter 9.

Hybrid cells have been examined for (a) general characteristics such as morphology, growth rate, contact inhibition, and senescence; (b) complex physiological or immunological properties, e.g., electrophysiological, nutritional, or antigenic characteristics; (c) specific gene products, e.g., enzymes, hormones, and immunoglobulins; (d) sensitivity to specific drugs.

Any one of these characteristics can be used as a marker if the two parental cells differ with respect to it. Obviously a distinction has to be made between properties common to all cells of a given organism (*constitutive markers*) and properties which are expressed only by certain determined cells as they enter the differentiated state (*facultative markers*).

A number of different gene expression patterns have been observed in multiplying mononucleate hybrid cells: *coexpression* of constitutive (Gershon and Sachs, 1963; Weiss and Ephrussi, 1966; Davidson *et al.*, 1966, 1968) and facultative markers (Bloom and Nakamura, 1974); *dominance* and *recessiveness* of drug resistance markers (M. Harris, 1974); *extinction* of facultative markers (Davidson *et al.*, 1966, 1968; Silagi, 1967; Schneider and Weiss, 1971; Sparkes and Weiss, 1973); and in some cases activation of new properties not expressed by either of the parental cells (Peterson and Weiss, 1972; McMorris and Ruddle, 1974; Malawista and Weiss, 1974; Darlington *et al.*, 1974). Extinction sometimes has been followed by reexpression of the extinguished markers as the hybrids undergo chromosome segregation (Weiss and Chaplain, 1971; Croce *et al.*, 1973). Although these observations are important for our understanding of how gene expression and cell differentiation are controlled in eukaryotic cells, they do not provide a clear picture of how gene activity is regulated. Still, it seems useful to emphasize some of the principles that have been illustrated by the results obtained in cell hybrid studies:

(a) The fact that extinguished markers can reappear (e.g., pigmentation in melanoma hybrids and liver-specific enzymes in hepatoma hybrids) after many cell cycles indicates that the epigenotype, that is, the programming for a certain type of cell differentiation, is quite stable and can be retained for long periods of time in the absence of its expression (see Chapter 9).

(b) There are no clear-cut examples of a complete reprogramming of a genome from one type of differentiation to another, but there are several observations which suggest that genes for individual facultative markers may be activated. The synthesis of *human*, as well as mouse, serum albumin in mouse hepatoma × human leukocyte hybrids is one example of this phenomenon (Darlington *et al.*, 1974).

(c) Gene dosage effects have been demonstrated in hepatoma (Peterson and Weiss, 1972) and melanoma (Fougère *et al.*, 1972; Davidson, 1972) hybrids since differentiated markers are extinguished in 1 : 1 hybrids but expressed in 2 : 1 hybrids where the input of chromosomes from the differentiated parental cell has been doubled.

(d) There are many examples of independent expression of one out of several markers characteristic of a specific form of cell differentiation. Extinction and reexpression of individual, liver-specific enzymes can occur independently of the expression of other facultative markers. At the same time it is evident from studies on neuroblastoma hybrids that groups of markers may be coordinately expressed (Minna *et al.*, 1972; McMorris and Ruddle, 1974).

Both interspecific and intraspecific hybrids have been used to study gene regulation and cell differentiation. The advantages and disadvantages of working with interspecific and intraspecific hybrids, respectively, can be summarized as follows: Interspecific crosses have the advantage that chromosome identification is facilitated. Furthermore, since homologous proteins from two different species usually differ slightly in amino acid composition and electrophoretic mobility, it is possible to determine whether genes in one or both parental genomes are expressed. The main drawback of the interspecific hybrids is the rapid elimination of chromosomes. It is often difficult or impossible to decide whether the loss of a given property, as for example the expression of an enzyme, is due to a normal regulatory event, to the elimination of the chromosome carrying the structural gene, or to loss of a chromosome with a regulatory gene. Another problem is the uncertainty as to whether regulatory molecules specified by one species will interact in a normal way with the genome of the other species.

Some of the problems with interspecific hybrids can be avoided by using intraspecific hybrids. This type of hybrid offers greater chromosomal stability and eliminates the possibility that one genome does not recognize the regulatory signals of the other because of a species barrier. On the other hand, chromosomal analysis and the identification of the gene products specified by each of the two genomes is very difficult.

In spite of all the uncertainties and difficulties of interpretation which are associated with cell hybridization studies, there is little doubt that this technique is a very valuable addition to the methodology of somatic cell genetics. Further analysis of the hybrid systems discussed on the preceeding pages will certainly throw light on the mechanisms regulating gene expression and cell differentiation in higher cells. Several of the systems appear very promising and a considerable amount of information has already been accumulated.

## F. Use of Somatic Cell Hybrids for Human Gene Mapping

### 1. Strategy in Isolating Hybrid Cells

Two different approaches have been used to obtain hybrid cells for gene mapping. One is based on fusing human cells with mouse cells deficient in some phenotypic character followed by cultivation of hybrids on media that select against the deficient cells. Thus, while unfused mouse cells die, any mouse $\times$ man hybrid cells which retain a human chromosome carrying genetic information that can overcome (complement) the defect survive. Phenotypic characters which may be used in this way are called *selective markers*. The specific retention of the human *HGPRT* gene in a $HGPRT^+$ $TK^-$ human + $HGPRT^-$ $TK^+$ mouse cell is an example of this approach. The presence of the human X chromosome or more specifically its *HGPRT* gene is an absolute requirement for survival on HAT medium, and this experimental system is therefore said to fix the retention of this human gene. Among man $\times$ mouse hybrids, there are at least five selective markers and corresponding selective media.

When the appropriate hybrids have been obtained on selective medium, it is sometimes possible to selectively remove the human chromosome by back selection on permissive medium. A man $\times$ mouse hybrid made by crossing $HGPRT^+$, $TK^-$ human cells with $HGPRT^-$, $TK^+$ mouse cells is, by mutual gene complementation, $HGPRT^+$, $TK^+$ and therefore survives on HAT medium. When such a hybrid cell is transferred to normal (permissive) medium and treated with 8-thioguanine, only those hybrids which lose the human *HGPRT* gene survive. This type of back-selective experiment makes it possible to check that markers assigned to a certain chromosome are lost when the chromosome is. The other method of obtaining hybrids for gene mapping is based on nonselective markers and random segregation of human chromosomes in a large number of human $\times$ mouse hybrids. By analyzing the expression of a variety of human markers in these hybrids it is possible to establish that some markers always occur together and that they are always lost as a group.

## 2. Assigning Gene Loci to Different Chromosomes

In order to assign gene loci to different chromosomes, phenotypic and chromosomal data from a large number of independent and hopefully randomly segregating clones are analyzed by making pairwise comparisons of the occurrence of markers and for a chromosome and an enzyme. If two human enzymes always occur together and are always lost together, it is likely that their structural genes belong to the same linkage group and map on the same chromosome. The establishment of enzyme–enzyme *linkage groups* has usually been the first step in assigning the corresponding genes to a specific chromosome. The next step has been to assign individual markers or linkage groups to a specific chromosome. If an enzyme marker and a specific chromosome show concordant segregation [that is, the presence or absence of the enzyme correlate with the presence or absence of a certain chromosome (Fig. 6)], then an enzyme–chromosome linkage is established. Thus, the structural gene for that enzyme is assigned to that chromosome.

## 3. Regional Mapping of Chromosomes

Chromosome mapping consists of establishing the linear order in which genes occur on the individual chromosomes, measuring distances between genes, and relating these data to the structure of the chromosomes. Chromosome mapping with hybrid cells is based either on the use of parental cells with rearranged chromosomes or on spontaneous or induced chromosomal alterations which may occur in the hybrid cells. Frequently hybrid cells are found containing broken chromosomes. If, after growth on selective medium, only a single broken chromosome is present, one may deduce that the gene for the selective marker is on this piece of chromosome. Then, by identifying what other markers have co-

**Fig. 6.** Gene assignment with hybrid cells is based on screening a large number of independent hybrid clones for a specific gene product, for instance an enzyme, and for the presence of specific chromosomes from the segregating genome. The different clones are then compared with each other. If the enzyme X, in a series of man–mouse hybrid clones, is correlated with the presence of a certain human chromosome, for instance chromosome 7, the structural gene for human enzyme X may be assigned to chromosome 7.

segregated, a number of genes can be mapped to a specific part of a chromosome—the part represented by the fragment. Moreover one can induce chromosome breaks at a high frequency with chemical agents or X-irradiation. This approach, the "disruptive strategy," can also be used to obtain information about the linear order of genes and distances between different gene loci (Goss and Harris, 1975). The greater the distance separating two genes on a chromosome, the higher the probability that chromosomal breaking agents will separate them, and similarly the smaller the distance the lower the probability.

Cell hybrids clearly represent a powerful tool which will make it possible to obtain detailed gene maps of the chromosomes of man and of other species. It should be pointed out, however, that it is not the only method used in gene mapping. A number of gene assignments have been made by family studies, nucleic acid annealing with cDNA or cRNA, and by studies of virus-induced changes on specific chromosomes. These techniques and the cell hybridization method complement each other in several ways and a large number of laboratories are now using these methods to study the organization of eukaryote genomes. Close to 100 genes have been assigned to the 22 different autosomal chromosomes in man. Cell hybridization has established or confirmed 56 of these assignments. Nineteen genes have been assigned to chromosome 1, and a considerable amount of information has also accumulated concerning the regional localization of individual genes on this chromosome. More than 90 different traits have been assigned to the X chromosome in family studies. Five of these assignments have been confirmed in cell hybrid studies. In the case of the X chromosome, cell hybridization has played a relatively small role in the assignment of new genes (only one has been added). It has been an important method, however, in the regional mapping of this chromosome. The information available is still fragmentary in the sense that the number of genes mapped is too small to see any correlations between the organization of the genome and more complex cell functions and their regulation. Some of the information is, however, already of great medical interest.

### G. Use of Hybrid Cells in the Analysis of Malignancy

The technique of fusing two nucleated cells into a hybrid cell has already gained wide applications in the analysis of genetic changes which result in malignant tumor cells. Malignancy is defined as the ability of a cell to multiply and kill the host animal into which it has been inoculated. Most fusions of normal and neoplastic mouse cells have produced hybrid cells which were malignant (Barski et al., 1961; Scaletta and Ephrussi, 1965; Defendi et al., 1967). Harris, Klein, and co-workers (for a review, see

Harris, 1972), have shown, however, that "normal" cells in some cases suppress the malignancy of a tumor cell so that the result is a nonmalignant hybrid cell. Hybrid cells which contain almost all the chromosomes from the normal and tumor cell parents rarely give rise to tumors, while segregant hybrid cells which have lost many chromosomes tend to be malignant. One interpretation of these results is that the malignant properties of tumor cells are due to chromosomal defects. These defects are *complemented* if the hybrid cells retain a complete set of normal chromosomes. When, however, some crucial chromosomes from the normal parental cell are lost, it is possible that the defect in the tumor chromosome is no longer compensated. It is, therefore, possible that a line of nonmalignant hybrid cells may produce malignant segregants which overgrow the nonmalignant hybrids.

This interpretation has been challenged by Croce and Koprowski (for references, see Croce *et al.*, 1975) who studied *in vitro* transformation by fusing SV40 virus-transformed human cells with normal contact inhibited mouse cells. The resulting hybrids were found to show a transformed phenotype *in vitro*. As chromosome segregation proceeded and human chromosomes were eliminated, it became possible to establish a positive correlation between human chromosome 7 on the one hand and transformed phenotype *in vitro*, malignancy *in vivo*, presence of viral genomes, and virus specified antigens on the other hand. These findings led Croce and Koprowski to conclude that viral DNA had been integrated on human chromosome 7 and that this DNA contained genes for transformation (malignancy) which were dominantly expressed.

Harris and Klein and Croce and Koprowski use a genetic nomenclature in discussing the properties of hybrids between malignant and normal cells. Harris and Klein interpret their results in terms of genetic defects, recessive expression of genes for malignancy, gene complementation, and chromosome segregation. Croce and Koprowski discuss the dominant expression of integrated viral genes as the cause for transformation and malignancy. At first these views would seem to be in conflict but there need not be a unitary hypothesis. It is unlikely that virus-induced tumors, chemically-induced tumors and radiation-induced tumors have undergone exactly the same genetic changes and that malignancy/transformation is due to a specific change in one single gene. It seems quite plausible that tumors may differ from each other with respect to their genetic defects and therefore that sometimes hybrids between such tumor cells or between tumor cells and normal cells will be nonmalignant because of gene complementation. This does not exclude the possibility that in other tumor cells there may be viral genes for transformation which are dominantly expressed. Clearly more work has to be done on a number of different types of tumors before these questions can be resolved. In addition to the alternatives discussed above epigenetic changes may also be involved in malignancy (Ephrussi, 1972). Epigenetic changes have been discussed mainly in connection with cell differentiation (see Chapter 9). Different types of specialized cells are viewed as self-maintaining regulatory states which are quite stable and can be inherited by daughter cells. Cell differentiation then can be viewed as a change from one regulatory equilibrium to another. The possibility that such mechanisms may also be involved in malignancy is stressed by recent results obtained by Mintz and Illmensee (1975). These authors injected malignant mouse teratoma cells into mouse blastocysts which were then implanted in the uterus of a foster mother. The newborn animals were chimeras (allophenic) animals made up of normal cells originating from both the teratoma and the blastocyst. Thus here is at least one type of malignant cell which can revert back to what appears to be a completely normal state. It should be pointed out, however, that these particular teratomas are unusual in that (a) they remain strictly diploid, and (b) that tumors consist of a malignant stemline from which a series of nonmalignant and highly differentiated cells can form within a teratoma.

Cell hybridization has also become an important tool in tumor virology. The most important applications have been in the mapping of virus integration sites (Croce et al., 1975), virus rescue and virus detection (Koprowski et al., 1967; Watkins and Dulbecco, 1967), analysis of factors determining the susceptibility of different cell types to viral infections (Basilico et al., 1970; Marin, 1970), and analysis of cellular mechanisms inhibiting viral replication or modifying viral gene expression (Koprowski and Knowles, 1974). For a discussion of these subjects, see Ringertz and Savage (1976).

## IV. FUSION WITH CELL FRAGMENTS

A new technique which promises to be of great interest in the study of gene regulation in animal cells is based on the fusion of cell fragments to reconstitute or reconstruct viable cells. The technique for generating cell fragments is based on the use of cytochalasin B, a metabolite produced by the fungus *Helminthosporium dematoideum*. In 1967 Carter observed that cells treated with this drug lost their nuclei. Cells exposed to cytochalasin undergo marked changes in shape and frequently the nuclei are extruded into the tip of long protrusions which are connected to the main cytoplasm via narrow stalks. These stalks often break spontaneously thus causing a loss of nuclei from some cells. The efficiency of *enucleation* has been increased to close to 100% by a technique developed by Prescott *et al.* (1972) and Wright and Hayflick (1972). In this technique cells adhering to glass or plastic slides are centrifuged in a cytochalasin-containing medium so that cytoplasmic stalks are broken by the centrifugal force (Fig. 7).

## A. Properties of Cell Fragments

### 1. Anucleate Cells (Cytoplasts)

After centrifugation, the enucleated cells remain on the glass slides and if returned to normal medium will resume the flattened appearance of normal cells. These anucleate cells retain several functional characteristics of the intact cells such as membrane ruffling, active cell movement, and endocytosis (Goldman and Pollack, 1974). While no RNA or DNA synthesis has been detected in autoradiographic studies of $^3$H-uridine and $^3$H-thymidine incorporation, protein synthesis continues. Immediately after enucleation the rate of $^3$H-amino acid incorporation is similar to that of the intact cell if the smaller size of the anucleate cell is taken into account (Prescott *et al.*, 1972; Goldman and Pollack, 1974). However, there is a gradual decrease in the rate of protein synthesis so that after 12–18 hours many anucleate cells fail to show amino acid incorporation at a level which can be detected by autoradiography. Most anucleate cells die between 16 and 30 hours after enucleation.

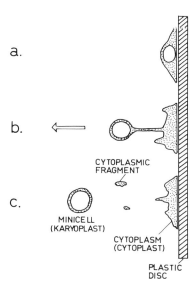

**Fig. 7.** Schematic illustration of the enucleation of cells adhering to plastic discs. The arrow indicates the direction of the centrifugal force (from Ringertz *et al.*, 1975).

## 2. Nuclei (Minicells or Karyoplasts)

Nuclei drawn out of cell monolayers during centrifugation in cytochalasin-containing media differ from detergent-isolated nuclei in that they retain a rim of cytoplasm equivalent to 10–20% of the original cytoplasm of the intact cell and are surrounded by an intact plasma membrane (Prescott and Kirkpatrick, 1973; Ege *et al.*, 1973, 1974a). Dry mass measurements on individual minicells show that their average weight is about one third of that of the average intact cell. The minicells can be collected from the bottom of the centrifuge tubes after enucleation. Usually the pellet also contains intact cells, which have been stripped off the plastic discs during enucleation, and small cytoplasmic fragments. The number of intact cells can be drastically reduced by precentrifugation in the absence of cytochalasin. This procedure removes the loosely attached cells.

Although some minicells are damaged, the majority exclude dyes in viability tests and synthesize RNA and DNA for some time after enucleation. When cultured under conditions in which intact cells thrive, minicells fail, however, to regenerate a cytoplasm and to multiply. Instead, they undergo a progressive decrease in dry mass, lose their ability to synthesize RNA and DNA, become permeable to dyes in viability tests and lyse. Most minicells are dead 24 hours after enucleation. The fact that minicells die may be exploited in cell genetic experiments where they are fused with anucleate or whole cells in an attempt to generate multiplying reconstituted or hybrid cells. No special efforts are required to eliminate the minicells; after a few days in culture they have eliminated themselves.

## 3. Microcells

Microcells are even smaller than minicells and differ from them by having only a fraction of the genome of the intact cell (Ege and Ringertz, 1974). The first step in generating microcells is to induce micronucleation. Several mitotic inhibitors (colchicine and other microtubular poisons) and X-irradiation are known to produce large numbers of micronucleated cells. In cells showing maximum micronucleation, each individual chromosome appears to form its own micronucleus. The frequency of cells which undergo micronucleation varies depending on which mitotic inhibitor is used, dose of inhibitor, time of exposure, and growth rate of the cell. Cells from different species vary in their sensitivity to mitotic inhibitors. Thus, it is necessary to work out optimal micronucleation conditions for each cell type examined. With a number of rodent cells it is possible to induce micronucleation in 70–90% of the cells but a minority of mononucleate cells always remains.

Formation of micronuclei is associated with a decondensation of mitotic chromosomes into interphase chromatin of a dispersed type. RNA synthesis is resumed and nucleoli reform.

Cytochalasin treatment and centrifugation of micronucleated cells results in the formation of microcells (Ege and Ringertz, 1974). The size and DNA content of individual microcells varies very much. Feulgen microspectrophotometry indicates that the smallest of the microcells have a DNA content equivalent to a single chromosome (Ege *et al.*, 1977, Sekiguchi *et al.*, 1977).

An alternative method of generating microcells has been described by Schor *et al.* (1975). HeLa cells were synchronized in metaphase and then subjected to a 9-hour cold shock. This treatment causes abnormal cleavage when the cells are returned to 37°C. A large number of budlike protuberances form and are pinched off to give microcells.

## B. Reconstructed Cells

### 1. Cytoplasmic Hybrids (Cybrids)

When an anucleate cell is fused with a nucleated cell the resulting cell is endowed with a hybrid cytoplasm (Fig. 8). Bunn *et al.* (1974) suggested the word "cybrid" as a convenient term for this kind of cytoplasmic hybrid cell. The first cybrids were reported by Poste and Reeve (1971, 1972). After fusion of the anucleate cell fractions with normal nucleated cells, the cybrids were identified as nucleated cells which had acquired markers characteristic of the cytoplasmic donor. Fusion of enucleated macrophages with L cells produced cybrids which carried macrophage receptors for sensitized red blood cells on their surface. These cells also resembled macrophages in that they remained attached to their substratum after trypsinization while L cells and L cell homokaryons were detached. The expression of these macrophage markers, however, was found to decrease with time after fusion.

The potential value of cybrids in studying cytoplasmic heredity is well illustrated by the study of Bunn *et al.* (1974). These authors fused anucleate mouse L cells carrying a mitochondrial gene marker in the form of chloramphenicol (CAP) resistance with a chloramphenicol sensitive but

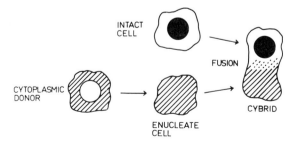

**Fig. 8.** Schematic illustration of the formation of cytoplasmic hybrids ("cybrids") by virus-induced fusion of anucleate cells with intact cells.

BUdR resistant subline of mouse L cells. The BUdR resistance of the latter cell type is due to nuclear gene mutation(s) which make(s) the cells deficient in the enzyme thymidine kinase. Fusion of the anucleate, CAP resistant cells with the BUdR resistant (but CAP sensitive) L cells gave cybrids which survived and multiplied in a selective medium containing both CAP and BUdR. Those L cells which did not fuse with anucleate cells died because of their sensitivity to CAP.

Wright and Hayflick (1975a) have developed another selection method for the isolation of cytoplasmic hybrids. This system is based on the ability of anucleate cells containing active enzymes to rescue cells poisoned with iodoacetate. The same principle can probably be used with other enzyme inhibitors. The intact cells are treated with the inhibitor at doses which kill all cells and then fused with anucleate but otherwise healthy cells. In this way the intact cell, or rather its nucleus, is rescued and can multiply to give a colony of cytoplasmic hybrids.

Wright and Hayflick (1975b) have used the iodoacetate selection method to test if anucleate young cells can rescue senescent cells which would normally stop dividing within a few cell generations. The negative results suggest that senescence is under nuclear rather than cytoplasmic control.

### 2. Reconstituted Cells

While cytoplasmic hybrids thus offer many new possibilities for cell genetic studies there are also experimental situations in which it would be more useful to combine the cytoplasm of one cell with a cytoplasm-free nucleus from another cell.

Although nuclei and cytoplasms have been successfully recombined in experiments with protozoa (for a review, see Goldstein, 1974) and amphibian oocytes (for a review, see Gurdon, 1974), there has been virtually no success with mammalian cells.

Attempts to reconstitute cells by fusing nucleated avian erythrocytes with enucleated mammalian cells have been made by several groups. The reasons these fusions approximate reconstitution is that erythrocytes lyse when exposed to Sendai virus. Ege *et al.* (1973, 1975) found that chick erythrocyte nuclei introduced into anucleate L cells or HeLa cells underwent an abortive reactivation. At first the erythrocyte nuclei grew in size as fast as during reactivation in heterokaryons, RNA synthesis began, and nucleoli developed. A small percentage of the nuclei even synthesized DNA. But after approximately 12 hours, the rate of $^3$H-uridine incorporation began to decrease and at 48 hours RNA synthesis was negligible. Nucleolar growth stopped, never reaching the size and level of organization it does in heterokaryons. The introduction of chick erythrocyte nuclei did not succeed in retarding the decay of protein synthesis in the anucleate cytoplasm. Although the reconstituted cells remained well attached to the substrate somewhat longer than did anucleate cells, nonetheless, after 3 days, practically all reconstituted cells had died. In more recent experiments (P. Elias and N. R. Ringertz, unpublished observations) cells reconstituted by fusing 8-day embryonic erythrocytes with enucleated chick cytoplasms have been found to survive for up to 8 days.

Because minicells are already active in RNA and DNA synthesis and because they are surrounded by a narrow protective layer of cytoplasm, they doubtless have a greater chance of producing reconstituted mammalian cells capable of long-term survival and cell multiplication than do isolated nuclei or nucleate erythrocyte ghosts. Retention of a plasma membrane containing virus receptors makes them susceptible to Sendai-induced fusion.

Recently, two groups independently reported reconstitution of mammalian cells (Fig. 9) after fusion of minicells with anucleate cells (Veomett *et al.*, 1974; Ege *et al.*, 1974b; Ege and Ringertz, 1975). Veomett *et al.* (1974) fused minicells (karyoplasts) prepared from L cells with anucleate L cells (cytoplasts). In order to distinguish between parental cells, reconstituted cells, and cybrids, an ingenious labeling procedure was used. The cells were allowed to ingest latex particles of specific sizes and these then acted as markers for the cytoplasms. The cytoplasm of the nuclear donor was labeled with *small* latex spheres and with $^3$H-thymidine in the nucleus. The cytoplasmic donor, on the other hand, was labeled only in the cytoplasm with *large* latex spheres. Reconstituted cells were then identified as cells which had $^3$H thymidine labeled nuclei and cytoplasm containing only large latex particles. Some of the reconstituted cells were found in different stages of mitosis suggesting that these cells are capable of proliferation.

The identification of viable reconstituted cells and cybrids is facilitated if enzyme deficient cells are used as nuclear or cytoplasmic donors. Ege

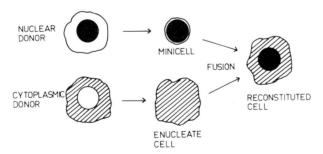

**Fig. 9.**   Schematic illustration of the formation of reconstituted cells by virus-induced fusion of minicells (karyoplasts) with anucleate cells (cytoplasts).

and Ringertz (1975) labeled HGPRT⁻ rat myoblasts with large latex beads according to Veomett *et al.* (1974). After enucleation the cytoplasms of these cells were fused with minicells prepared from normal L6 rat myoblasts. Before enucleation these cells had been labeled with a large number of small latex beads. Some of these remained in the minicells and served as a measure of the amount of cytoplasm introduced by the minicells into the reconstituted cells. At different time points after fusion the cells were exposed to ³H-hypoxanthine, a metabolite which can only be used by HGPRT⁺ cells. In autoradiograms of the fused preparations reconstituted cells were identified as cells having ³H-hypoxanthine-labeled nuclei and large latex beads in the cytoplasm. Cybrids were radioactively labeled and had both large and small latex beads. Reconstituted cells and cybrids were also easily distinguished from contaminating intact nuclear and cytoplasmic donor cells.

Little is yet known about the properties and viability of reconstituted cells. The evidence which exists can be summarized as follows (for further references, see Ringertz *et al.*, 1975; Ringertz and Savage, 1976):

1. The nucleus of reconstituted cells appears to be normally integrated into the cytoplasm. Electron microscopic studies of chick erythrocyte nuclei introduced into anucleate rat cells show that the outer layer of the nuclear envelope is in direct contact with the rat cytoplasm and that the Golgi apparatus assumes its normal position relative to the nucleus.

2. Normal nucleocytoplasmic relations are established with respect to the exchange of macromolecules between the nuclear and the cytoplasmic compartments. In reconstituted cells containing a chick erythrocyte nucleus and a rat cytoplasm, the erythrocyte nucleus takes up and concentrates rat nuclear antigens. In cells reconstituted from L6 minicells and L6 cytoplasms newly synthesized RNA is transported from the nucleus into the cytoplasm.

3. The nucleus responds to regulatory signals from the surrounding cytoplasm. Inactive chick erythrocyte nuclei introduced into the cytoplasm of active cells resume RNA synthesis, form nucleoli and in a few cases also synthesize DNA.

4. The reconstituted cells are viable in the sense that they synthesize macromolecules and survive longer than the cell fragments from which they were made.

5. Some types of reconstituted cells are capable of multiplication. Recently Krondahl *et al.* (1977) reported that cells reconstituted from *rat* myoblast (L6) nuclei and *mouse* fibroblast (A9) cytoplasms divide and form colonies which can be cloned and subcultured. Clones derived from reconstituted cells formed myotubes which produced myosin and developed the cross-striated pattern typical of skeletal muscle (Ringertz *et al.*, 1977). The myogenic program of the rat myoblast thus can persist through

the enucleation and reconstitution procedures, and is not obviously altered by a period of exposure to mouse fibroblast cytoplasm.

These results clearly establish the feasibility of using cell reconstitution to analyze the relative roles of nucleus and cytoplasm in regulating the expression of those specific phenotypes which characterize different cell types. Identification of reconstituted cells and their progeny would be very much facilitated if new types of nuclear and cytoplasmic markers could be found. As indicated in the previous section on cybrids, mitochondrial gene mutations for chloramphenicol resistance are the only form of cytoplasmic gene markers known so far. Since chloramphenicol resistance has been successfully used in the selection of multiplying cybrids, there are reasons to believe that the same principle could be used to isolate the progeny of reconstituted cells.

### 3. Microcell Heterokaryons and Hybrids

The main application of microcells would be as a vehicle by which partial genomes from one cell could be introduced into another (Fig. 10). Although it has been possible to introduce isolated chromosomes or

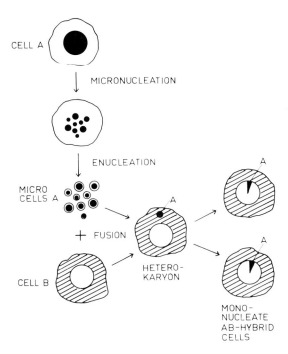

**Fig. 10.** Schematic illustration of the preparation of microcells and fusion of microcells with intact cells. The resulting heterokaryons may give rise to mononucleate hybrid cells with only a few chromosomes from the micronucleate parental cell.

chromosome fragments into living cells by microsurgery or endocytosis (for references, see Ringertz and Savage, 1976), it has been difficult to exploit this technique to its full potential since in most cases the gene material appears to be rapidly degraded or eliminated. In microcells the genetic material is protected by membranes and a small amount of cytoplasm. Microcells can be fused with intact cells to generate microcell heterokaryons (Ege *et al.*, 1974b). Some of these give rise to mononucleate hybrid cells which, from the beginning, contain only a few chromosomes from a microcell and a complete chromosome set from the intact "recipient" cell (Fournier and Ruddle, 1977). Although microcells with different combinations of chromosomes will fuse randomly with the intact cells, it will be possible to control which hybrids survive by using mutant "recipient" cells and culture conditions which select for hybrids in which a microcell chromosome complements for the mutation in the recipient cell (see Section III,F,1).

## V. CONCLUDING REMARKS

Clearly, somatic cell genetics is already too vast a field to be fully covered in a single review article. In the preceding pages we have tried to indicate how three types of altered somatic cells can be generated and used in the study of genetic mechanisms in somatic cells.

*Mutant cells* can be used for a variety of purposes. In those cases where the mutation or the altered gene product can be identified, the mutant cells provide insights into the molecular architecture of eukaryotic cells and can be used to examine metabolic pathways and complex functions. Mutants of cells capable of differentiating *in vitro* and which are defective with respect to one or several characteristics in the differentiated phenotype can be used to examine metabolic pathways and complex functions. Fibroblasts and lymphocytes from individuals suffering from genetic diseases can be cultured *in vitro* and studied with cell genetic and other methods. The information obtained is of direct medical, as well as cell biological, interest.

*Hybrid cells* show great promise in the study of gene regulation, differentiation, malignancy, and virus–host cell interaction, and can also be used as a tool in gene mapping.

The biological information obtained with heterokaryons stresses the role of the cytoplasm in the control of nuclear activity. When a $G_1$ nucleus is brought into contact with the cytoplasm of a S phase cell, the $G_1$ nucleus is stimulated to synthesize DNA. If an interphase cell is combined with a mitotic cell, the chromatin of the interphase nucleus forms prematurely condensed chromosomes. The fact that it is possible to visualize chromosomes at all stages of interphase is of great interest in the analysis of how mutagenic substances and radiation damages the genetic material. Gene complementation tests with heterokaryons can be used to define and analyze genetic defects in hereditary diseases.

Mononucleate hybrid cells (synkaryons) have been used to study mechanisms which control phenotypic expression in mammalian cells. The simultaneous expression of homologous genes of both parental cells has been described as codominance while loss of markers has been interpreted as recessiveness. In some respects the use of a genetic terminology is unfortunate since the analysis of hybrid cell phenotypes is different from the analysis of progenies arising from sexual crosses. For this reason many workers prefer to use a descriptive "nongenetic" terminology. Hybrids exhibiting a parental marker are simply said to show expression while failure to express the marker is referred to as extinction.

Extinction of a marker may be due to regulatory phenomena, to loss of the chromosome carrying the structural gene for the marker, or to disturbances caused by the loss of regulatory genes. In many experimental systems it is at present difficult to distinguish between these alternatives. In some cases, however, the extinguished marker has been reexpressed after prolonged culture of the hybrids. Reexpression makes it possible to conclude that the initial extinction was due to regulatory phenomena rather than to loss of structural genes. In a few cases hybrids have expressed new properties which have not been present in the parental cells. These observations indicate that hybrid cells can be used to study gene regulation. Much more information is needed, however, before the different patterns of gene expression can be interpreted in more specific terms. This is true also for the analysis of malignancy.

Another main application of cell hybrids is as an instrument in gene mapping. A large number of research groups are now engaged in the mapping of the human genome. In view of the effort and money which now goes into this it may be appropriate to ask why genes should be mapped. First, it is clear that gene maps may improve our understanding of genetic disease in man. Hybrids made by crossing cells from patients can be used in complementation analysis to establish if two patients with similar symptoms carry the same or different gene defects. Second, genetic maps will be of importance in the study of many cellular processes, in particular, regulatory phenomena. The knowledge of whether the synthesis of a specific protein is controlled by one or several genes and whether control genes and structural genes are located on the same or different chromosomes may be fundamental to our understanding of gene regulation. Third, as gene maps of many different species become available new perspectives on the evolution of the human gene material and of animal species will be gained.

Reconstituted cells, cybrids, and other types of reconstructed cells made by fusing cell fragments with each other or with intact cells have not yet been studied in any detail. The main areas of application are likely to be in experiments designed to analyze nucleocytoplasmic interactions, gene regulation and cell differentiation, mitochondrial genetics, and virus–host cell interactions.

## REFERENCES

Appels, R., and Ringertz, N. R. (1975). Chemical and structural changes within chick erythrocyte nuclei introduced into mammalian cells by cell fusion. *Curr. Top. Dev. Biol.* **9,** 137–166.

Barski, G. S., Sorieul, S., and Cornefert, F. (1960). Production dans des culture *in vitro* de deux souches cellulaires en association, de cellules de caractère "hybride." *C. R. Hebd. Seances Acad. Sci.* **251,** 1825–1827.

Barski, G. S., Sorieul, S., and Cornefert, F. (1961). "Hybrid" type cells in combined cultures of two different mammalian cell strains. *J. Natl. Cancer Inst.* **26,** 1269–1291.

Basilico, C., and Meiss, H. K. (1974). Methods for selecting and studying temperature sensitive mutants of BHK-21 cells. *Methods Cell Biol.* **8,** 1–22.

Basilico, C., Matsuya, Y., and Green, H. (1970). The interaction of polyoma virus with mouse-hamster somatic hybrid cells. *Virology* **41,** 295–305.

Bloom, A. D., and Nakamura, F. T. (1974). Establishment of a tetraploid immunoglobulin-producing cell line from the hybridization of two human lymphocyte lines. *Proc. Natl. Acad. Sci. U.S.A.* **71,** 2689–2692.

Bunn, C. L., Wallace, D. C., and Eisenstadt, J. M. (1974). Cytoplasmic inheritance of chloramphenicol resistance in mouse tissue culture cells. *Proc. Natl. Acad. Sci. U.S.A.* **71,** 1681–1685.

Carlsson, S.-A., Luger, O., Ringertz, N. R., and Savage, R. E. (1974). Phenotypic expression in chick erythrocyte × rat myoblast hybrids and in chick myoblast × rat myoblast hybrids. *Exp. Cell Res.* **84,** 47–55.

Carter, S. B. (1967). Effects of cytochalasins on mammalian cells. *Nature (London)* **213,** 261–266.

Croce, C. M., Litwack, G., and Koprowski, H. (1973). Human regulatory gene for inducible tyrosine aminotransferase in rat-human hybrids. *Proc. Natl. Acad. Sci. U.S.A.* **70,** 1268–1272.

Croce, C. M., Huebner, K., Girardi, A. J., and Koprowski, H. (1975). Genetics of cell transformation by Simian Virus 40. *Cold Spring Harbor Symp. Quant. Biol.* **39,** 335–343.

Darlington, G. J., Bernard, H. P., and Ruddle, F. H. (1974). Human serum albumin phenotype activation in mouse hepatoma-human leukocyte cell hybrids. *Science* **185,** 859–862.

Davidson, R. L. (1972). Regulation of melanin synthesis in mammalian cells: Effect of gene dosage on the expression of differentiation. *Proc. Natl. Acad. Sci. U.S.A.* **69,** 951–955.

Davidson, R. L. (1974). Gene expression in somatic cell hybrids. *Annu. Rev. Genet.* **8,** 195–218.

Davidson, R. L., Ephrussi, B., and Yamamoto, K. (1966). Regulation of pigment synthesis in mammalian cells as studied by somatic hybridization. *Proc. Natl. Acad. Sci. U.S.A.* **56,** 1437–1440.

Davidson, R. L., Ephrussi, B., and Yamamoto, K. (1968). Regulation of melanin synthesis in mammalian cells as studied by somatic hybridization. I. Evidence for negative control. *J. Cell. Phys.* **72**, 115–127.

Defendi, V., Ephrussi, B., Koprowski, H., and Yoshida, M. C. (1967). Properties of hybrids between polyoma-transformed and normal mouse cells. *Proc. Natl. Acad. Sci. U.S.A.* **57**, 299–305.

deWeerd-Kastelein, E. A., Kleijzer, W. J., Sluyter, M. L., and Kleijzer, W. (1973). Repair replication in heterokaryons derived from different repair–deficient *Xeroderma pigmentosum* strains. *Mutat. Res.* **19**, 237–243.

Ege, T., and Ringertz, N. R. (1974). Preparation of microcells by enucleation of micronucleate cells. *Exp. Cell Res.* **87**, 378–382.

Ege, T., and Ringertz, N. R. (1975). Viability of cells reconstituted by virus induced fusion of minicells with anucleate cells. *Exp. Cell Res.* **94**, 469–473.

Ege, T., Zeuthen, J., and Ringertz, N. R. (1973). Cell fusion with enucleated cytoplasms. *Chromosome Ident., Proc. Nobel Symp., 23rd, 1972* pp. 189–194.

Ege, T., Hamberg, H., Krondahl, U., Ericsson, J., and Ringertz, N. R. (1974a). Characterization of minicells (nuclei) obtained by cytochalasin enucleation. *Exp. Cell Res.* **87**, 365–377.

Ege, T., Krondahl, U., and Ringertz, N. R. (1974b). Introduction of nuclei and micronuclei into cells and enucleated cytoplasms by Sendai virus-induced fusion. *Exp. Cell Res.* **88**, 428–432.

Ege, T., Zeuthen, J., and Ringertz, N. R. (1975). Reactivation of chick erythrocyte nuclei after fusion with enucleated cells. *Somatic Cell Genet.* **1**, 65–80.

Ege, T., Ringertz, N. R., Hamberg, H., and Sidebottom, E. (1977). Preparation of microcells. *In* "Methods in Cell Biology" (D. M. Prescott, ed.), Vol. XV, pp. 339–357. Academic Press, New York.

Ephrussi, B. (1972). "Hybridization of Somatic Cells." Princeton Univ. Press, Princeton, New Jersey.

Fougère, C., Ruiz, F., and Ephrussi, B. (1972). Gene dosage dependence of pigment synthesis in melanoma × fibroblast hybrids. *Proc. Natl. Acad. Sci. U.S.A.* **69**, 330–334.

Fournier, R. E. K., and Ruddle, F. H. (1977). Microcell-mediated transfer of murine chromosomes into mouse, Chinese hamster, and human somatic cells. *Proc. Natl. Acad. Sci. U.S.A.* **74**, 319–323.

Gershon, D., and Sachs, L. (1963). Properties of a somatic hybrid between mouse cells with different genotypes. *Nature (London)* **198**, 912–913.

Goldman, R. D., and Pollack, R. (1974). Use of enucleated cells. *Methods Cell Biol.* **8**, 123–143.

Goldstein, L. (1974). Movement of molecules between nucleus and cytoplasm. *In* "The Cell Nucleus" (H. Busch, ed.), Vol. I, pp. 387–438. Academic Press, New York.

Gordon, S., and Cohn, Z. (1970). Macrophage-melanocyte heterokaryons. I. Preparation and properties. *J. Exp. Med.* **131**, 981–1003.

Gordon, S., and Cohn, Z. (1971). Macrophage-melanoma cell heterokaryons. IV. Unmasking the macrophage specific membrane receptor. *J. Exp. Med.* **134**, 947–962.

Goss, S. J., and Harris, H. (1975). New method for mapping genes in human chromosomes. *Nature (London)* **255**, 680–684.

Gurdon, J. B. (1974). The genome in specialized cells, as revealed by nuclear transplantation in amphibia. *In* "The Cell Nucleus" (H. Busch, ed.), Vol. I, pp. 471–489. Academic Press, New York.

Harris, H. (1970). "Cell Fusion." Harvard Univ. Press, Cambridge, Massachusetts.

Harris, H. (1972). The Croonian Lecture 1971. Cell fusion and the analysis of malignancy. *J. Natl. Cancer Inst.* **48**, 851–864.

Harris, H. (1974). "Nucleus and Cytoplasm." Oxford Univ. Press (Clarendon), London and New York.

Harris, H., and Watkins, J. F. (1965). Hybrid cells derived from mouse and man. Artificial heterokaryons of mammalian cells from different species. *Nature (London)* **205,** 640–646.

Harris, H., Watkins, J. F., Ford, C. E., and Schoefl, G. I. (1966). Artificial heterokaryons of animal cells from different species. *J. Cell Sci.* **1,** 1–30.

Harris, M. (1964). "Cell Culture and Somatic Variation." Holt, New York.

Harris, M. (1971). Mutation rates in cells at different ploidy levels. *J. Cell. Phys.* **78,** 177–184.

Harris, M. (1974). Mechanisms of *de novo* variation in mammalian cell cultures. *In* "Somatic Cell Hybridization" (R. L. Davidson and F. de la Cruz, eds.), pp. 221–226. Raven, New York.

Johnson, R. T., and Rao, P. N. (1971). Nucleocytoplasmic interaction in the achievement of nuclear synchrony in DNA synthesis and mitosis in multinucleate cells. *Biol. Rev. Cambridge Philos. Soc.* **46,** 97–155.

Kao, F. T., and Puck, T. T. (1967). Genetics of somatic mammalian cells. IV. Properties of Chinese hamster cell mutants with respect to the requirement for proline. *Genetics* **55,** 513–524.

Kao, F. T., and Puck, T. T. (1974). Induction and isolation of auxotrophic mutants in mammalian cells. *Meth Cell Biol.* **8,** 23–39.

Kit, S., Leung, W. C., and Trkula, D. (1973). Properties of mitochondrial thymidine kinases of parental and enzyme-deficient HeLa cells. *Arch. Biochem. Biophys.* **158,** 503–513.

Koprowski, H., and Knowles, B. (1974). Viruses, immune functions and antigenic determinants in heterokaryons and hybrids. *In* "Somatic Cell Hybridization" (R. L. Davidson and F. de la Cruz, eds.), pp. 71–100. Raven, New York.

Kopropwski, H., Jensen, F. C., and Steplewski, Z. (1967). Activation of production of infectious tumor virus SV40 in heterokaryon cultures. *Proc. Natl. Acad. Sci. U.S.A.* **58,** 127–133.

Krondahl, U., Bols, N., Ege, T., Linder, S., and Ringertz, N. R. (1977). Cells reconstituted from cell fragments of two different species multiply and form colonies. *Proc. Natl. Acad. Sci. U.S.A.* **74,** 606–609.

Littlefield, J. W. (1964). Selection of hybrids from matings of fibroblasts *in vitro* and their presumed recombinants. *Science* **145,** 709–710.

McMorris, F. A., and Ruddle, F. H. (1974). Expression of neuronal phenotypes in neuroblastoma cell hybrids. *Dev. Biol.* **39,** 226–246.

Malawista, S. E., and Weiss, M. C. (1974). Expression of differentiated functions in hepatoma cell hybrids. High frequency of induction of mouse albumin production in rat hepatoma × mouse lymphoblast hybrids. *Proc. Natl. Acad. Sci. U.S.A.* **71,** 927–931.

Marin, G. (1970). Somatic cell hybridization and problems of viral oncogenesis. *In Vitro* **5,** 94–108.

Minna, J. D., Glazer, D., and Nirenberg, M. (1972). Genetic dissection of neural properties using somatic cell hybrids. *Nature (London), New Biol.* **235,** 225–231.

Minna, J. D., and Coon, H. G. (1974). Human × mouse hybrid cells segregating mouse chromosomes and isozymes. *Nature (London)* **252,** 401–404.

Mintz, B., and Illmensee, K. (1975). Normal genetically mosaic mice produced from malignant teratocarcinoma cells. *Proc. Natl. Acad. Sci. U.S.A.* **72,** 3585–3589.

Mitchell, C. H., England, J. M., and Attardi, G. (1975). Isolation of chloramphenicol-resistant variants from a human cell line. *Somatic Cell Genet.* **1,** 215–234.

Naha, P. M. (1974). Isolation of temperature sensitive mutants of mammalian cells. *Methods Cell Biol.* **8,** 41–46.

Okada, Y., and Murayama, F. (1965). Multinucleated giant cell formation by fusion between cells of two different strains. *Exp. Cell Res.* **40**, 154–158.

Okada, Y., Suzuki, I., and Hosaka, Y. (1957). Interaction between influenza virus and Ehrlich's tumor cells. III. Fusion phenomenon of Ehrlich's tumor cells by the action of HVJ Z-strain. *Med. J. Osaka Univ.* **7**, 709–717.

Peterson, J. A., and Weiss, M. C. (1972). Expression of differentiated functions in hepatoma cell hybrids: Induction of mouse albumin production in rat hepatoma-mouse fibroblast hybrids. *Proc. Natl. Acad. Sci. U.S.A.* **69**, 571–575.

Pontecorvo, G. (1971). Induction of directional chromosome elimination in somatic cell hybrids. *Nature (London)* **230**, 367–369.

Poste, G., and Reeve, P. (1971). Formation of hybrid cells and heterokaryons by fusion of enucleated and nucleated cells. *Nature (London), New Biol.* **229**, 123–125.

Poste, G., and Reeve, P. (1972). Enucleation of mammalian cells by cytochalasin B. II. Formation of hybrid cells and heterokaryons by fusion of anucleate and nucleated cells. *Exp. Cell Res.* **73**, 287–294.

Prescott, D. M., and Kirkpatrick, (1973). Mass enucleation of cultured animal cells. *Methods Cell Biol.* **7**, 189–202.

Prescott, D. M., Myerson, D., and Wallace, J. (1972). Enucleation of mammalian cells with cytochalasin B. *Exp. Cell Res.* **71**, 480–485.

Rao, P. N., and Johnson, R. T. (1974). Regulation of cell cycle in hybrid cells. *In* "Control of Proliferation in Animal Cells" (B. Clarkson and R. Baserga, eds.), pp. 785–800. Cold Spring Harbor Lab., Cold Spring Harbor, New York.

Ringertz, N. R., and Bolund, L. (1974). Reactivation of chick erythrocyte nuclei by somatic cell hybridization. *Int. Rev. Exp. Pathol.* **13**, 83–116.

Ringertz, N. R., and Savage, R. E. (1976). "Cell Hybrids." Academic Press, New York.

Ringertz, N. R., Ege, T., Elias, P., and Sidebottom, E. (1975). Reconstruction of cells from cell fragments. "Proceedings Federal European Biochemical Society Meeting, 10th " (G. Bernardi and F. Gros, ed.), Vol. 38, pp. 235–244. North-Holland, Amsterdam

Ringertz, N. R., Krondahl, U., and Coleman, J. R. (1977). Reconstitution of cells by fusion of cell fragments. I. Myogenic expression after fusion of minicells from rat myoblasts (L6) with mouse fibroblast (A9) cytoplasm. *Exp. Cell Res.* (in press).

Röhme, D. (1974). Prematurely condensed chromosomes of the Indian muntjac: A model system for the analysis of chromosome condensation and banding. *Hereditas* **76**, 251–258.

Ruddle, F. H., and Creagan, R. P. (1975). Parasexual approaches to the genetics of man. *Annu. Rev. Genet.* **9**, 407–486.

Sager, R. (1972). "Cytoplasmic Genes and Organelles." Academic Press, New York.

Scaletta, L. J., and Ephrussi, B. (1965). Hybridization of normal and neoplastic cells *in vitro*. *Nature (London)* **205**, 1169–1171.

Schneider, J. A., and Weiss, M. C. (1971). Expression of differentiated functions in hepatoma cell hybrids. I. Tyrosine aminotransferase in hepatoma–fibroblast hybrids. *Proc. Natl. Acad. Sci. U.S.A.* **68**, 127–131.

Schor, S. L., Johnson, R. T., and Mullinger, A. M. (1975). Perturbation of mammalian cell division. II. Studies on the isolation and characterization of human minisegregant cells. *J. Cell Sci.* **19**, 281–304.

Sekiguchi, T., and Shelton, K., and Ringertz, N. R. (1977). Preparation of microcells by cytochalasin enucleation of micronucleated animal cells. *Expt. Cell Res.* (in press).

Silagi, S. (1967). Hybridization of a malignant melanoma cell line with L-cells *in vitro*. *Cancer Res.* **27**, 1953–1960.

Sparkes, R. S., and Weiss, M. C. (1973). Expression of differentiated functions in hepatoma cell hybrids: Alanine aminotransferase. *Proc. Natl. Acad. Sci. U.S.A.* **70**, 377–381.

Sperling, K., and Rao, P. N. (1974). The phenomenon of premature chromosome condensation: Its relevance to basic and applied research. *Humangenetik* **23**, 235–258.

Spolsky, C. M., and Eisenstadt, J. M. (1972). Chloramphenicol-resistant mutants of human HeLa cells. *FEBS Lett.* **25**, 319–324.

Thompson, E. B., and Gelehrter, T. D. (1971). Expression of tyrosine aminotransferase activity in somatic cell heterokaryons: Evidence for negative control of enzyme expression. *Proc. Natl. Acad. Sci. U.S.A.* **68**, 2589–2593.

Thompsom, L. H., and Baker, R. M. (1973). Isolation of mutants of cultured mammalian cells. *Methods Cell Biol.* **6**, 209–281.

Veomett, G., Prescott, D. M., Shay, J., and Porter, K. R. (1974). Reconstruction of mammalian cells from nuclear and cytoplasmic components separated by treatment with cytochalasin B. *Proc. Natl. Acad. Sci. U.S.A.* **71**, 1999–2002.

Watkins, J. F., and Dulbecco, R. (1967). Production of SV40 virus in heterokaryons of transformed and susceptible cells. *Proc. Natl. Acad. Sci. U.S.A.* **58**, 1396–1403.

Watkins, J. F., and Grace, D. M. (1967). Studies on the surface antigens of interspecific mammalian cell heterokaryons. *J. Cell Sci.* **2**, 193–204.

Weiss, M. C., and Chaplain, M. (1971). Expression of differentiated functions in hepatoma cell hybrids: Reappearance of tyrosine aminotransferase inducibility after the loss of chromosomes. *Proc. Natl. Acad. Sci. U.S.A.* **68**, 3026–3030.

Weiss, M. C., and Ephrussi, B. (1966). Studies of interspecific (rat × mouse) somatic hybrids. III. Lactate dehydrogenase and β-glucuronidase. *Genetics* **54**, 1111–1122.

Weiss, M. C., and Green, H. (1967). Human-mouse hybrid cell lines containing partial complements of human chromosomes and functioning human genes. *Proc. Natl. Acad. Sci. U.S.A.* **58**, 1104–1111.

Wiblin, C. N., and Macpherson, I. (1973). Reversion in hybrids between SV40-transformed hamster and mouse cells. *Int. J. Cancer* **12**, 148–161.

Wright, W. E., and Hayflick, L. (1972). Formation of anucleate and multinucleate cells in normal and SV40-transformed WI38 by cytochalasin B. *Exp. Cell Res.* **74**, 187–194.

Wright, W. E., and Hayflick, L. (1975a). Use of biochemical lesions for selection of human cells with hybrid cytoplasms. *Proc. Natl. Acad. Sci. U.S.A.* **72**, 1812–1816.

Wright, W. E., and Hayflick, L. (1975b). Nuclear control of cellular aging demonstrated by hybridization of anucleate and whole cultured normal human fibroblasts. *Exp. Cell Res.* **96**, 113–121.

# 6

# Cytogenetics

*E. D. Garber*

## I.  INTRODUCTION

The new biology began with the formulation and acceptance of the cell theory which focused attention on the cell as the basic organizational unit of life. Fortunately, the nucleus and chromosomes of many plant and animal species were amenable to detailed study by light microscopy so that cellular and organismal reproduction could be related to nuclear and chromosomal observations. Three principles emerged: (1) cells arise from preexisting cells; (b) nuclei arise from preexisting nuclei; and (3) chromosomes arise from preexisting chromosomes. Once the sequence of nuclear and chromosomal events during mitosis, meiosis, and fertilization was established, the cell or specialized nucleus could be viewed as the structural unit of heredity. By the end of the nineteenth century, cytologists had implicated the chromosomes in heredity.

The basic principles of genetics were formulated from the analyses of breeding experiments which involved the hybridization of individuals with contrasting characteristics, and the scoring of individuals with one or the other characteristic in successive generations. These experiments implicated sex as an essential component of heredity. Sex can be operationally defined as the pooling of genetic information from different sources for eventual distribution to progeny. To account for the characteristically constant chromosome number of a species, Weismann (1887) predicted an unusual type of nuclear division responsible for the precise reduction of the somatic chromosome number in each generation.

Early cytological studies of germinal cells revealed two nuclear divisions preceded the production of gametes in animals and of pollen and embryo sac in flowering plants. A detailed examination of the first nuclear division indicated that the chromosomes behaved differently when compared with their behavior during the second nuclear division or in mitosis. In 1883, van Beneden reported the significant observation that the sperm and eggs of *Parascaris equorum* (the horse threadworm) had two chromosomes while the zygote and cells derived from the zygote by mitosis had four chromosomes. By coining the term "meiosis," Farmer and Moore (1905) emphasized the basic difference between the mitotic process and

the process resulting in the precise reduction of the chrc
in germinal cells. Strasburger (1905) proposed "haploid
as terms for the gametic and somatic chromosome numb
thereby relating gametogenesis and sporogenesis with fe
two major events with respect to chromosome numbers i
life cycle.

Regardless of the diversity of life cycles, sexual reproduction involves
the production of haploid nuclei by meiosis and the means to deliver these
nuclei for fertilization to restore the diploid number. Plant and animal
species exhibit the basic pattern of meiosis. In animals, the immediate
products of meiosis are gametes: spermatogenesis for sperm and oogenesis
for eggs. In higher plants, the immediate products of meiosis are spores:
microsporogenesis for microspores and megasporogenesis for megaspores.
The microspore develops into the haploid male gametophyte after a
limited number of mitotic divisions, which produces the gametic nucleus.
The megaspore nucleus undergoes a limited number of mitotic divisions
yielding the haploid female gametophyte or embryo sac, which contains
the egg. Only one of the four products of oogenesis and megasporogenesis
functions directly or indirectly as the egg nucleus; the other three products
degenerate.

## II. MEIOSIS

Light microscopists first described the nuclear and chromosomal events
during mitosis. Consequently, similar events during meiosis were assigned
previously coined terms: prophase, metaphase, anaphase, and telophase.
The original terminology of reductional and equational divisions to distinguish the two nuclear divisions was abandoned on genetic grounds and the
Roman numerals I and II were assigned to the first and second division,
respectively. In the reductional division, homologous chromosomes separate and go to opposite poles during meiosis I; in the equational division,
sister chromatids separate and proceed to opposite poles during meiosis
II.

Mitosis is involved in the vegetative multiplication of nuclei, cells or
organisms. Meiosis is concerned with sexual reproduction and occurs in
specific cells or tissues at the appropriate stage of the life cycle. Mitosis is
complete after one nuclear division and meiosis after two nuclear divisions. Mitosis yields two nuclei, each identical to the other and to the
parental nucleus in chromosome number and genotype. In diploid species,
meiosis yields four haploid nuclei from one diploid nucleus, and each
nucleus may have a different chromosomal constitution, particularly in

nimal species, and different genotypes, depending on the level of heterozygosity. Finally, chromosomal behavior during prophase I has no counterpart during mitotic prophase.

Cytogenetics correlates breeding data and chromosome number, morphology and behavior during meiosis and mitosis. The basic tools of cytogenetics are chromosomal aberrations. In favorable cytological material, structural and numerical chromosomal aberrations can be assigned or attributed to specific chromosomes or regions of these chromosomes. Although meiotic chromosomes generally yield more detailed information on numerical and structural aberrations than mitotic chromosomes, considerable information on aberrations can be obtained from mitotic metaphase chromosomes which have been processed to yield characteristic banding patterns.

The generalized scheme of meiosis and particularly of prophase I to be presented does not consider dissenting minority opinions (Grell, 1969). Unusual or anomalous meiosis is not uncommon in insect species and plant species. Consequently, the genetics of these species would not follow basic patterns of heredity in species with the usual type of meiosis. Detailed studies of such species could shed light on the genetic control of the meiotic process.

Meiosis will be described in terms of an idealized cell termed a meiocyte. Cytological criteria signaling the onset of meiosis are essentially subjective and presumably reflect earlier molecular events. The premeiotic nucleus enlarges, and thready structures emerge from the structurally incoherent nucleus.

## A. Prophase I

This stage is the arena for controversies concerned with chromosome strandedness, synapsis, chiasma formation, and crossing-over. The different chromosomal events have been assigned to substages identified by terms associated with the morphology or behavior of chromosomes within the nuclear membrane: leptotene (thin threads), zygote (yoked threads), pachytene (thick threads), diplotene (double threads), and diakinesis, the terminus of both diplotene and prophase I.

### 1. Leptotene

The emergence of extended, apparently single-stranded chromosomes with obvious chromomeres from the chromatinic network present in the interphase nucleus constitutes the first visual marker for the onset of meiosis (Fig. 1). The chromomeres occur as irregularly shaped and spaced, densely straining chromatinic material giving the chromosomes a beaded appearance. During this substage, the ends of the chromosomes of animal and some plant species are appressed to a restricted area of the inner surface of the nuclear envelope opposite the centriole to form a bouquet. In many plant species, the chromosomes at this or the next substage form an irregular mass near the nucleolus termed the synizetic knot.

### 2. Zygotene

The lateral association of chromosomal segments marks this substage. This pairing or synapsis is a visible expression but not *prima facie* evidence of homology. In favorable material, the chromomeres are precisely matched in the synapsed segments. Pairing continues throughout zygotene in zipperlike fashion for segments proximal to synapsed segments (Fig. 2). The occasional lateral association of nonhomologous segments of structurally aberrant chromosomes may result from the zipper effect.

### 3. Pachytene

The complete pairing of homologous chromosomes, which are shorter and thicker than the chromosomes of the previous substage, indicates the pachytene substage (Fig. 3). Chromosomes are homologous when all of their segments are homologous. In a number of plant species with numerous genetic markers, such as maize (Fig. 4) and tomato (Fig. 5), the excellent pachytene chromosomes can be distinguished by their relative length, arm ratio (long arm/short arm) and such topological features as heavily staining segments, prominent chromomeres or knobs, and constrictions. These cytological advantages are responsible for the extensive cytogenetic studies of maize and tomato. The nucleolus-organizing chromosome or chromosomes are identified by their association with the nucleolus at a specific site, the nucleolus-organizing region, on the chromosome.

### 4. Diplotene

This substage commences when the synapsed chromosomes separate, usually at the centromere (Fig. 6). As the chromosomes continue to separate, they clearly exhibit two strands or chromatids and are relationally twisted (Fig. 7). Finally, the chromosomes are completely separate, except where chiasmata appear to hold the chromosomes together as a bivalent (Fig. 8). A chiasma is the site where the mutual switching of nonsister chromatids has occurred. Chiasmata presumably represent visual expressions of crossing-over (Zohary, 1955).

While each bivalent, regardless of chromosome length, has by definition at least one chiasma, the number of chiasmata in each chromosome arm is

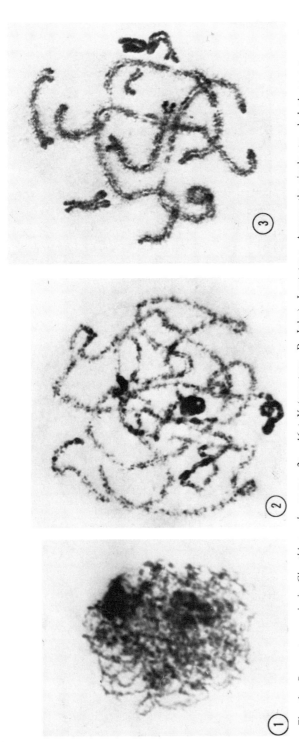

**Fig. 1.** Spermatogenesis in *Chorthippus brunneus*, $2n = 16 + X$ (courtesy, B. John). Leptotene. Apparently single-stranded chromosomes can be seen at the periphery of the chromosome mass. The X chromosome is the large heterochromatic body.

**Fig. 2.** Spermatogenesis in *Chorthippus brunneus*, $2n = 16 + X$ (courtesy, B. John). Zygotene. Synapsis is complete but the chromosomes remain extended.

**Fig. 3.** Spermatogenesis in *Chorthippus brunneus*, $2n = 16 + X$ (courtesy, B. John). Pachytene. Each chromosome appears to have one chromatid but the formerly obvious chromomere patterning is obscured by the contracting of the chromosomes.

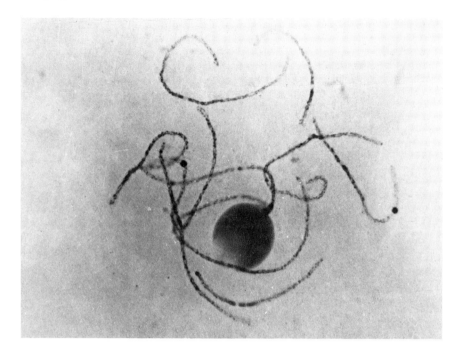

**Fig. 4.**   Pachytene. *Zea mays*, $2n = 20$ (courtesy, D. T. Morgan, Jr.).

usually proportional to the length of the arm and can be genetically regulated. The failure to produce at least one chiasma results in two univalents whose distribution to daughter nuclei during later stages cannot be predicted. In species such as *Drosophila melanogaster*, the male is apparently achiasmatic but meiosis is regular, indicating that chiasmata are not absolutely necessary for the proper distribution of chromosomes during meiosis. A satisfactory explanation for the situation in achiasmatic species is not yet available. As diplotene progresses, the coiling of the chromosomes increases so that the chromosomes become shorter and thicker. Furthermore, the total number of chiasmata per nucleus diminishes by the displacement of chiasmata from their original site toward and off the ends of the chromosomes, a phenomenon termed "terminalization."

### 5. Diakinesis

At this substage, the bivalents are short, thick and dispersed throughout the nuclear volume (Fig. 9), signaling the end of both diplotene and prophase I. The number of bivalents is readily determined and the presence of univalents easily detected.

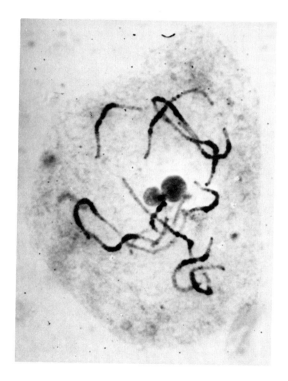

**Fig. 5.** Pachytene. Tomato, *Lycopersicon esculentum*, $2n = 24$ (courtesy, D. W. Barton.)

## B. Metaphase I

The loss of the nuclear envelope and, in higher plant species, the absence of the spindle apparatus characterizes the brief prometaphase I stage (Fig. 10). In many protists and fungi, the nuclear membrane does not seem to disappear during mitosis or meiosis. The association between centromeres and spindle fibers aligns the bivalents in an equatorial plane which need not coincide with the diameter of the cell (Fig. 11). The spindle apparatus during oogenesis is formed nearer the cell envelope to facilitate the extrusion of one nucleus into a polar body. The centromere of each chromosome in the bivalents is associated with spindle fibers from only one pole; the centromere of mitotic metaphase chromosomes is associated with fibers from both poles. These observations suggest that the replication of the centromere is not synchronous with chromatid replication or that replicated centromeres behave as a single centromere during the first meiotic division. Terminalization continues throughout metaphase I.

**Fig. 6.**   Spermatogenesis in *Chorthippus brunneus*, $2n = 16 + X$ (courtesy, B. John). Early diplotene. Homologous chromosomes separate at the centromeres.

## C. Anaphase I

The onset of this stage is detected by the first complete separation of homologous chromosomes, usually occurring first in the small bivalents (Fig. 12). As the separation for other bivalents is detected, it is obvious that the homologous chromosomes are proceeding toward opposite poles (Fig. 13). Finally, all of the homologous chromosomes separate, moving toward the poles as a group within the confines of the spindle apparatus (Fig. 14). The chromatid arms are separated except at the centromere and each chromosome is clearly a dyad.

## D. Telophase I

Nuclear envelopes appear synchronously around the two groups of chromosomes before they reach the pole. The spindle apparatus gradually disappears, the nucleolus reappears, and the chromosomes elongate. The nuclei do not return to the state of highly dispersed chromatin present in the premeiotic nucleus.

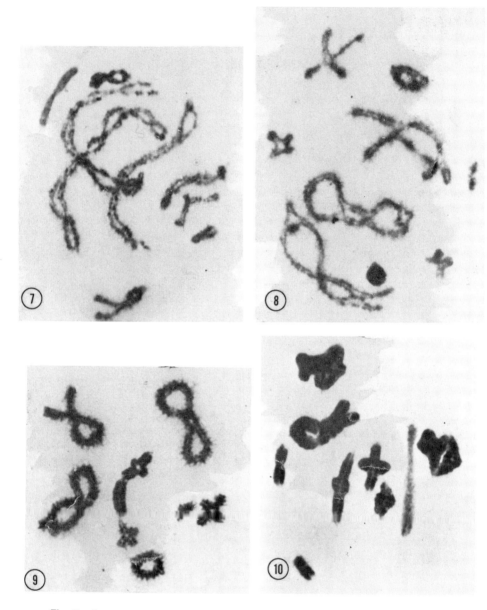

**Fig. 7.**   Spermatogenesis in *Chorthippus brunneus*, $2n = 16 + X$ (courtesy, B. John). Early diplotene. Twisted double-stranded chromosomes separate.

**Fig. 8.**   Spermatogenesis in *Chorthippus brunneus*, $2n = 16 + X$ (courtesy, B. John). Diplotene. Chiasmata formed by the mutual switching of chromatids hold together the obviously double-stranded chromosomes to produce the X-configurations between homologous chromosomes forming bivalents.

**Fig. 9.**   Spermatogenesis in *Chorthippus brunneus*, $2n = 16 + X$ (courtesy, B. John). Diakinesis. Increasing spiralization yields shorter, thicker chromosomes and bivalents.

**Fig. 10.**   Spermatogenesis in *Chorthippus brunneus*, $2n = 16 + X$ (courtesy, B. John). Prometaphase I. While some centromeres are associated with spindle fibers, the bivalents have not yet formed in the equatorial plane midway between the poles.

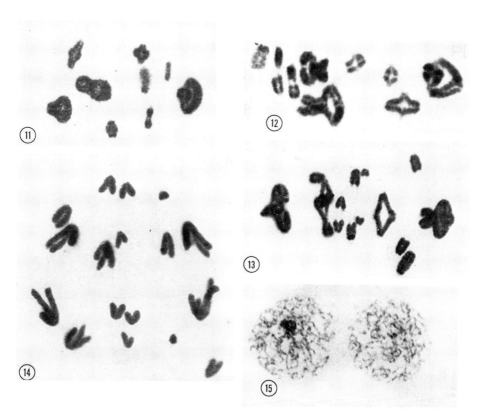

**Fig. 11.** Spermatogenesis in *Chorthippus brunneus*, $2n = 16 + X$ (courtesy, B. John). Metaphase I. Coorientation of centromeres establishes the equatorial plane for the bivalents in the plane of the photograph.

**Fig. 12.** Spermatogenesis in *Chorthippus brunneus*, $2n = 16 + X$ (courtesy, B. John). Early anaphase I. Homologous chromosomes of smallest bivalents separate to one pole at the top and one at the bottom of the photograph.

**Fig. 13.** Spermatogenesis in *Chorthippus brunneus*, $2n = 15 + X$ (courtesy, B. John). Middle anaphase I. Homologous chromosomes of larger bivalents separate.

**Fig. 14.** Spermatogenesis in *Chorthippus brunneus*, $2n = 16 + X$ (courtesy, B. John). Anaphase I. All homologous chromosomes separate and each haploid group proceeds to the respective pole.

**Fig. 15.** Spermatogenesis in *Chorthippus brunneus*, $2n = 16 + X$ (courtesy, B. John). Interkinesis. Each group of extended chromosomes is enclosed by a nuclear membrane. One nucleus includes the heterochromatic X chromosome.

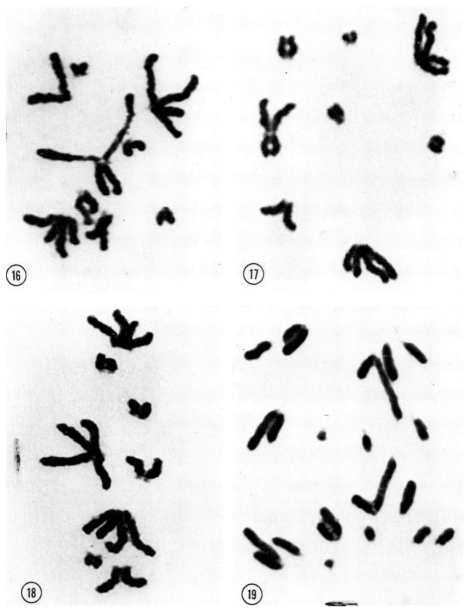

**Fig. 16.** Spermatogenesis of *Chorthippus brunneus*, $2n = 16 + X$ (courtesy, B. John). Only one of the two meiotic nuclei is presented to illustrate the second meiotic division. Prophase II. The extended chromosomes are relatively short and thick, and the sister chromatids are held together at the centromere.

**Figs. 17 and 18.** Spermatogenesis of *Chorthippus brunneus*, $2n = 16 + X$ (courtesy, B. John). Only one of the two meiotic nuclei is presented to illustrate the second meiotic division. Metaphase II. The chromosomes are viewed at the equatorial plane from a pole.

### E. Interkinesis

The interval between the two meiotic nuclear divisions is termed "interkinesis" (Fig. 15) and not "interphase," a term which is only applied to the interval between successive mitotic divisions. At interkinesis, the primary meiocyte has yielded either one secondary meiocyte with two haploid nuclei or two secondary meiocytes, each with a haploid nucleus, depending on the species.

### F. Prophase II

The second meiotic division is essentially mitotic in nature. Chromosomal events (except for chromosome condensation) observed during prophase I do not occur at prophase II. Both haploid nuclei are usually synchronized in the second meiotic division. The chromosomes emerge from the interkinesis nuclei and become shorter and thicker by coiling (Fig. 16). The nucleolus is still present.

### G. Metaphase II

The loss of the nuclear envelope and the alignment of the centromeres at an equatorial plane midway between the spindle poles characterize this stage (Figs. 17 and 18). The nucleolus is not present. Spindle fibers from each pole are associated with each centromere, indicating that the centromere has been replicated or that each replicated centromere has acquired centromeric activity.

### H. Anaphase II

The centromeres of sister chromatids separate and proceed to the respective poles (Fig. 19). Each haploid group of apparently single stranded chromosomes is derived from the four chromatids of the two homologous chromosomes that formed the bivalents of diplotene. Each group of chromosomes is confined within the volume of the spindle in their progress towards the respective poles.

### I. Telophase II

Each of the four haploid groups of chromosomes is enclosed by a nuclear envelope; the nucleolus appears, and the chromosomes progres-

---

**Fig. 19.** Spermatogenesis of *Chorthippus brunneus*, $2n = 16 + X$ (courtesy, B. John). Only one of the two meiotic nuclei is presented to illustrate the second meiotic division. Anaphase II. Haploid groups of one-stranded chromosomes proceed to the respective pole.

sively uncoil and disappear as discernible individuals to form the chromatinic network seen in strained material. At the end of this stage, the nuclei are processed by appropriate cellular divisions and developmental events to become gametes in animals and spores in higher plants.

## III. UNUSUAL CHROMOSOMES

Plant and animal chromosomes share a sufficient number of characteristics to define a usual or standard chromosome: a linear organelle with one centromere, two nonhomologous arms, and unique allotment of Mendelian genes arranged in a specific sequence. Unusual or nonstandard chromosomes are found in individuals, species, genera and occasionally larger taxonomic groups or encountered as chromosomal aberrations. Such chromosomes provide valuable cytogenetic tools and, in some cases, useful materials for the molecular geneticist.

The most prevalent unusual chromosomes are concerned with sex determination in animal species and relatively few dioecious plant species. Supernumerary or B chromosomes occur in plant and animal species, do not appear as a rule to have alleles for genes in standard chromosomes, and are usually heterochromatic. Chromosomes with no specific site for the centromere, or with multisites or with neocentric sites of centromeric activity have been found in plant and animal species. The polytene chromosomes and lampbrush chromosomes are usually restricted to certain large groups of animal species. The nonstandard chromosomes derived from standard chromosomes are discussed in the section on cytogenetics.

### A. Sex Chromosomes

Sex determination in animal and some plant species is directly or indirectly related to specific chromosomes in the complement. Usually one sex has a pair of homologous or X chromosomes and the other sex, a pair of heteromorphic chromosomes, the X and Y chromosomes (Fig. 20). The sex with the homologous chromosomes is termed homogametic, and the sex with heteromorphic chromosomes, heterogametic. This chromosomal association with sex determination constitutes a permanent test cross for the contrasting phenotypes of femaleness and maleness and is responsible for the approximately equal frequencies of each sex in most animal species. Many insect and some plant species have more than two sex chromosomes (Fig. 21) whose association and behavior during meiosis are somewhat bizarre when compared with the autosomes. An extensive literature

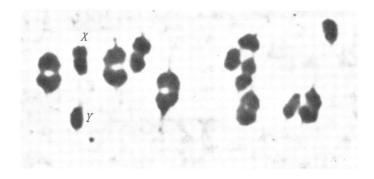

**Fig. 20.** *Humulus lupulus* (hop), metaphase I, $2n = 18 + $ XY (courtesy, B. John).

on the cytogenetics of the sex chromosomes has accumulated (Ohno, 1967).

Although no acceptable explanation to account for the origin of sex chromosomes has been proposed, animal evolution is presumably intimately related to the production of well-defined sexes in approximately equal numbers in each generation. Genes are responsible for diverting developmental processes to yield a male or a female. Consequently, a reasonably precise method is needed to ensure the distribution of such genes during meiosis. In *D. melanogaster*, maleness and femaleness genes occur in the X chromosome and the autosomes, but one chromosome, the X chromosome, assumes the dominant role in sex determination. Two X chromosomes determine femaleness and one X chromosome, maleness. Another solution to this problem involves sex determination by only one chromosome so that its presence is responsible for maleness and its absence for femaleness, for example, the Y chromosome in mammals.

Animal species seem to tolerate the loss or gain of sex chromosomes to a greater extent than autosomes. In man, individuals with up to five X

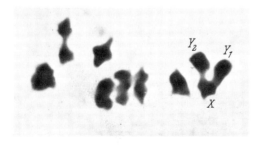

**Fig. 21.** *Humulus japonicus* (hop), metaphase I, $2n = 14 + $ XY, $Y_2$ (courtesy, B. John).

chromosomes or with $3X + 2Y$ chromosomes have been found. The inactive X chromosome hypothesis accounts for this situation in man and other mammalian species (Lyon, 1971). After commitment to normal sexual development during early embryogenesis, one or the other X chromosome in cells of the XX female is inactivated and appears as a small heterochromatic (Barr) body in resting somatic nuclei. Clones from cells with an inactive X chromosome retain this chromosome in an inactive form. In females or males with two or more X chromosomes, all but one X chromosome are inactivated. All X chromosomes in female or male germinal tissue presumably remain active.

## B. Supernumerary Chromosomes

One or more extra chromosomes, which are not sex chromosomes, standard chromosomes, or derived from standard chromosomes, have been found in plant and animal species. These supernumerary or B chromosomes are not necessary for viability or the normal development and functioning of the individual. These unusual chromosomes share a number of characteristics: usually they are smaller than the smallest standard chromosome of the complement (Fig. 22), are mostly or completely heterochromatic, are not homologous with any standard chromosome of the complement, and apparently lack alleles for any genes in standard chromosomes. In many plant species, which have yielded most of the information on supernumerary chromosomes (Battaglia, 1964), the chromosomes exhibit a high frequency of mitotic nondisjunction. Although supernumerary chromosomes appear to be genetically inert, each species tolerates its own quota of these chromosomes. The most common phenotype associated with increasing number of supernumerary chromosomes in plant species is sterility.

The origin of supernumerary chromosomes has not yet been satisfactorily explained. Moens (1965) isolated an isochromosome (2S.2S) with the short arms of chromosome 2 in tomato which had two homologous, heterochromatic arms and which simulated a supernumerary chromosome except for its normal mitotic disjunction. Dhillon and Garber (1960) reported an extra, standard chromosome in *Collinsia heterophylla* which simulated a supernumerary chromosome by high mitotic nondisjunction during gametophyte development and by the lack of morphological changes in plants with up to five such chromosomes.

## C. Chromosomes with Unusual Centromeric Characteristics

The standard chromosome has a single centromere, usually identifiable as an achromatic segment or the site of association with spindle fibers at a

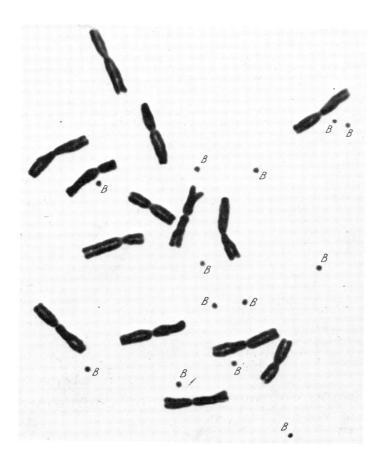

**Fig. 22.**   *Allium cernuum*, $2n = 14 + 13B$ (courtesy, B. John).

specific locus in the chromosome. Morphological and cytogenetic studies indicate that the centromere is a complex structure (Fig. 23) which can be broken into at least two parts (Fig. 24), each retaining centromeric activity (McClintock, 1938a). No acceptable explanation is available to account for the origin of the centromere.

A dicentric chromosome resulting from a structural aberration must cope with the possibility that each centromere could proceed to opposite poles during a nuclear division. In *Triticum aestivum* a dicentric chromosome was transmissible because the spindle fibers were associated with the same centromere in each dicentric chromosome in a bivalent during the first meiotic division (Sears and Camara, 1952). Either one of the centromeres was active during the second meiotic division or mitotic divisions.

**Fig. 23.** Centromere (arrow) of metaphase chromosome of *Allium Cepa* (courtesy, B. John).

### D. Multicentric Chromosomes

The horse threadworm, *Parascaris equorum*, includes one chromosomally defined race with two bivalents and a second race with one bivalent, and in each race, the chromosomes appear to have one centromere per chromosome. During early embryogenesis, the chromosomes in cells fated to be somatic tissue undergo spontaneous fragmentation. These chromosomes behave as standard chromosomes, each with one centromere during subsequent mitotic divisions. One explanation for this unusual situation assumes multicentric chromosomes in germinal cells and a dominant median centromere. Spontaneous chromosome breakage at predetermined loci in cells committed to somatic tissue produces centromere-containing "fragments" with healed chromosome ends.

### E. Diffuse Centromeric Activity

Chromosomes of the plant genus *Luzula*, of lepidopteran, homopteran, and hemipteran species of insects, and of euglenoids are associated with spindle fibers along the entire length of the chromosomes. During metaphase, the chromosomes seem to be stiff at the equatorial plane, and during anaphase sister chromosomes are parallel until bent by conforming to the shape of the spindle. Chromosome breakage by X-rays in *L. purpurea* yields fragments with diffuse centromeric activity which are not lost in successive mitotic divisions as in the case for acentric fragments of standard chromosomes (Malheiros and de Castro, 1947).

**Fig. 24.** Misdividing centromere of univalent during anaphase I in a triploid plant of *Scilla* sp. The centromere is associated with spindle fibers from opposite poles and each chromosome arm is progressing to different poles (courtesy, B. John).

## F. Neocentric Activity

The centromere of standard chromosomes is the only site of association between chromosome and spindle fibers, indicating that no other chromosomal region has centric activity. Sites of neocentric activity have been observed in standard chromosomes of a few grass species and of *Lilium formosanum*. Neocentric activity has been extensively studied in *Zea mays* (Rhoades, 1955). In the presence of abnormal chromosome 10 with a terminal heterochromatic segment, the prominent knobs of standard chromosomes have neocentric activity. A chromosome arm with a

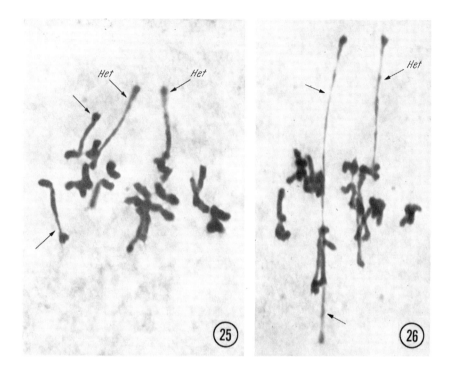

**Figs. 25 and 26.** Metaphase II and anaphase II in maize showing sites of neocentric activity at knob sites (arrows) in one or both homologous chromosomes in *Zea mays* (courtesy, B. John).

knob precedes the centromere towards the pole during meiotic and mitotic nuclear division (Figs. 25 and 26).

### G. Polytene Chromosomes

Giant interphase chromosomes are found in certain tissues of dipteran and some ciliated protozoan species and suspensor cells of the embryo of some plant species. In the familiar salivary gland chromosomes, endomitotic chromosomal replications yield a bundle of parallel, greatly extended strands or chromonemata which is at least one hundredfold longer and has a ten-thousandfold greater diameter than the standard metaphase chromosomes. The *appression* of corresponding chromomeres in the approximately thousand chromonemata produces the banding effect in polytene chromosomes. Furthermore, the somatic pairing of the polytene chromosomes simulates pachytene, a feature long exploited by Drosophila cytogeneticists to detect structural aberrations with far greater resolution than possible in pachytene chromosomes.

## H. Lampbrush Chromosomes

Lampbrush chromosomes (Fig. 27) are observed at diplotene during oogenesis in a number of vertebrate and invertebrate species with greatly

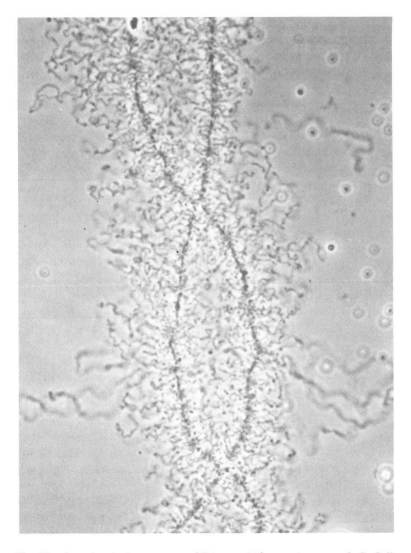

**Fig. 27.** Lampbrush chromosomes of *Triturus viridescens* (courtesy, J. G. Gall).

increased nuclear and cytoplasmic volume at prophase I. These chromosomes are obviously longer than standard chromosomes of other species at diplotene and display paired but opposite lateral loops of different dimensions along the homologs. Each loop appears to originate and to terminate in a chromomere; opposite loops of a pair are approximately the same length. Although lampbrush chromosomes provide notable experimental material for molecular biology, cytogeneticists have not exploited these chromosomes.

## IV. CYTOGENETICS

The correlation of breeding data and cytological observations constitutes cytogenetics. A species has a specific chromosome number, and each chromosome has its unique allotment of genic and chromosomal material arranged in a specific sequence. Furthermore, chromosome distribution during mitotic and meiotic nuclear division is regular, except for species with anomalous meiosis. Aberrations in the structure or number of chromosomes can be expected to alter the pattern of inheritance for the pertinent genic markers. Meiotic mutants are usually discovered by unexpected breeding data. The more common intrachromosomal structural aberrations, interchromosomal structural aberrations, and numerical aberrations involving the addition or loss of chromosomes or sets of chromosomes are presented here. Detailed presentations have been given by Brown (1972), Garber (1972), and Khush (1973).

## V. INTRACHROMOSOMAL STRUCTURAL ABERRATIONS

The standard chromosome with its unique allotment of genic and chromosomal material in a specified sequence can undergo a structural aberration involving the reorientation, loss, or gain of a chromosomal segment. Intrachromosomal structural aberrations result as primary events from the spontaneous or induced breakage of one or two chromosomes and the fusion of appropriate broken ends. Only broken chromosome ends fuse. The cytogenetic analysis of these and the other types of chromosomal aberrations depends on the transmissibility of the aberrations by gametophytes or gametes.

An inversion requires two breaks in one chromosome and the fusion of the broken ends of the three segments so that the segment with two broken ends is 180° out of phase with the standard arrangement. For inversions, chromosomal material is neither added nor subtracted. A deficiency results from the loss of an intercalary (internal) or a terminal segment. Two breaks are needed to produce an intercalary deficiency and one break, a terminal deficiency. The addition of a chromosomal segment already present in the chromosome is termed a duplication and requires two breaks in the donor chromosome and one break in the recipient homolog, followed by the translocation of the segment with two broken ends and the appropriate fusions. Duplications are relatively uncommon when compared with the frequency of inversions and intercalary deficiencies. The recovery of the different types of intrachromosomal structural aberration is determined by a number of factors: (1) the probability of one or two breaks in one or two chromosomes, (2) the probability of the pertinent fusions of broken ends, (3) the fate of the aberrant chromosome during meiosis and mitosis, and (4) the impact of the aberrant chromosome on the viability or functioning of the products of meiosis, zygotes, and embryos.

Individuals with intrachromosomal structural aberrations were first encountered during genetic studies. For example, heterozygotes with a standard and inversion homolog were considered to have a "crossover suppressor gene" for a number of linked genes in the pertinent chromosome. Many so-called "lethal" genes and a number of mutant genes were later found to correspond to specific deficiencies. The famous *Bar* mutation in *Drosophila melanogaster* was eventually attributed to a tandem duplication.

Ring chromosomes represent a special type of nonstandard chromosome resulting from the loss of terminal segments, creating two deficiencies, and the subsequent fusion of the broken ends of the segment containing the centromere. In several cases (McClintock, 1938a; Schwartz, 1953), a ring chromosome was implicated in variegation for contrasting phenotypes when the dominant allele was in the ring and the recessive allele in the standard homolog. Ring chromosomes undergo structural alterations during meiosis and mitosis so that the segment with the dominant allele may be lost and only the recessive allele remains (see Garber, 1972).

## A. Inversions

Inversions are classified by the absence (paracentric) or presence (pericentric) of the centromere in the inverted segment, regardless of the length of the segment. Depending on the site and length of the inverted segment, a pericentric inversion can grossly alter the arm ratio of a standard chromosome and provides one means to alter chromosome morphology in related species. Both types of inversion can displace such topological markers as heterochromatic segments or prominent chromomeres from their standard location.

### 1. Inversion Heterozygosity

The length of the inverted segment is important in determining chromosome configuration at pachytene. The synapsis of a long inverted segment and standard segment can produce a linear element with asynapsed nonhomologous terminal segments. The synapsis of an intermediate length inverted segment and standard segment yields a loop which includes the inverted segment (Fig. 28). In species with favorable pachytene chromosomes, the site and location of the inversion loop can be determined. A very small inverted segment may or may not be appressed to the homologous standard segment.

A single crossing-over within the loop formed by synapsis between a standard and an inverted segment of homologous chromosomes yields visual evidence to distinguish a paracentric from a pericentric inversion and accounts for the sterility often associated with inversion heterozygosity (Fig. 29). While the single crossing-over yields two deficient and two

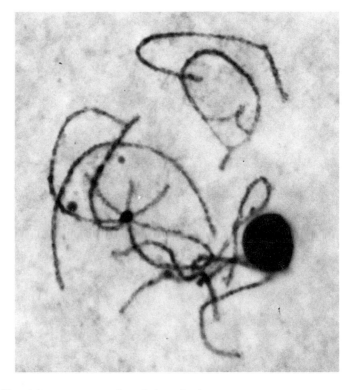

**Fig. 28.**   A heterozygous pericentric inversion in chromosome 2 of *Zea mays*, appearing as a looped chromosome pair at 12 o'clock (courtesy, M. M. Rhoades).

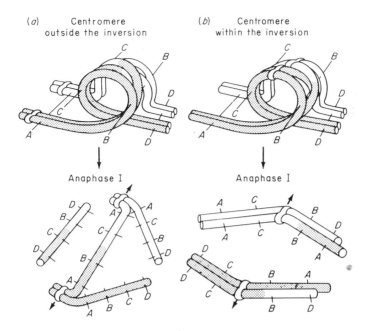

**Fig. 29.** Anaphase I configurations resulting from a crossover within a paracentric (a) or pericentric (b) inversion. (From "General Genetics," 2nd ed., by A. M. Srb, R. D. Owen, and R. S. Edgar. W. H. Freeman and Company. Copyright © 1965.)

intact chromosomes, the paracentric but *not* the pericentric inversion produces a dicentric chromatid bridge and an acentric fragment at anaphase I (Fig. 30). The site and length of the deficiencies determine the viability or functioning of spores and gametophytes and of zygotes receiving a deficient chromosome.

For heterozygous inversions in plants which include not more than one chiasma, the percentage of aborted pollen grains is correlated with the frequency of crossing-over within the heterozygous inversion. Consequently, factors influencing the frequency of crossing-over can be readily studied in plants with such heterozygous inversions. The percentage of aborted ovules, however, is related to the frequency of crossing-over in the pericentric but *not* paracentric heterozygous inversion. The high ovule fertility in both plant and animal species reflects the directed orientation of the dicentric chromatid bridge during the first meiotic division so that the surviving megaspore or the egg nucleus receives a standard or inversion chromosome and not a deficient chromosome. Relatively long inversions that accommodate two chiasmata are valuable tools in studying the cytological consequences of 2-strand, 3-strand, and 4-strand double

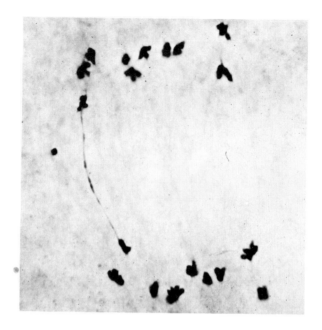

**Fig. 30.** Dicentric chromatid bridge and acentric fragment (to the right of the bridge) at anaphase I resulting from a crossing-over within a heterozygous paracentric inversion in *Zea mays* (courtesy, M. M. Rhoades).

crossing-over. McClintock (1938b) detected three unique chromosomal configurations at anaphase I and II each derived from specific types of double crossing over in a long paracentric inversion in the distal segment of the long arm of chromosome 9 of maize.

The crossover–suppressor effect for linked genes in a heterozygous inversion is a misnomer. The recombinants are in the deficient chromosomes, which give rise to inviable products and, therefore, are not present in the scorable progeny. The crossover–suppressor effect is not found for inversion homozygotes. The linkage relationships among loci in the inverted and in the adjacent standard segments provide one genetic means to determine the site and length of the inversion in terms of the linkage map.

## B. Deficiency

The terminally deficient chromosome results from a single break, yielding a centric and an acentric fragment with broken ends. Terminally deficient chromosomes are usually not transmissible because the two chromatids after replication are joined by the fusion of their broken ends to produce a dicentric chromosome. When the centromeres proceed to opposite poles, the chromosome breaks and the cycle is repeated in each mitotic division. Transmissible, terminally deficient chromosomes can be generated by crossing over in the appropriate segment of a chromosome with a double included inversion as the nonstandard homolog (McClintock, 1941). A double included inversion results from two separate events: one large inversion and later, a second smaller inversion involving a segment within the first inversion. The broken end of the terminally deficient chromosome is healed when the chromosome is included in the zygotic nucleus and behaves as a "normal" chromosome.

The site and length of an intercalary deficiency may be established by detecting a buckled (looped) segment in the pachytene chromosomes of a deficiency heterozygote. Heteromorphic bivalents often indicate a deficiency for the shorter chromosome. The detailed banding of salivary gland chromosomes permits the detection of deficiencies too small to be noted in pachytene chromosomes.

In plants, the female gametophyte is more likely to tolerate deficiencies than the male gametophyte. The missing genes could interfere with the specialized functions of the male gametophyte which produces the pollen tube in effecting fertilization. Deficient gametes of animal species are likely to function, and zygotes with a heterozygous deficiency often develop, producing a scorable individual. As a rule, long deficiencies are more apt to be lethal than short deficiencies. Deficiencies of equal length, however, may or may not be lethal, depending on their site in specific chromosomes. Such observations suggest that "vital" genes may not be randomly distributed throughout one or all of the chromosomes.

Deficiencies can produce mutant phenotypes and by their transmission simulate genic mutations. Such gene simulants are usually lethal when homozygous in the zygotic nucleus. Minute deficiencies in *D. melanogaster* can be detected by the careful examination of the salivary gland chromosomes in deficiency heterozygotes where the absence of relatively few bands is obvious. Pachytene chromosomes do not provide suitable material for such resolution. McClintock (1944) investigated two different but overlapping terminal deficiencies in maize and demonstrated mutant phenotypes for different combinations of these deficiencies. If the relationships between certain mutant phenotypes and specific deficiencies had not been established by a direct examination of the pertinent chromosomes, an unorthodox explanation involving genic mutations with unusual properties might have been proposed.

Deficiencies are used to assign mutant genes to specific chromosomal sites. Homozygotes for dominant alleles are exposed to ionizing radiation or radiomimetic chemicals to break chromosomes and then crossed with homozygotes for recessive alleles. A few progeny exhibit the unexpected recessive phenotype, pseudodominance, when the missing segment includes the corresponding dominant allele. Depending on the site and length of the missing segment, closely linked recessive alleles may exhibit pseudodominance. While deficiencies should be established cytologically, minute deficiencies beyond visual resolution can be postulated to account for pseudodominance. Electron microscopy has been used to detect deficiencies by melting and annealing DNA from wild-type and presumably deficient strains of coliphage (Westmoreland et al., 1969). In the reconstituted double helix, a loop in the wild-type polynucleotide chains represents the site and length of the missing segment in the polynucleotide chain from the deficient mutant strain.

The precise cytological localization of a gene requires overlapping deficiencies responsible for pseudodominance and sharing a common missing site. Slizynska (1938) and Judd et al. (1972) have successfully used this technique in D. melanogaster, employing light and electron microscopy and increasing cytological resolution by examining the salivary gland chromosomes. Overlapping deficiencies each producing an rII mutation of coliphage T4 were detected by the absence of wild-type progeny from crosses between these deficient mutants (Benzer, 1961). Furthermore, a collection of sequenced but nonoverlapping deficiencies facilitated the mapping of gene mutant sites within the rII locus. The absence of wild-type progeny from crosses between a genic mutant and a specific deficiency indicated that the mutant site resides in the missing segment. These studies represent one of the early examples of prokaryotic cytogenetics.

## C. Duplication

The Bar duplication in D. melanogaster will be presented because the cytogenetic analysis of this duplication had a significant impact on genetic theory. The Bar mutation occurred spontaneously and was first identified as a "genic" mutation drastically altering the number of eye facets. The mutation was assigned a locus in the X chromosome and viewed as a dominant allele. When Sturtevant (1925) demonstrated crossing-over within the Bar locus, the integrity or indivisibility of the classical gene was threatened. Bridges (1936) correlated the presence of a tandem duplication of the 16A segment in the salivary gland X chromosome with the Bar phenotype and one 16A segment with the wild-type phenotype (Fig. 31). Although the integrity of the classical gene had been restored, intragenic crossing over was later demonstrated by the increased resolution of

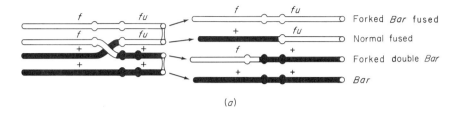

(a)

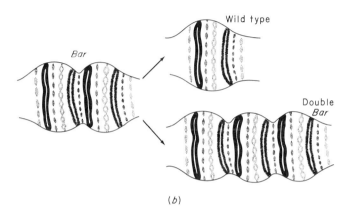

(b)

**Fig 31** Diagrams illustrating (a) the origin of normal and double *Bar* (extremely narrow eyes) by crossing-over in females homozygous for *Bar* (narrow eyes) and heterozygous for the flanking markers forked (*f*) and fused (*fu*) and (b) the number of 16A segments in *Bar*, double-*Bar*, and normal salivary gland X chromosomes (after White, 1973).

crossing-over products recovered in microbial eukaryotic and prokaryotic species.

## VI. INTERCHROMOSOMAL STRUCTURAL ABERRATIONS

The translocation of segments between chromosomes is responsible for interchromosomal structural aberrations. The most common example of such aberrations is the mutual exchange of segments which is not a consequence of crossing-over. One break in each of two chromosomes yields four segments each with one broken end, and appropriate fusions produce two chromosomes each with segments from different chromosomes and one centromere. Reciprocal translocations may occur between homologous chromosomes but generally involve nonhomologous chromosomes. The reciprocal translocation of intercalary segments, requiring two breaks in two chromosomes, is possible but rarely encountered.

## A. Cytology

During pachytene (Fig. 32), the four chromosomes of a heterozygous reciprocal translocation forms a cross-configuration when the homologous segments of these chromosomes synapse. In good cytological material, the "intersection" of the cross indicates the site of the break in each chromosome and the length of each translocated segment. Furthermore, the two nonhomologous chromosomes involved in the reciprocal translocation can be identified. Chromosome breaks yielding reciprocal translocations in *D. melanogaster* must occur in the euchromatic regions and not in the darkly staining heterochromatic regions adjacent to the centromere to detect the cross-configuration in the salivary gland chromosomes. The pooling of the heterochromatin adjacent to the centromeres in each chromosome produces a darkly staining body, the "chromocenter," and each chromosome arm extends outward from the chromocenter. Consequently a reciprocal translocation with the intersection in the heterochromatin would be cytologically indistinguishable from individuals without a reciprocal translocation.

At diakinesis (Fig. 33) and metaphase I (Figs. 34, 35, and 36), a heterozygous reciprocal translocation yields an "interchange complex," a chain or ring of four chromosomes when at least one chiasma occurs in three or four "arms" of the cross-configuration, respectively. The interchange complex at metaphase I appears as an open (Fig. 34) or alternate (Fig. 35) configuration. The frequency of sporocytes, meiocytes, with one or the other configuration is responsible for the fertility of the individual and for the transmission of genes in the two nonhomologous chromosomes of the interchange complex. For the open configuration, two adjacent chromosomes proceed to the same pole at anaphase I and two types of open configuration are possible: *adjacent*-1, homologous centromeres go to one pole, and *adjacent*-2, nonhomologous centromeres go to one pole. For the alternate configuration, two alternate chromosomes with nonhomologous centromeres proceed to the same pole during anaphase I. Two alternate configurations are also possible (Endrizzi, 1974). While chromosome distribution from an interchange complex is usually 2:2, a 3:1 distribution is not uncommon and gives spore or gametic nuclei with $n + 1$ or $n - 1$ chromosomes.

The fertility of individuals with a reciprocal translocation is directly related to the frequency of the alternate configuration of the interchange

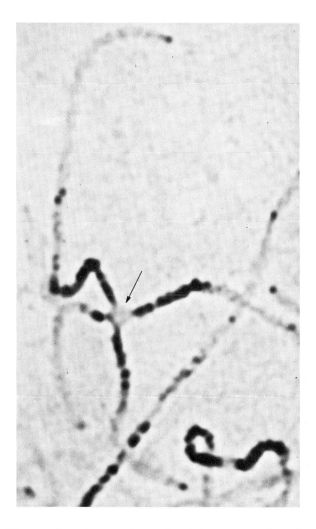

**Fig. 32.**   The cross-figuration at pachytene for a heterozygous reciprocal translocation in tomato. The site of breakage (arrow) in each nonhomologous chromosome occurred in or very close to each centromere so that the intersection of the cross appears at the centromeres. The cross results from the pairing of homologous segments of standard and translocation chromosomes (courtesy, B. Snoad).

complex at metaphase I (Fig. 37). Spore or gametic nuclei from cells from the alternate configuration include both standard chromosomes or both translocation chromosomes of the interchange complex. Therefore, these nuclei do not have a deficient chromosome. The spore or gametic nuclei

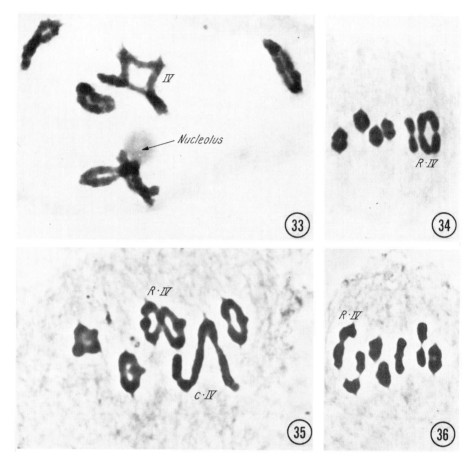

**Fig. 33.**   Interchange complex (IV) at diakinesis in a heterozygous reciprocal transloca-
tion of rye, *Secale cereale* (courtesy B. John)
**Fig. 34.**   Adjacent orientation of interchange complex ring at metaphase I.
**Fig. 35.**   Alternate orientation for a ring and a chain interchange complex at metaphase I.
**Fig. 36.**   Discordant adjacent orientation of interchange complex ring at metaphase I.

from cells with an adjacent configuration include one standard and one
translocation chromosome and these nuclei have a deficient chromosome.
In plants, spores with a deficiency usually abort; in animals, deficient
gametes function but deficient zygotes or embryos are generally inviable.
The site and length of the deficiency determine whether spores, zygotes or
embryos will be nonfunctional.

In many plant species, individuals with a reciprocal translocation are
usually semisterile (50% pollen and ovule abortion) and 50% of the meiocytes display the alternate configuration of the interchange complex at

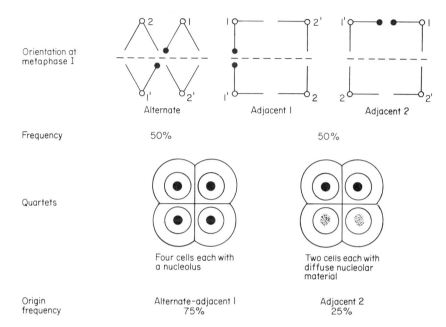

**Fig. 37.** Scheme illustrating the origin of spore-quartets with one nucleolus or diffuse nucleolar material in two adjacent nuclei for a translocation heterozygote involving a nucleolus-organizing chromosome. [From E. D. Garber (1972). "Cytogenetics: An Introduction." McGraw-Hill Book Company. Copyright © 1972.]

metaphase I. In several plant species, plants with a reciprocal translocation are highly fertile and significantly more than 50% of the meiocytes have the alternate configuration. The site of the chiasma formation and of the break in each chromosome may be responsible for the different frequencies of the alternate configuration of the interchange complex in these species (Garber and Dhillon, 1961).

A direct examination of the interchange complex at metaphase I usually does not distinguish the *adjacent*-1 and *adjacent*-2 configurations or the two types of alternate configuration. Garber (1948) demonstrated the two types of adjacent configuration in *Sorghum versicolor* for a reciprocal translocation involving the nucleolus-organizing chromosome (Fig. 37). Endrizzi (1974) distinguished the two types of alternative configuration in an interchange complex involving chromosomes with obviously different morphology.

## B. Fertility

Progeny from a self-pollinated semisterile, translocation heterozygote or from a cross between a semisterile location heterozygote and a fertile standard plant include approximately equal numbers of semisterile and fertile plants (Fig. 37). The functional spores and gametophytes include both standard or both translocation chromosomes from the alternate configuration of the interchange complex. The fertile progeny from a self-pollinated translocation heterozygote include approximately equal numbers of standard plants and translocation homozygotes which are fully fertile. The transmission of the interchange complex in animals is somewhat complicated by the functioning of gametes with deficient chromosomes from the adjacent configurations and the lethality of deficiencies for the zygotes or embryos. The translocation homozygotes can be identified cytologically when the chromosome morphology of two bivalents obviously differs from two standard bivalents or by obtaining only translocation heterozygotes in progeny from a cross with standard plants.

### C. Genetics

The semisterility of translocation heterozygotes behaves as a genic simulant in that crosses with standard plants yield semisterile or fully fertile progeny in approximately equal numbers. Consequently, genes can be assigned to chromosomes in the interchange complex by making a test cross: a translocation heterozygote with heterozygous markers crossed with a homozygous recessive standard individual (Fig. 38). When a gene is *not* in the interchange complex, the progeny includes four categories of phenotypes in approximately equal frequencies: (1) semisterile dominant, (2) fully fertile dominant, (3) semisterile recessive, and (4) fully fertile recessive phenotype. Deviations from the expected frequencies for each category indicate that the gene is located in one of the interchange complexes. Unless it were known that the test cross also involved a translocation heterozygote, the breeding data would show a linkage for genes known to be in nonhomologous chromosome, a phenomenon termed "pseudolinkage." The recombinant genotypes in the chromosomes of the adjacent configurations are not recovered because the meiotic products include deficient chromosomes. When a collection of reciprocal translocations is available, genes can be assigned to one of the translocated chromosomes by linkage with one reciprocal translocation but not to another reciprocal translocation for a common chromosome.

### VII. NUMERICAL ABERRATIONS

### A. Aneuploidy

Species are characterized by a standard number of chromosomes in the gametic ($n$) or somatic ($2n$) nuclei. In many animal species, males and females have different chromosome numbers, which can be related to the

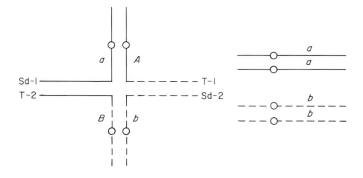

|  | \| Semisterile | | | | Fertile | | | |
|---|---|---|---|---|---|---|---|---|
|  | AB | Ab | aB | ab | AB | Ab | aB | ab |
|  | Standard heterozygote | | | | | | | |
|  | – | – | – | – | I | I | I | I |
|  | Translocation heterozygote | | | | | | | |
| No crossing over |  |  |  |  |  |  |  |  |
| A–a  or  B–b | I | O | O | O | O | O | O | I |
| Crossing over |  |  |  |  |  |  |  |  |
| A–a | High | O | Low | O | O | Low | O | High |
| B–b | High | Low | O | O | O | O | Low | High |
| A–a and B–b | High | Low | Low | Low | Low | Low | Low | High |

**Fig. 38.** Relative frequencies of fertile or semisterile progeny from a test cross between a standard plant homozygous for recessive alleles and a plant with a reciprocal translocation and heterozygous for the corresponding alleles. Crossing-over may or may not occur in the chromosomal segment between the pertinent locus and site of interchange for each chromosome. [From E. D. Garber (1972). "Cytogenetics: An Introduction." McGraw-Hill Book Company. Copyright © 1972.]

sex chromosomes rather than the autosomes. The loss or gain of chromosomes in gametes, gametophytes, and in diploid, somatic cells or individuals is termed aneuploidy. The cytogenetics of aneuploidy in plants is discussed in detail by Khush (1973).

In trisomy ($2n + 1$), an extra standard or structurally aberrant chromosome is added to the diploid complement, and in monosomy ($2n - 1$), one standard chromosome is missing. The additional chromosome is termed a trisome and the individual, a trisomic. The remaining chromosome in a monosomic individual is termed a monosome. A gametic nucleus with an added chromosome is disomic ($n + 1$). Diploid species are more likely to tolerate an added chromosome than the loss of a standard chromosome. Sex chromosomes are unusual in that added sex chromosomes are more often tolerated than added autosomes, and the loss of one X chromosome in XX females, or the Y chromosome is more likely to be tolerated than the loss of an autosome.

In tetrasomy $(2n + 1^{II})$, a bivalent is added so that one chromosome occurs four times and in nullisomy $(2n - 1^{II})$, a bivalent is lost. These aneuploid types are usually restricted to hexaploid plant species such as wheat which have three genomes with homologous chromosomes in the different genomes. Homoeologous chromosomes include a large number of homologous genes and chromosome segments.

Mitotic nondisjunction is responsible for the loss of one chromosome and usually the addition of this chromosome to daughter nuclei. Depending on viability and selective forces, the aneuploid cells can yield aneuploid clones which become part of the germinal tissue. The meiotic products would then be $n$, $n + 1$ or $n - 1$ gametes nuclei, and when the aneuploid gametes participate in fertilization, the zygotes are diploid $(2n)$, trisomic $(2n + 1)$ or monosomic $(2n - 1)$. Meiotic nondisjunction also produces $n$, $n + 1$ or $n - 1$ gametic nuclei. Monosomic progeny are more likely to be found in tetraploid species with two genomes than in diploid species with one genome. Self-pollinated or crossed trisomic individuals can produce diploid, trisomic, or tetrasomic progeny, but the impact of a duplicated bivalent is usually lethal for diploid species compared with tetraploid and hexaploid species. Self-pollinated or crossed monosomic individuals can produce diploid, monosomic, or nullisomic progeny, but the impact of a missing bivalent is usually lethal for diploid and tetraploid species compared with hexaploid species.

## B. Monosomy

The loss of a standard chromosome represents a gross deficiency for a basic diploid species. In plants, $n - 1$ gametophytes are almost always inviable or malfunctional so that monosomic individuals do not result from fertilization. In animals, the absence of an X chromosome does not interfere with gametic function and XO progeny can be obtained. Monosomics can be obtained in tetraploid plant species with bivalents, and $n - 1$ female gametophytes are more likely than $n - 1$ male gametophytes to be functional in fertilization. Tetraploid species with two related genomes have numerous genic duplications which compensate for the loss of a chromosome. Consequently, the cytogenetics of monosomy has been most extensively investigated in tetraploid and hexaploid plant species.

The number of different, possible monosomics for a species equals the gametic chromosome number. The various types of monosomic individuals usually have unique morphological characteristics and can be distinguished from the diploid and each other. Clausen and Cameron (1944) used monosomy to assign mutant genes of the tetraploid species *Nicotiana tabacum* to specific linkage groups and these, in turn, to particular chromosomes. Earlier genetic studies in this and other tetraploid species had been frustrated by the relatively few detectable, monogenically determined, contrasting phenotypes and a relatively large number ($n$ = 14) of morphologically nondescript chromosomes. For example, the recessive allele (*wh*) for white flower in *N. tabacum* was assignable to chromosome C when the progeny from a cross between a monosomic C seed parent with the dominant allele (*Wh*, carmine flower) and a diploid recessive (*wh wh*) pollen parent include monosomics with white flowers (*wh*O). In this experiment, the recessive phenotype appears as expected for a deficiency for the dominant allele (pseudodominance).

## C. Nullisomy

In hexaploid plant species, a relatively low percentage of $n - 1$ pollen grains can function to fertilize eggs. Consequently, nullisomics ($2n - 1^{II}$) have been obtained by self-pollinating or hybridizing monosomics ($2n - 1$). Sears (1953, 1954) obtained all of the possible nullisomics ($n = 21$) in the hexaploid species *Triticum aestivum* and developed several methods for assigning mutant genes to specific chromosomes. Although nullisomic analysis is almost completely restricted to hexaploid species, many agronomically important species are hexaploids.

## D. Trisomy

The most common type of aneuploid is a trisome ($2n + 1$), usually a third standard chromosome. Because the extra chromosome may be either a standard chromosome or a structurally aberrant chromosome, the cytogenetics of trisomy has become a specialized area of investigation and has a voluminous literature covering plant and animal species. The early literature on trisomy recognized three categories determined by the chromosomal contribution of the trisome: (1) primary, an additional standard chromosome, (2) secondary, an additional chromosome with homologous arms (isochromosome), and (3) tertiary, an additional chromosome with segments from two nonhomologous chromosomes. The discovery of other categories of trisomy was responsible for a relatively complicated terminology (Khush, 1973). Primary and tertiary trisomy is more common than the other categories and will be presented to illustrate the cytogenetic consequences of trisomy.

## 1. Primary Trisomy

In a number of species, the impact of an extra standard chromosome has a marked phenotypic effect, and each primary trisome can be related to a distinctive syndrome of physiological and morphological alterations. The number of detectable different primary trisomes in a species is determined by the viability, development and functioning of $n + 1$ gametophytes from plant trisomics and of trisomic zygotes and embryos in animals. *Datura stramonium*, jimsonweed (Avery *et al.*, 1959), and *Lycopersicon esculentum*, tomato (Khush, 1973) are remarkably responsive to chromosomal duplication, and the cytogenetics of trisomy in these species provides models for other species.

Primary trisomes offer one means to extend the range of heterozygosity for alleles in the extra chromosome (*AAa*, simplex; *Aaa*, duplex) and provide a method to assign these genes to a specific chromosome and, in turn, to a particular linkage group. Simplex individuals from a cross between a homozygous dominant trisomic (*AAA*) and a homozygous recessive diploid (*aa*) are detected cytologically or by their characteristic syndrome of altered characters. The ratio of wild-type and mutant progeny from the cross between the simplex individual (*AAa*) and a homozygous recessive diploid (*aa*) is 2 to 5 $A : 1a$, which can be readily distinguished from the $1 : 1$ ratio for a test cross involving mutant genes *not* in the primary trisome.

## 2. Tertiary Trisomy

The $3 : 1$ distribution of chromosomes from an interchange complex can yield $n + 1$ gametic nuclei with both standard chromosomes and either one of the translocation chromosomes. The tertiary trisome with segments from nonhomologous chromosomes may be cytologically distinguishable from a primary trisome by its characteristic morphology at metaphase or by forming an association of five rather than three chromosomes at diakinesis or metaphase I. By suitable crosses to obtain simplex trisomics for test crossing to homozygous recessive diploids, mutant genes can be assigned to specific segments of particular chromosomes.

Down's syndrome usually results from primary trisomy for chromosome 21. A tertiary trisome with a segment of the long arms of chromosomes 15 and 21 is also associated with this syndrome (Carter *et al.*, 1960). This observation indicates that the responsible duplication involves a specific number of genes in the long arm of chromosome 21. Furthermore, the loss of the short arms of chromosomes 15 and 21 in the other translocation chromosome of the reciprocal translocation is not lethal, suggesting that these arms do not carry "vital genes."

## E. Euploidy

In basic diploid species, gametophytic or gametic nuclei have one genome and the sporophytic or somatic nuclei, two homologous genomes. To contrast chromosome numbers in gametic and somatic nuclei, Strasburger (1905) coined the terms haploid and diploid, respectively. Individuals with one genome or three or four homologous genomes also occur. The terminology for these numerical chromosomal aberrants considers the number of *homologous* genomes in the somatic nuclei: one genome, monoploidy ($n$); three genomes, autotriploidy ($3n$); and four genomes, autotetraploidy ($4n$). Individuals with greater than four homologous genomes are relatively uncommon. Stebbins (1950) presents a detailed discussion of autopolyploidy from an evolutionary point of view.

### 1. Monoploidy (n)

Monoploids regularly occur in the life cycle of the lower plants and hymenopteran insects. Monoploids differ from haploids in that a monoploid has a single genome while a haploid may have one or more genomes, depending on the species. Consequently, a haploid with a single genome would also be a monoploid. Monoploid sporophytes from basic diploid plant species are relatively uncommon, generally differing from diploid individuals by their reduced stature, slender proportions, smaller organs, and almost complete sterility. A deliberate search for monoploids requires the availability of plants with homozygous recessive alleles in different chromosomes. Progeny with recessive phenotypes (pseudodominance) from crosses between a homozygous dominant pollen parent and a homozygous recessive parent are likely candidates for monoploids, pending cytological confirmation.

### 2. Autotriploidy (3n)

Spontaneous autotriploids are occasionally found in progeny, presumably products of a fertilization between haploid and diploid gametic nuclei. The usual procedure, however, requires a hybridization between an autotetraploid seed parent and a diploid pollen parent. Cytogenetic interest in autotriploidy generally stems from a need to obtain primary trisomics. The random distribution of three homologous chromosomes during meiosis, the malfunctioning or inviability of gametophytes or gametes with one or more extra chromosomes, and the aberrant development or inviability of zygotes or embryos with one or more extra chromosomes are responsible for the occurrence of primary trisomics and diploids in the progeny of poorly fertile autotriploids. In some species, the autotriploid is used as seed or pollen parent in crosses with the diploid.

### 3. Autotetraploidy (4n)

The occasional failure of a mitotic nucleus to complete nuclear division at anaphase is responsible for an autotetraploid cell. When this event happens during early development, the incorporation of autotetraploid cells into germinal tissue can yield plants with autotetraploid flowers. Seed from such self-fertilized flowers will produce autotetraploid individuals. Colchicine is often used to increase the frequency of autotetraploid nuclei and to facilitate the production of autotetraploid plants.

Plant cytogeneticists and plant breeders are interested in autotetraploidy for several reasons. Autotetraploids indirectly provide a source of primary trisomics by yielding autotriploids as progeny from crosses with diploids. The failure of crosses between a tetraploid and a diploid species may be overcome when the tetraploid species is hybridized with an autotetraploid. Except for an identical chromosome number in each parent, there is no obvious reason for reported successes from crossing a tetraploid species with an autotetraploid from a related diploid species. The hybrid usually has a tetraploid chromosome number. Finally, autotetraploids offer useful characteristics for horticulturists seeking more slowly maturing, robust, and larger varieties of diploid species. Nuclei and cells of autotetraploids generally are larger than those of the corresponding diploids and the greater volume can be reflected in the dimensions of the plant or its organs. In comparing diploid, autotriploid, and autotetraploids of one species, the increase in stature and in size of certain organs of autotriploids is usually intermediate between diploid and autotetraploid. The fertility of autotetraploids is relatively high but not approaching that of diploids. The random distribution of four homologs in a quadrivalent or trivalent and univalent for each chromosome during meiosis is responsible for a characteristic frequency of progeny with more or less than the autotetraploid chromosome number.

### F. Allopolyploidy

Chromosome number is as important a datum for a species as any other characteristic deemed significantly stable to merit taxonomic significance. Plant genera commonly include species with different chromosome numbers which are multiples of a basic number. Winge (1917) proposed the first explanation for these observations. Related diploid species can cross to produce an interspecific hybrid with greatly reduced fertility or complete sterility. The spontaneous doubling of the chromosomes in somatic or germinal tissue can yield fertile flowers. Seed from such flowers produce fertile plants with the tetraploid chromosome number and these tetraploid populations may evolve into a tetraploid species. This process can be repeated for interspecific hybrids between diploid and tetraploid species to produce hexaploids which eventually yield hexaploid species.

The cytogenetic analysis of interspecific hybrids and of polyploid species was responsible for a confusing terminology to account for chromosome associations in the hybrids and species. The genome or set of chromosomes designated as X in the gametic nucleus of the basic diploid organism (2X) of a genus was used as the unit in considering the origin of the polyploid species of the genus. For example, *Triticum* includes species with a gametic chromosome number of 7, 14, or 21 and in this genus, one genome has seven chromosomes (X = 7). Furthermore, the diploid (2X), tetraploid (4X) and hexaploid (6X) species have bivalents at diakinesis and metaphase I, suggesting that the tetraploid species have two different genomes and the hexaploid species three different genomes in the gametic nucleus.

Stebbins (1950) proposed an idealized scheme for classifying polyploid species by considering the origin and relationship of the genomes (Fig. 39): autopolyploids, segmented allopolyploids, true or genomic allopolyploids or autoallopolyploids. While this scheme represents an acceptable

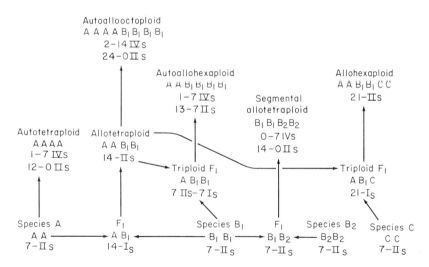

**Fig. 39.** An idealized diagram to illustrate the genomic constitution of four basic diploid species, the chromosome associations in their interspecific hybrids and the interrelationships and chromosome associations in the polyploid species which evolved from populations derived from interspecific hybrids with a doubled chromosome number. In the autoallopolyploid species, two different genomes are present and at least one genome is represented four times. [From G. L. Stebbins, Jr. (1950). "Variation and Evolution in Plants" by G. L. Stebbins, Jr. Columbia University Press, Copyright © 1950.]

theoretical construct, genome analysis for interspecific hybrids may not permit the tidy assignment of polyploid species unequivocally to one of the four polyploid categories.

Amphiploids include the diploid chromosome number of each species in an interspecific hybrid. This type of polyploid is usually the product of an interspecific hybridization, eventually followed by a chromosome doubling in the interspecific hybrid. Amphiploid cell populations are now available for a number of mammalian species which cannot hybridize (Davidson, 1973). For example, cells from man ($2n = 46$) and mouse ($2n = 40$) are adapted to tissue culture, mixed in the presence of inactivated Sendai virus to enhance the formation of intercellular bridges and then added to an appropriate medium favoring the growth of heterokaryons or amphiploid cells ($2n = 86$). Each human and mouse chromosome can be identified by morphological characteristics and unique banding patterns after treatment with certain compounds and staining. In the course of establishing clones, the mouse chromosomes are retained and human chromosomes are lost, until stable clones with 40 mouse and 1–15 human chromosomes are present. When homologous enzymes with different electrophoretic mobilities are present in man and mouse cells, it is possible to assign the gene to a specific human chromosome by noting the presence or absence of the human enzyme in extracts from clones with different arrays of human chromosomes. This ingenious use of amphiploid cells to construct human linkage groups is essentially a modification of the parasexual cycle first discovered in fungi and of nullisomic analysis first used in hexaploid plant species.

## REFERENCES

Avery, A. G., Satina, S., and Rietsema, J. (1959). "Blakeslee: The genus *Datura*." Ronald Press, New York.

Battaglia, E. (1964). Cytogenetics of B-chromosomes. *Caryologia* 17, 245–299.

Benzer, S. (1961). On the topography of the genetic fine structure. *Proc. Natl. Acad. Sci. U.S.A.* 47, 403–415.

Bridges, C. B. (1936). The Bar "gene," a duplication. *Science* 83, 210–211.

Brown, W. V. (1972). "Textbook of Cytogenetics." Mosby, St. Louis, Missouri.

Carter, C. O., Hamerton, J. L., Polani, P., Gunlap, A., and Weller, S. D. V. (1960). Chromosome translocation as a cause of familial mongolism. *Lancet* 2, 678–680.

Clausen, R. E., and Cameron, D. R. (1944). Inheritance in *Nicotiana tabacum*. XVIII. Monosomic analysis. *Genetics* 29, 447–477.

Davidson, R. (1973). "Somatic Cell Hybridization: Studies on Genetics and Development." Addison-Wesley, Reading, Massachusetts.

Dhillon, T. S., and Garber, E. D. (1960). The genus *Collinsia* X. Aneuploidy in *C. heterophylla*. *Bot. Gaz. (Chicago)* 121, 125–133.

Endrizzi, J. E. (1974). Alternate-1 and alternate-2 disjunctions in heterozygous reciprocal translocations. *Genetics* **77**, 55–60.

Farmer, J. B., and Moore, J. E. S. (1905). On the meiotic phase (reduction division) in animals and plants. *Q. J. Microsc. Sci.* [N.S.] **48**, 489–557.

Garber, E. D. (1948). A reciprocal translocation in *Sorghum versicolor* Anderss. *Am. J. Bot.* **35**, 295–297.

Garber, E. D. (1972). "Cytogenetics: An Introduction." McGraw-Hill, New York.

Garber, E. D., and Dhillon, T. S. (1961). The genus *Collinsia.* XVIII. A cytogenetic study of radiation-induced reciprocal translocations in *C. heterophylla. Genetics* **47**, 461–467.

Grell, R. F. (1969). Meiotic and somatic pairing. *Genet. Organ.* **I,** 361–492.

Judd, B. H., Shen, M. W., and Kaufman, T. C. (1972). The anatomy and function of a segment of the X chromosome of *Drosophila melanogaster. Genetics* **71**, 139–156.

Khush, G. S. (1973). "Cytogenetics of Aneuploids." Academic Press, New York.

Lyon, M. F. (1971). Possible mechanisms of X-chromosome inactivation. *Nature (London), New Biol.* **232**, 229–232.

McClintock, B. (1938a). The production of homozygous deficient tissues with mutant characteristics by means of the aberrant behavior of ring-shaped chromosomes. *Genetics* **23**, 315–376.

McClintock, B. (1938b). The fusion of broken ends of sister half-chromatids following chromatid breakage at meiotic anaphases. *M., Agric. Exp. Stn., Res. Bull* **290**.

McClintock, B. (1941). The stability of broken ends of chromosomes in *Zea mays. Genetics* **26**, 234–282.

McClintock, B. (1944). The relation of homozygous deficiencies to mutations and allelic series in maize. *Genetics* **29**, 478–502.

Malheiros, N., and de Castro, D. (1947). Chromosome numbers and behavior in *Luzula purpurea* Link. *Nature (London)* **160**, 156.

Moens, P. B. (1965). The transmission of a heterochromatic isochromosome in *Lycopersicon esculentum. Can. J. Genet. Cytol.* **7**, 296–303.

Morgan, T. H. (1919). "The Physical Basis of Heredity." Lippincott, Philadelphia, Pennsylvania.

Ohno, S. (1967). "Sex Chromosomes and Sex-Linked Genes." Springer-Verlag, Berlin and New York.

Rhoades, M. M. (1955). The cytogenetics of maize. *Agronomy* **5**, 123–219.

Schwartz, D. (1953). The behavior of an X-ray induced ring chromosome in maize. *Am. Nat.* **87**, 19–28.

Sears, E. R. (1953). Nullisomic analysis in common wheat. *Am. Nat.* **87**, 245–252.

Sears, E. R. (1954). The aneuploids of common wheat. *M., Agric. Exp. Stn., Res. Bull.* **572**.

Sears, E. R., and Camara, C. A. (1952). A transmissible decentric chromosome. *Genetics.* **37**, 129–135.

Slizynska, H. (1938). Salivary chromosome analysis of the white-facet region in *Drosophila. Genetics* **23**, 291–299.

Srb, A. M., Owen, R. D., and Edgar, R. S. (1965). "General Genetics," 2nd ed. Freeman, San Francisco, California.

Stebbins, G. L. (1950). Variation and Evolution in Plants." Columbia Univ. Press, New York.

Strasburger, E. (1905). Typische und allotypische Kernteilung. Ergebnisse und Erörterungen. *Jahrb. Wiss. Bot.* **42**, 1–71.

Sturtevant, A. H. (1925). The effects of unequal crossing over at the Bar locus in *Drosophila. Genetics* **10**, 117–147.

van Beneden, E. (1883). Recherches sur la maturation de l'oeuf et la fecondation. *Arch. Biol.* **4**, 265–638.

Weismann, A. (1887). ``Über die Zahl des Richtungskörper und ihr Bedeutung für die Verer-
    bung.'' Jena.
Westmoreland, B., Szybalski, W., and Ris, H. (1969). Mapping of deletions and substitutions
    in heteroduplex DNA molecules of bacteriophage lambda by electron microscopy.
    *Science* **163,** 1343–1348.
White, M. J. D. (1973). ``Animal Cytology and Evolution.'' 3rd. ed. Cambridge Univ.
    Press, Cambridge.
Winge, O. (1917). The chromosomes: Their numbers and general importance. *C. R. Trav.
    Lab. Carlsberg* **13,** 131–276.
Zohary, D. (1955). Chiasmata in a pericentric inversion in *Zea mays. Genetics* **40,** 874–877.

# 7

# Cytoplasmic Inheritance

*Ruth Sager*

## I. INTRODUCTION

The first evidence for the existence of cytoplasmic genes as part of the permanent genomic constitution of eukaryotic organisms was reported in the very early days of genetics by Correns (1909), who described maternal inheritance of genetic factors influencing chloroplast development in the four o'clock flower *Mirabilis*, and by Baur (1909), who described biparental but non-Mendelian inheritance of similar traits in another flowering plant, *Pelargonium*. This evidence was based on two properties of cytoplasmic genes that are still useful in identifying cytoplasmic genes today: (1) deviations from Mendelian ratios, favoring transmission from one of the two parents, and (2) somatic segregation during vegetative growth of cytoplasmic alleles that originated in the two parents.

Despite a substantial number of well-documented instances of non-Mendelian inheritance, there was wide skepticism among geneticists concerning the reality of cytoplasmic genes until the discovery of high molecular weight, double-stranded cytoplasmic DNA's with unique nucleotide sequences located in chloroplasts (Sager and Ishida, 1963; Chun *et al.*, 1963) and in mitochondria (Luck and Reich, 1964). Following these publications, geneticists finally came to accept the validity, not only of the physical evidence, but also of the genetic evidence, which had previously been doubted. Cytochemical identification of chloroplast (Ris and Plaut, 1962) and mitochondrial (Nass and Nass, 1962) DNA's alone was not convincing to geneticists, since no evidence was presented that these DNA's were unique rather than amplified copies of nuclear DNA sequences.

This chapter will present a brief summary of current knowledge about chloroplast and mitochondrial DNA's, primarily from the viewpoint of genetic analysis: classes of mutants, methods of genetic analysis, correlations of physical and genetic evidence about these genomes, and finally some speculations about the significance of the organelle genomes in the overall genetic organization of the eukaryotic organism.

This subject matter was rather extensively reviewed in "Cytoplasmic Genes and Organelles" (Sager, 1972) and many aspects of the subject have since been updated in three symposium volumes (Birky *et al.*, 1975; Saccone and Kroon, 1976; Bücher *et al.*, 1976) and in several review articles (e.g., Schatz and Mason, 1974; Sager and Schlanger, 1976; Sager, 1977; Gillham, 1974). In this chapter the emphasis will be primarily upon the underlying ideas and upon key results, rather than upon specific documentation. The reader is referred to the books and reviews cited above for further details and references to individual research papers, as well as to other chapters in this volume.

Before turning to the principal subject matter of this chapter, i e , or ganelle genomes, it is appropriate to consider other possible mechanisms of nonMendelian heredity. Three kinds of phenomena will be considered: (1) other classes of cytoplasmic DNA's; (2) long-lived messenger RNA's; and (3) non-nucleic acid-based heredity.

First and most proximate to our present knowledge is the possibility that additional classes of cytoplasmic DNA's exist with specific and unique genetic functions. Suggestions of this sort have come from the electron microscope identification of small circular double-stranded DNA's in cell fractions from yeast and from mammalian cells, and from genetic evidence in yeast of non-Mendelian genes that are apparently also nonmitochondrial (Guérineau et al., 1974; Griffiths *et al.*, 1975; Aigle and Lacroute, 1975). These preliminary findings are quite exciting, since they raise the possibility of new and as yet unknown classes of genetic determinants. If further classes of DNA's do exist in eukaryotic cells, we need to find them (cf. Chapter 8). In this effort, physical evidence obviously needs to be coupled with evidence of mutationally altered phenotypes; and to identify such phenotypes, hypotheses need to be developed concerning the functions of these DNA's.

Second, a different phenomenon needs to be considered: namely, the effect of long-lived messenger RNA's or even proteins of maternal origin present in fertilized eggs during early development. Examples of such pseudomaternal inheritance, i.e., "maternal effects," need to be identified and distinguished from phenotypes attributable to actual cytoplasmic genes.

A third category of phenomena, which is very obscure in terms of mechanism, has never been rigorously eliminated or demonstrated: namely, non-nucleic acid-based heredity. Phenomena such as cortical inheritance in *Paramecium* and cell wall inheritance in *Chlamydomonas* have been presented as possible examples of hereditary determinants consisting of protein structure and architecture, rather than of nucleic acid-determined information. This point of view is presented by Dr. Beisson in Chapter 9. Here I wish merely to outline my view of this concept.

Two mechanisms are known which could account for phenomena such as cortical and cell wall inheritance. One is the self-assembly of macromolecules into complex aggregates, as studied in phage $T_4$ morphogenesis (cf. Casjens and King, 1975) and ribosome assembly (Nomura and Held, 1974). The other is the use of primers (e.g., in starch and in DNA synthesis) to add on to preexisting molecules in a structured way.

In both of these mechanisms, no new genetic information is contributed beyond that already present in the DNA-coded primary sequence of amino acids in the proteins. All the complex, subcellular structures arise from the architectonic properties of the building blocks themselves. Until the potentialities inherent in these known mechanisms have been tested, it seems unnecessary to view heredity beyond the framework of nucleic acid-based mechanisms with all their ramifications.

## II. ORGANELLE DNA'S

### A. Properties

Since 1963 (Sager and Ishida, 1963; Chun *et al.*, 1963) density gradient centrifugation in CsCl has been used routinely to identify and isolate organelle DNA's in most organisms in which the average base compositions of nuclear and organelle DNA's are sufficiently different to permit use of this easy and unambiguous method. The separation of nuclear from chloroplast DNA of *Chlamydomonas* by CsCl density gradient centrifugation is shown in Fig. 1. Buoyant density data for the DNA's of a number of organisms are given in Table I. It is noteworthy that the buoyant densities of chloroplast DNA's (chlDNA's) of all higher plants and green algae so far reported lie in the range of 1.695–1.697 gm/cm³ and mitochondrial DNA's (mitDNA's) of higher plants and most animals lie in the range of 1.701–1.707 gm/cm³. MitDNA's of lower eukaryotes, however, vary widely in average base composition and, consequently, in buoyant density in CsCl.

Organelle DNA's, with some exceptions among the lower eukaryotes, are circular molecules that exhibit remarkable uniformity in size (Table II). It is difficult to establish the length of a linear DNA molecule with assurance, since enzymes might cleave it or add on to it during extraction procedures. Covalently closed, circular double-stranded DNA molecules, however, are stable structures whose contour length can be measured with precision in electron microscope preparations. Thus, the native circularity of organelle DNA's has greatly simplified size determinations.

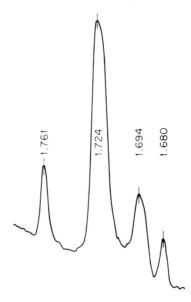

**Fig. 1.**    Principal DNA's of *Chlamydomonas*. Microdensitometer tracing of DNA's from gametes (mating type plus) centrifuged to equilibrium in CsCl density gradient. Bands seen are nuclear DNA at 1.724 gm/cm³ and chloroplast DNA at 1.694 gm/cm³, as computed from markers at 1.761 gm/cm³ (SP-15 phage DNA, from Dr. Marmur) and 1.680 gm/cm³ [crab poly(dAT), from Dr. Sueoka]. In gametes, chloroplast DNA is 7% of total DNA, based on calibration with known amount of SP-15 DNA. (From Sager, 1972.)

**TABLE I**

**Buoyant Densities in CsCl of DNA's of Selected Eukaryotic Microorganisms, Higher Plants and Animals**[a]

| | Nuclear | Chloro-plast | Mito-chondrial |
|---|---|---|---|
| Microorganisms | | | |
| Algae | | | |
| *Acetabularia* | 1.696 | 1.702–1.704 | 1.714–1.715 |
| *Chlamydomonas* | 1.724–1.725 | 1.693–1.695 | 1.706 |
| *Euglena* | 1.707 | 1.685 | 1.690–1.692 |
| *Polytoma* | 1.710 | 1.683 | 1.715 |
| *Porphyra* | 1.720 | 1.696 | unknown |
| Other | | | |
| *Neurospora* | 1.712–1.713 | — | 1.692–1.702 |
| *Paramecium* | 1.689 | — | 1.702 |
| *Tetrahymena* | 1.683–1.692 | — | 1.683–1.686 |
| Yeast | 1.698–1.699 | — | 1.683–1.684 |
| *Physarum* | 1.700 | — | 1.686 |
| Plants | | | |
| Broad bean | 1.695 | 1.697 | 1.705 |
| Mung bean | 1.695 | 1.697 | 1.706 |
| Lettuce | 1.694 | 1.697 | 1.706 |
| Spinach | 1.694–1.695 | 1.696–1.697 | 1.706 |
| Animals | | | |
| Sea urchin | 1.694 | — | 1.704 |
| *Drosophila* | 1.699–1.702 | — | 1.680–1.685 |
| Fish | 1.697–1.699 | — | 1.703–1.708 |
| Birds | 1.698–1.701 | — | 1.707–1.711 |
| Amphibia | 1.702 | — | 1.700–1.704 |
| Mammals (except man) | 1.698–1.704 | — | 1.698–1.704 |
| Human | 1.700 | — | 1.706–1.707 |

[a] From Borst and Kroon (1969), Nass (1976), and Sager and Schlanger (1976).

ChlDNA's of higher plants are about 40 $\mu$m in diameter, corresponding to a molecular weight of about $10^8$ daltons. Circular mitDNA's of animals are about one-tenth this size, measuring 4.5–5.5 $\mu$m in different species, a length that corresponds to about $10^7$ daltons.

In the lower eukaryotes, however, a surprising variety of sizes and base compositions of organelle DNA's are found, and the significance of this variability is not known. In *Paramecium*, *Tetrahymena*, and *Physarum*, for example, only linear mitDNA molecules have been found; their uniform size supports the view that they are linear *in vivo*, rather than circles that have been cleaved during extraction.

The mitDNA of the yeast *Saccharomyces cerevisiae* is five times the size found in animal cells. Circles have been reported, although they are dif-

**TABLE II**

**Physical and Genomic Sizes of Organelle DNA's**[a]

|  | Physical | Genomic | Circular |
|---|---|---|---|
| Chloroplast DNA |  |  |  |
| *Chlamydomonas* | $4 \times 10^9$ daltons[b] | $2 \times 10^8$ (major component)[b] | unknown |
| *Acetabularia* | unknown | $1.1–1.5 \times 10^9$ | unknown |
| *Euglena* | $9.2 \times 10^7$ | $1.8 \times 10^8$ | yes |
| Lettuce | $9.6 \times 10^7$ | $1.2 \times 10^8$ | yes |
| Pea | $8.9 \times 10^7$ | $9.5 \times 10^7$ | yes |
| Spinach | $\sim 9.5 \times 10^7$ | $\sim 1 \times 10^8$ | yes |
| Mitochondrial DNA |  |  |  |
| *Saccharomyces cerevisiae* | $\sim 5 \times 10^7$ | $5 \times 10^7$ | probably |
| *Kluyveromyces lactis*[c] | $\sim 2.2 \times 10^7$ | unknown | yes |
| *Tetrahymena* | $\sim 3 \times 10^7$ | $3 \times 10^7$ | no |
| *Neurospora* | $\sim 4 \times 10^7$ | unknown | yes |
| *Euglena*[d] | unknown | $4 \times 10^7$ | unknown |
| Sea urchin | $\sim 9 \times 10^6$ | unknown | yes |
| Amphibia | $\sim 1.1 \times 10^7$ | $\sim 1 \times 10^7$ | yes |
| Chick | $\sim 1.1 \times 10^7$ | $\sim 1 \times 10^7$ | yes |
| Mammals | $\sim 1 \times 10^7$ | $\sim 1 \times 10^7$ | yes |

[a] See Borst (1972), Sager and Schlanger (1976), and Nass (1976) for references and discussion.

[b] See Sager (1975) and Section II,A for discussion.

[c] Sanders *et al.* (1974).

[d] Talen *et al.* (1974).

ficult to recover because of their large size. At 25 $\mu$m, yeast mitDNA is among the largest mitDNA's known. Yeast mitDNA is also remarkable for its base composition, which is about 82% adenine + thymine. As shown in Table III, the expected correlations between base composition and buoyant density in CsCl (Schildkraut *et al.*, 1962) and between $T_m$ (temperature at which 50% of the hydrogen bonds are broken) and buoyant density (Marmur and Doty, 1962) do not hold for DNA's with such low guanine + cytosine content.

The genomic size of organelle DNA's is of particular interest, since we would like to know their capacity for genetic information. Reannealing kinetic studies (Britten and Kohne, 1968) of mitDNA's have thus far indicated that the analytical size and genomic size correspond well (Borst, 1972). In yeast, however, the very high A + T content has introduced the possibility that some simple sequence DNA may be present as poly(dAT)-

TABLE III

Guanine–Cytosine Content of Mitochondrial DNA's from
Two Strains of Yeast[a]

| Strain | Molar ratio G–C calculated from | | |
|---|---|---|---|
| | Chemical analysis | Buoyant density | $T_m$ |
| A | 17.4 | 23.5 | 13.2 |
| B | 16.8 | 23.5 | 10.7 |

[a] From Bernardi et al. (1970).

like material. Recent studies of fragments formed by degradation of yeast mitDNA with micrococcal nuclease, which preferentially attacks A + T-rich stretches (Prunell and Bernardi, 1974), and by restriction enzymes (Bernardi et al., 1975) have supported Bernardi's hypothesis that yeast mitDNA contains "spacer" regions of very high A + T (more than 96%). Such sequences probably do not carry genetic coding information but are interspersed with "genes," i.e., regions with a relatively high G + C content (averaging 32%). (Bücher et al., 1976).

The simple relationship between genomic size and physical size seen with mitDNA's is also found in the chlDNA's of higher plants. Here too, covalently closed circles seen by electron microscopy (Kolodner and Tewari, 1975) have a contour length corresponding to about $10^8$ daltons, a value that coincides well with the genomic size, determined by reannealing kinetic analysis for numerous higher plants and for *Euglena* (cf. Table II and Sager and Schlanger, 1976).

The situation appears to be more complicated in *Chlamydomonas* and in the macroscopic alga *Acetabularia*. It is not known whether the chlDNA's of *Chlamydomonas* and *Acetabularia* are circular, although preparations from both organisms have been examined by numerous investigators over the past few years. Linear pieces of broken chlDNA molecules of over 50 $\mu$m (corresponding to more than $2 \times 10^8$ daltons) have been described. In *Acetabularia*, Green et al. (1975) reported a genomic size determined by reannealing kinetic analysis of about $1 \times 10^9$ daltons, which could correspond to the physical size.

In *Chlamydomonas*, Wells and Sager (1970) found two components by reannealing kinetic analysis: a fast but minor component in the size range of $10^6$–$10^7$ daltons and a major component of about $1$–$2 \times 10^8$ daltons.

If these two components are the only ones present in chloroplast DNA, then the $10^8$ dalton component would be reiterated about 40 times, since the total chlDNA per cell in *Chlamydomonas* is about $4 \times 10^9$ daltons.

This result is in disagreement with the genetic data, to be presented below, which establishes that the chloroplast DNA of *Chlamydomonas* is genetically diploid, i.e., present in two copies per cell. The 40-fold reiteration is also in disagreement with the cytochemical findings of Ris and Plaut (1962) who described two Feulgen-positive, acridine-fluorescent bodies in the chloroplast of *Chlamydomonas*. For these reasons, we have postulated the presence of a third kinetic component in the chlDNA of *Chlamydomonas*, a slow annealing component not yet identified by reannealing kinetics because of technical difficulties. If this slow annealing single-copy component can be found, then all the data will be consistent with the description of chlDNA of *Chlamydomonas* as a multicomponent molecule with both single-copy and reiterated sequences, present in two copies per cell. No electron microscopic evidence of circularity or linearity has yet been reported, but pieces of chloroplast DNA larger than $10^8$ daltons have been seen (W. G. Burton, unpublished).

Considerable advances in the understanding of organelle DNA structure have been made in the past year (1976) with the availability of many well-characterized restriction enzymes that cut DNA molecules at specific sites. Physical maps of chloroplast and mitochondrial DNA's have been generated from the DNA fragments produced by specific restriction enzymes, and organelle RNA's have been mapped by hybridization to restriction fragments. In the near future it seems likely that many genes may be mapped with the use of potent antisera directed against specific proteins, used to fractionate the corresponding mRNA by precipitating nascent polypeptide chains on polysomes of organelle origin (cf., Bücher *et al.*, 1976).

## B. Functions of Organelle DNA's

The functions that have been clearly established for both chloroplast and mitochondrial DNA's include transcription of ribosomal and transfer RNA's that are part of the organelle protein synthesizing apparatus and of mRNA's that are translated to produce proteins coded by the respective organelle DNA's (cf. Schatz and Mason, 1974; Wildman *et al.*, 1975; Kroon and Saccone, 1974; Sager and Schlanger, 1976; Bücher *et al.*, 1976). In HeLa cell mitDNA (Attardi *et al.*, 1975) and by inference in all animal mitDNA's, about 20% of the $10^7$ dalton molecule is used for transcription of ribosomal and transfer RNA's. The remaining $8 \times 10^6$ daltons correspond to about 12,500 nucleotide pairs, which can then code for a maximum of 2080 amino acids, or 20 proteins with an average content of 100 amino acids each. ChlDNA's of higher plants, with a genomic size of about $10^8$, have the capacity to code for about 200 proteins, and in *Chlamydomonas* and *Acetabularia* the number may be even 10-fold higher.

Beyond numerology, however, is the intriguing problem of identifying these proteins. Fundamentally, there are three experimental approaches: (1) to obtain organelle mutants or protein isoenzymes from different strains and identify the gene product; (2) to use radioisotopic labeling of proteins synthesized in the presence of an inhibitor of cytosol protein synthesis (e.g., cycloheximide or emetine) to identify proteins synthesized in chloroplasts or mitochondria; and (3) to identify proteins of organelle origin directly by *in vitro* transcription and translation of the corresponding chlDNA or mitDNA, or restriction fragments thereof. The first method, can, in principle, lead to the identification of every organelle gene product, but in practice a combination of the three approaches is much more efficient. Within the next several years, it should be possible to identify most of the chloroplast and mitochondrial gene products.

What functions of organelle-coded proteins are known or have been suggested by indirect evidence? Four classes of proteins coded by organelle DNA's were proposed a few years ago (Sager, 1972): (1) proteins that function in electron transport, coupled phosphorylation, and photosynthesis; (2) other membrane-bound proteins unique to the inner mitochondrial or chloroplast lamellar membranes; (3) proteins of the organelle protein synthesizing apparatus, e.g., ribosomal proteins; and (4) regulatory proteins that function as transmitters of signals between organelle genomes and the nucleus (Blamire *et al.*, 1974).

In the search for proteins altered by mitochondrial mutations, yeast mutants have provided important clues. The cytoplasmic petites originally described by Ephrussi and his students were deficient in respiratory activity and in cytochromes of electron transport (reviewed in Ephrussi, 1953; Sager, 1972). However, these petite mutants, long studied both genetically and biochemically, turned out to contain large deletions, duplications, and gross rearrangements of mitDNA, leading to loss of many, sometimes all mitochondrial functions. Since yeast is a facultative anaerobe, this class of mutant could be grown, but could not be used as a source of point mutations to identify individual mitochondrial genes.

The first mitochondrial mutants of yeast which were *not* petites *were* carrying mutations to antibiotic resistance (Thomas and Wilkie, 1968; Linnane *et al.*, 1968). Subsequently many more mutant strains were recovered resistant to one or more of the antibiotics: chloramphenicol, erythromycin, spiramycin, mikamycin, and paromomycin. In bacteria (Nomura and Held, 1974) and in chloroplasts (Sager and Schlanger, 1976) resistance to these antibiotics results from mutationally-induced alterations in ribosomal proteins and, in a few instances, ribosomal RNA. On this basis, yeast geneticists expected that mitochondrial mutations to antibiotic resistance would also result from changes in mitochondrial ribosomal components. As yet, however, no altered mitochondrial ribosomal proteins or

RNA's have been reported, despite extensive research (e.g., Grivell *et al.*, 1973). Although the gene products have not been identified, the genes responsible for antibiotic resistance have been exceedingly useful in genetic analysis.

Mutations in the mitDNA of yeast that alter electron transport or oxidative phosphorylation have been sought especially with the use of drugs such as venturicidin, antimycin A, aurovertin, rhodamine 6G, and valinomycin (Griffiths, 1975). four mutant genes (see Section IV,B for genetic analysis), each conferring resistance to the oxidative phosphorylation inhibitor oligomycin, have been identified. These probably correspond to the four proteins of the oligomycin-sensitive ATPase complex (Tzagaloff *et al.*, 1973), whose biosynthesis is blocked by chloramphenicol (inhibitor of mitochondrial protein synthesis) but not by cycloheximide (inhibitor of cytosol protein synthesis). This complex also contains six proteins, presumably nuclear-coded, whose synthesis are blocked by cycloheximide but not by chloramphenicol. Analogous inhibitor studies have identified three out of seven components of the cytochrome oxidase complex as mitochondrial in origin, and mutant genes corresponding to these proteins have been located (Saccone and Kroon, 1976; Bucher *et al.*, 1976). Biochemical and genetic evidence also suggest that one or two subunits of the cytochrome $bc_1$ complex are mitochondrial in origin. (Saccone and Kroon, 1976).

It is not yet established whether every protein synthesized within an organelle is coded by the organelle DNA. Alternatively, mRNA molecules from the cytosol might be translated in the organelle. As yet there is no evidence for this process. As more and more organelle genes are identified, it becomes increasingly likely that all proteins synthesized in organelles are coded by organelle DNA's.

ChlDNA has a vastly larger coding potential than mitDNA, but few protein products of the chloroplast genome have yet been identified. One reason is that very few investigators have concentrated on selection of chloroplast mutants. In both systems, vastly more nuclear genes than organelle genes contribute to the total composition of the organelle, and one needs special methods to select for mutations in the organelle genome. In *Chlamydomonas* the antibiotic streptomycin acts as a specific mutagen for chlDNA (Sager, 1962), and we have used streptomycin mutagenesis to select mutations in chlDNA affecting chloroplast ribosomes (Sager, 1972; Schlanger and Sager, 1974; Section IV,A.). Unfortunately, this method has not been used by others for the selection of photosynthetic mutants (Levine and Goodenough, 1970), and consequently those that have been characterized are of nuclear origin.

Recently Wildman and his students have used electrophoretic differences in the fraction I protein RUdP carboxylase, a major chloroplast protein (Fig. 2), to follow their inheritance in the $F_1$ progeny of crosses between various varieties of tobacco. The carboxylase consists of two subunits: the larger subunit contains three major peptides, and the small subunit contains two (Kung *et al.*, 1974). Maternal inheritance of the genes coding for the large subunit proteins has been inferred from studies of the pattern of transmission of the proteins themselves in interspecies crosses of plants in the genus *Nicotiana* (Wildman *et al.*, 1973, 1975). Isozymes of the large subunit proteins show maternal inheritance and are presumably coded by chlDNA, while those of the small subunit, examined in the same material, show Mendelian inheritance, and are therefore presumably coded by nuclear DNA.

Evidence that at least two of the chloroplast ribosomal proteins are

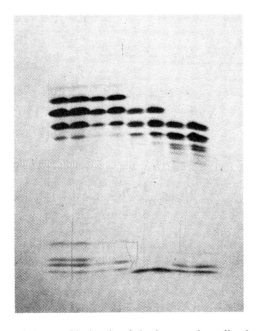

**Fig. 2.** Position of the peptide bands of the large and small subunits of Fraction I proteins isolated from four species of *Nicotiana* after electrofocusing in polyacrylamide gels containing 8 $M$ urea. Left to right: 1, 2, *N. excelsior*; 3, 4, *N. gossei*; 5, 6, *N. glauca*; 7, 8, *N. tabacum*. For each species, two concentrations of Fraction I protein were used, 20 $\mu$g to left, 30 $\mu$g to right. Top bands (pH 7 region), large subunit. The lowermost band of *N. tabacum* seems to split in two but not consistently. For details on electrofocusing technique and evidence that the minor bands arise by decomposition of urea and oxidation of thio groups, see Kung *et al.* (1974). Lower bands (pH 6 region), small subunit. (From Sakano *et al.*, 1974; reproduced with the permission of Dr. S. G. Wildman.)

coded by chlDNA comes from genetic and biochemical studies of drug-resistant mutants. Mutations that map in chlDNA (Sager and Ramanis, 1976a,b) in *Chlamydomonas* confer resistance to the antibiotics streptomycin, spectinomycin, neamine, carbomycin, and cleocin. The resistance is at the ribosome level as shown in *in vitro* studies of the protein synthesizing capacity of ribosomes from mutant and wild-type cells in the presence and absence of each of these drugs (Schlanger *et al.*, 1972; Schlanger and Sager, 1974). In the streptomycin-resistant mutant, we (Ohta *et al.*, 1975) were able to identify a single protein of the 30 S subunit as altered or missing in a chromatographic study of proteins extracted from chloroplast ribosomes of a streptomycin-resistant strain, compared with those from wild-type cells (Fig. 3). In a companion study of ribosomal proteins from a spectinomycin-resistant strain, no difference in mutant and wild-type profiles was found. However, Burton (1972) demonstrated the loss of spectinomycin binding capacity by chloroplast ribosomes from the mutant strain, suggesting a protein alteration responsible for resistance to the drug.

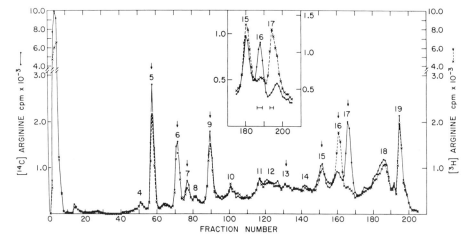

**Fig. 3.** Chromatography on a CM-cellulose column of 30 S ribosomal subunit proteins from *Chlamydomonas*. The mixture of 30 S subunits from [$^{14}$C] arginine-labeled wild-type and [$^{3}$H]arginine-labeled quadruple mutant was combined with 10–15 mg of nonlabeled 70 S ribosomes, digested by ribonucleases, and layered onto a column (5 mm × 30 cm) of CM-cellulose. The proteins were eluted at a flow rate of 5–6 ml/hour by a gradient of 0.05 to 0.5 M sodium acetate containing 6 M urea at pH 5.6 in a total volume of 400 ml. An aliquot of 0.5 ml from each fraction was assayed for radioactivity (cpm). •: $^{14}$C-labeled 21 gm (wild-type) protein; ×: $^{3}$H-labeled 11,925-1 (quadruple mutant) protein. In the insert the labeling was reversed. ×: $^{3}$H-labeled 21 gm (wild-type) protein; •: $^{14}$C-labeled 11,925-1 (quadruple mutant) protein. (From Ohta *et al.* 1975.)

Mets and Bogorad (1971, 1972) described several erythromycin-resistant mutants in *Chlamydomonas*, one coming from a mutation of cytoplasmic origin (presumably chloroplast) and the others of nuclear origin. All of them have been shown to alter one or more chloroplast proteins, indicating that both nuclear and chloroplast genes code for chloroplast ribosomal proteins. Which of the ribosomal proteins are coded by each of the genomes remains for future studies to elucidate.

At an even more fundamental level lies the tantalizing and unanswered question of why these organelles possess a unique protein synthesizing machinery that is distinctive from that of the cytosol of the same organism. An important consequence of this distinctiveness is a differential response to inhibitors. In general, organelle ribosomes are sensitive to the same classes of inhibitors as bacterial ribosomes, including aminoglycosides (e.g., streptomycin), macrolides (e.g., erythromycin), and chloramphenicol. Organellar ribosomes are resistant to inhibitors of cytosol ribosomes, such as cycloheximide. These differences in susceptibility to drugs have been used to good advantage experimentally to block protein synthesis in one cell compartment but not the other, but as yet no one has established what use the cell makes of this property. Are there intracellular signals that turn off protein synthesis in organelles while permitting it to continue in the cytosol and vice versa?

The remarkable generality that emerges from the evidence summarized in this section is that all organelle-coded and organelle-synthesized proteins so far identified are parts of functional complexes which include both nuclear-coded and organelle-coded components. This is true not only of the membrane bound cytochromes and $F_1$–ATPase but also of the soluble RUdP-carboxylase, which comprises about 50% of the total chloroplast protein in photosynthesizing cells.

## III. GENERAL FEATURES OF CYTOPLASMIC GENETIC SYSTEMS

### A. Maternal Inheritance

Classically, cytoplasmic genes have been identified by their non-Mendelian pattern of inheritance. In lower eukaryotes, e.g., in *Chlamydomonas*, yeast, and *Neurospora*, in which complete tetrads are recovered from zygotes after meiosis, nuclear genes show 2:2 haploid Mendelian inheritance, while cytoplasmic genes typically show 4:0 inheritance, in which the mutant gene is transmitted to all or none of the progeny (Fig. 4). This pattern is formally similar to the classical maternal inheritance of non-Mendelian genes in higher plants first described by Correns (1909).

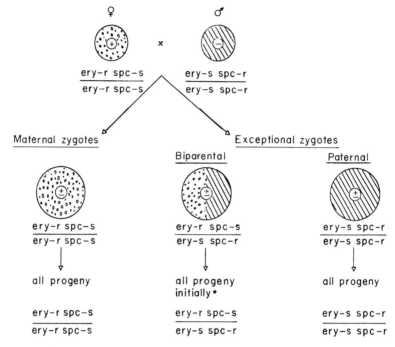

**Fig. 4.** Classes of zygotes based on transmission of chloroplast genes. Female parent (cytoplasm dotted) carries chloroplast markers ery–r and spc–s. Male parent (cytoplasm striped) carries ery–s and spc–r. Nucleus labeled by mating type alleles *mt*⁺ (+) and *mt*⁻ (−). *Recombination occurs later, during multiplication of progeny clones.

In animal systems maternal inheritance of mitDNA has been demonstrated at the molecular level. In *Xenopus*, Dawid *et al.* (1974) showed maternal transmission of mitDNA by taking advantage of the difference in buoyant density between mitDNA's of two interbreeding species. In the F₁ progeny of reciprocal crosses, the animals contained only the mitDNA inherited from the female parent. Hutchison *et al.* (1974) utilized another molecular method. They took advantage of the different cleavage patterns seen by gel electrophoresis after digestion with specific restriction endonucleases of mitDNA's from horse and from donkey. They examined mitDNA's from horse, from donkey and from the F₁ progeny resulting from reciprocal crosses: mule (from horse × donkey) and hinney (from donkey × horse). The result was unambiguous; the mitDNA's of the progeny were of maternal origin in both crosses.

On purely logical grounds, it is surprising that organelle DNA's, usually present in many copies per cell, exhibit maternal (i.e., uniparental) inheritance, instead of some form of biparental inheritance, as one might have predicted *a priori*. In higher plants both maternal and biparental patterns of transmission of nonMendelian genes influencing chloroplast development have been described, although maternal inheritance is by far the commoner pattern. In order to "explain" maternal inheritance, most geneticists have considered it the consequence of a physical barrier to male transmission of cytoplasm by sperm or pollen. Rare instances of biparental transmission have been considered to result from occasional successes of male factors in penetrating the barrier. On these assumptions, the otherwise surprising simplicity of maternal inheritance was not considered remarkable; and maternal inheritance was taken as THE hallmark of nonnuclear or cytoplasmic inheritance.

Within the last few years, however, these assumptions have been questioned (Sager, 1972, 1975; Tilney-Bassett, 1975; Laughnan and Gabay, 1975) and the mechanism of maternal inheritance has become a very interesting subject of research at the molecular level. The new evidence is principally (1) that biparental inheritance of cytoplasmic genes is more frequent than previously thought; and (2) that physical exclusion of cytoplasmic DNA's from zygotes does not occur, e.g., in *Chlamydomonas*. Recent studies of the mechanism of organelle inheritance in *Chlamydomonas* (reviewed in Sager, 1975, 1977) and in higher plants (reviewed in Tilney-Bassett, 1975) have provided new information and new perspectives on the problem that will be summarized here.

In *Chlamydomonas*, mating type is determined by a pair of nuclear alleles, $mt^+$ and $mt^-$. Chloroplast genes are transmitted from the $mt^+$ parent to all progeny in regular zygotes and the homologous genes from the $mt^-$ parent are not transmitted. We call this pattern *maternal* inheritance by analogy with higher plants and animals, and refer to the $mt^+$ strain as the *female* (Fig. 4).

This maternal pattern can be altered by UV irradiation (at a dose that permits survival of all zygotes) of the $mt^+$ parent immediately before mating (Sager and Ramanis, 1967) to give biparental zygotes. At higher UV doses, replication of chloroplast DNA from the $mt^+$ irradiated gamete is inhibited and paternal inheritance of chloroplast genes is the result.

In addition to UV irradiation, a nuclear gene mutation, *mat-1*, has been described (Sager and Ramanis, 1974) that has a similar effect, giving rise to about 50% biparental zygotes.

We postulate that the loss of chloroplast genes from the $mt^-$ parent results from destruction of the chlDNA of $mt^-$ origin after zygote formation (Sager, 1972). The direct demonstration of disappearance of chlDNA originating from the $mt^-$ parent in the presence of surviving chlDNA from the $mt^+$ parent was achieved by two methods, differential prelabeling of the parental DNA's with $^{15}$N and with $^{14}$N (Sager and Lane, 1972), and with $^3$H- and $^{14}$C-adenine (Schlanger and Sager; 1974; Schlanger et al., 1977). A typical preparative gradient is shown in Fig. 5A. Following UV irradiation (Fig. 5B), the pattern of elimination is reversed, with some or all of the $mt^+$ chlDNA being lost while that from the $mt^-$ is retained. Similar results were found with zygotes from crosses involving the mat-1 gene (Fig. 5C).

In addition to the loss of chlDNA of $mt^-$ origin in the zygote, we found that the chlDNA of $mt^+$ origin undergoes a density shift, becoming about 0.005 gm/cm$^3$ lighter in buoyant density in CsCl gradients than the corresponding chlDNA of gametes and vegetative cells.

On the basis of these results, we have postulated a molecular mechanism to account for maternal inheritance in Chlamydomonas (Sager, 1972, 1975; Sager and Kitchin, 1975). On this hypothesis, the chlDNA of male origin is attacked by a restriction endonuclease present only in the zygote, while the chlDNA of female origin is protected from destruction by a modification enzyme that adds protective side chains (perhaps methyl groups), resulting in the shift in buoyant density that occurs immediately after zygote formation. Both UV irradiation and the mat-1 mutation interfere with the action (probably block formation) of both enzymes. This hypothesis has led to the discovery and characterization of the first known eukaryotic restriction endonuclease (Burton et al., 1976, 1977).

Modification and restriction of chlDNA, analogous to modification and restriction of bacterial DNA (Boyer, 1974), might account for maternal inheritance in Chlamydomonas, but is there evidence that a similar mechanism is also present in other organisms?

Most of the evidence concerning the control of plastid transmission in Oenothera and in Pelargonium has been summarized recently (Sager, 1972, 1975; Tilney-Bassett, 1975). The non-Mendelian pattern of inheritance of the plastid abnormalities has been clearly established in both genera. However, they both show a high frequency of transmission of non-Mendelian, i.e., chloroplast genes, from the male as well as female parent. In addition, nuclear genes have been shown to influence transmission ratios of maternal and paternal chloroplast genes, analogous to the influence of mat-1 in Chlamydomonas.

In these higher plants, the basic method of studying plastid transmission is by making crosses between green (G) and white (W) parents (Fig. 6). The white parent comes from white sectors containing germ cells present on sectored green and white plants. In calculating the transmission ratios of green and white plastids, not only are the numbers of pure green and pure white seedlings determined, but also the relative amounts of green and white tissue are evaluated on variegated (i.e., striped) seedlings.

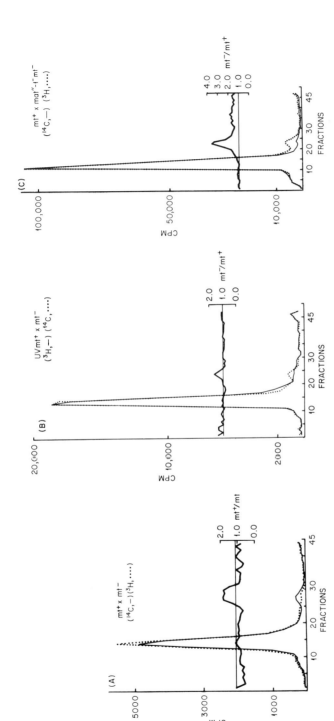

**Fig. 5.** (A) Preparative CsCl gradient of DNA extracted from zygotes 6 hours after mating. $Mt^+$ cells were pregrown with $^{14}$C-adenine and $mt^-$ cells with $^3$H-adenine to prelabel DNA. (B) Same procedure as in (A) except radioisotope labels were reversed and $mt^+$ gametes were irradiated with UV before mating. (C) Same as in (A) except $mt^-$ parental strain carried $mat-l$ mutation which affects zygotes similarly to UV, giving biparental and paternal inheritance of chloroplast genome (from Schlanger *et al.*, 1977).

**Fig. 6.** Patterns of chloroplast inheritance in higher plants. (a) *Maternal*: All progeny resemble the female parent; no transmission of chloroplast genes from the male parent. (b) *Biparental*: Both female and male chloroplast genes are transmitted; ratios are non-Mendelian, variable, influenced by genotype and environment (from Sager, 1972).

In *Oenothera*, the progeny from G♀ × W♂ crosses are green and variegated, but no pure white seedlings are recovered, and in the reciprocal crosses, W♀ × G♂, no pure green seedlings are found. Thus, maternal chloroplasts are always transmitted, as well as a variable number of plastids from the male. The relative rates of plastid multiplication were evaluated by Stubbe (1964) on a series of different nuclear genetic backgrounds;

he found that five distinct plastome (i.e., plastid genome) types could be identified by their characteristic relative percent recovery as green or white tissue. More recently, Schotz (1970) made crosses between female plants containing the fastest multiplying plastomes and males with the slowest ones, and found that the progeny showed maternal inheritance of multiplication rate. In crosses between females with the slowest plastomes and males with the fastest, however, biparental transmission of the multiplication rate occurred, indicating that male transmission was occurring here, but could not fully overcome the maternal advantage seen in the reciprocal cross. These results are important in demonstrating a certain amount of plastid genome autonomy in the control of plastid (presumably chloroplast DNA) replication rates.

In *Pelargonium*, there is much less maternal effect per se. Green, white, and variegated seedlings are recovered from reciprocal crosses between isogenic parents differing only in their plastid constitution. However, there are strong parental genetic effects upon $G:W$ ratios, but they are not sex specific as in *Oenothera*. In crosses between particular strains, the plastids of male origin may predominate, regardless of the polarity of the cross, and in other crosses, plastids of female origin predominate. Apparently, genes present in both parents influence the plastid DNA transmission ratio. Tilney-Bassett (1975) has studied one gene in particular called *Pr* that has a strong enough effect to be identified and followed in crosses. One allele, $Pr_1$, determines a mainly maternal pattern of chloroplast inheritance called Type I. When the male parent is homozygous $Pr_1Pr_1$, then all progeny show Type I inheritance. Another allele, $Pr_2$, seems to be lethal when homozygous, but heterzygotes ($Pr_1Pr_2$), when used as male parents, give rise to progeny, half of which show Type I inheritance and the other half show the Type II pattern, in which about equal numbers of green and white seedlings are produced.

Thus, *Pelargonium* shows clear-cut inheritance of characteristic and different plastid transmission ratios regulated by single nuclear genes. This system has not yet been examined experimentally at the molecular level, but the formal similarity to chloroplast gene inheritance in *Chlamydomonas* is striking. The $G:V:W$ (i.e., green:variegated:white) plastid ratios in *Pelargonium* are analogous to the maternal:biparental:paternal ratios in *Chlamydomonas*. The excess of green plastids, like the excess of maternal zygotes, reflects maternal transmission, and the excess of white plastids, like paternal zygotes, reflects transmission of cytoplasmic genes from the male. We have proposed (Sager, 1975; Sager and Kitchin, 1975) that both phenomena may reflect the same underlying molecular mechanism, namely, the modification and restriction of chloroplast DNA's from the two parents, regulated by particular nuclear genes. The similar patterns seen in *Oenothera* and in other plants may be regulated in the same way.

In summary, this section has discussed the identification of organelle genes by their non-Mendelian pattern of inheritance, and has stressed that this pattern may be maternal, biparental, or even paternal, depending upon molecular mechanisms that regulate the pattern. Since organelle genomes are usually present in multiple copies per cell, specific mechanisms are required to distribute their copies at meiosis. Biparental inheritance would be the rule, unless a particular mechanism is present to eliminate one parental genome. The mechanism may be as simple as physical exclusion, but the complicated and precise patterns of transmission seen in *Pelargonium*, *Oenothera*, and *Chlamydomonas* suggest other mechanisms of control, perhaps by differential enzymatic mechanisms, i.e., modification and restriction of DNA.

Thus far, we have only considered the patterns of cytoplasmic inheritance as determined in the sexual cycle, i.e., at fertilization in higher plants and animals and in zygote formation in *Chlamydomonas*. However, a second aspect of cytoplasmic inheritance is the segregation and recombination that occurs in somatic cells during vegetative growth. Under conditions of strict maternal inheritance, the progeny are of course homozygous, and therefore neither recombination nor somatic segregation can be detected. In biparental inheritance, however, with progeny that are initially heterozygous, it is possible to study segregation of single pairs of alleles and recombination with two or more pairs of alleles (cf. Section IV).

## B. Stability and Mutability of Organelle Genes

One powerful reason why the existence of cytoplasmic genes was doubted for so long by most geneticists was the rarity with which cytoplasmically inherited mutants were found. Even Correns, who tried vigorously to establish the generality of non-Mendelian inheritance (Correns, 1937), found few examples, in contrast to the large numbers of Mendelian mutations that were found in the formative years of genetics, 1919–1940. Even now cytoplasmic mutants of higher plants are relatively rare, leading investigators of organelle biogenesis to study nuclear genes that influence organelle development, rather than organelle genes themselves (e.g., von Wettstein *et al.*, 1971).

In view of the small size of mitDNA, we can now understand why mitochondrial mutations are rare. Studies of forward mutation and reversion frequencies of the few mitochondrial mutations known in yeast and in animal cells have indicated that mutations occur with frequencies similar to those of nuclear genes, but that careful regimes of selection are required to establish and maintain mutant strains, owing presumably to the large number of copies of mitDNA per cell.

The high frequency of spontaneous mutations to cytoplasmic *petites* in yeast (cells that have lost the capacity to respire, but can grow anaerobically in the presence of $O_2$) has led to the view of mitDNA as highly mutable. In fact, most petites result from large deletions in mitDNA (Locker *et al.*, 1974) in contrast to the usual "point" mutations, which result from base substitutions or frame shifts in other DNA's.

In the chloroplast genome of *Chlamydomonas*, extensive studies of mutation to streptomycin resistance (Sager, 1962; Gillham and Levine, 1962) have shown that this mutation occurs rarely if ever spontaneously, but can be induced by growth in the presence of toxic but nonlethal concentrations of streptomycin. No reversions have ever been reported or seen in our laboratory. Similarly, chloroplast mutations conferring resistance to other drugs have been induced by growth in the presence of streptomycin as a mutagen (Sager, 1962) followed by selection with the appropriate drug, e.g., spectinomycin, erythromycin, neamine, cleocin, and others (Sager, 1972, 1977). Mutations conferring temperature sensitivity and acetate requirement have shown similar inducibility by streptomycin and similar absence of revertants.

The paucity of organelle mutants and the difficulty in identifying gene products of organelle DNA's may result from the essential nature of these gene products, such that mutations are lethal. If so, an intensive search for temperature-sensitive and other classes of conditional lethals might prove fruitful, as it was in phage T4 and lambda genetics.


## IV. REVIEW OF SPECIFIC CYTOPLASMIC GENETIC SYSTEMS

### A. Chloroplast Genome of *Chlamydomonas*

The phenotypic classes of mutants identified as nonMendelian and found to be located in a single linkage group in chlDNA include (1) acetate-requirement, as a result of loss of ability to grow photosynthetically; (2) resistance to antibiotics that inhibit chloroplast protein synthesis; and (3) temperature sensitivity (growth at 25° but not at 35°C). In addition, 4:0 inheritance of other traits such as tiny colony formation and weak cell-wall formation has been observed.

Genetic studies have been carried out with biparental zygotes (Fig. 4) recovered after UV irradiation of the $mt^+$ gametes (see Section III,A). The

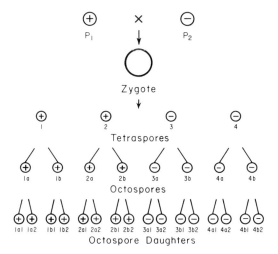

**Fig. 7.** Procedure for pedigree analysis. After germination, zoospores are allowed to undergo one mitotic doubling and then the eight cells (octospores) are transferred to a fresh Petri plate and respread. After one further doubling, each pair of octospore daughters is separated and allowed to form colonies. The sixteen colonies, derived from the first two doublings of each zoospore are then classified for all segregating markers. (From Sager and Ramanis, 1970.)

four products of meiosis (zoospores) can be recognized by the segregation of three pairs of unlinked nuclear markers. Progeny have been recovered after two mitotic doublings of zoospores, under conditions in which pairs of daughter cells can be identified, so that we recover a pedigree of four cells from each zoospore, or 16 progeny per zygote (Fig. 7). These pedigrees have proven essential for genetic analysis because recombination of chloroplast genes occurs very rapidly beginning at the first doubling after meiosis. Contrary to the behavior of nuclear genes, little if any recombination of chloroplast genes occurs during meiosis in our strains. In distantly related strains of the same species studied by Gillham (1969, 1974), frequent recombination of chloroplast genes does occur during meiosis, as well as in mitosis.

By means of pedigree analysis, segregation and recombination data have been acquired from a large number of multi-factor crosses, each involving four to seven cytoplasmic markers as well as three pairs of nuclear markers. The principal findings from these crosses follow (cf. Sager and Ramanis, 1970, 1976a,b; Singer et al., 1976, for details).

1. Zoospores from biparental zygotes are usually heterozygous for all the cytoplasmic markers involved in the cross. This evidence shows that segregation is rare in zygotes.

2. Segregation of cytoplasmic genes begins at the first mitotic doubling after meiosis and results in the appearance of homozygous clones. On the average, each heterozygous pair segregates 1 : 1 in each clone, demonstrating that the zoospore initially contained equal numbers of copies from the two parents.

3. For each chloroplast gene pair, two different patterns of segregation occur, reciprocal and nonreciprocal. The nonreciprocal events, resembling gene conversion, produce one daughter cell that is a pure parental type, and one that is still heterozygous. This pattern is called Type II, and the reciprocal pattern, producing two daughter cells, each carrying one of the parental alleles, is called Type III. Our conception of how these patterns may arise is shown in Fig. 8. On the basis of this model, all segregation events are the result of recombination, i.e., strand exchanges located between the segregated gene and the attachment point $(ap)$.

4. Different chloroplast gene pairs in the same DNA molecule (i.e., from the same parent) may undergo Type II and Type III segregation events in the same round of replication.

5. Strand changes seen as segregation or recombination occur at each cell division beginning with the first mitotic doubling after meiosis, and can be shown to continue as long as any heterozygous markers remain to detect the process (Sager and Ramanis, 1968; Singer *et al.*, 1976). They occur with a constant probability per doubling.

6. Zoospores from biparental zygotes are haploid for the nuclear genome. They are diploid for the chloroplast genome, as shown by (1) the occurrence of segregation at a constant rate per doubling; and (2) detailed analysis of the patterns of reciprocal and nonreciprocal segregation (Sager and Ramanis, 1976a,b).

Two different procedures have been principally used to map the chloroplast genome: (1) gene cosegregation frequencies; and (2) frequencies of Type III segregation events. The frequencies with which two or more genes cosegregate at the same doubling, and therefore presumably in the same recombinational event, provide unambiguous mapping data, so long as the distances between the genes relate to the cosegregation distances well enough to generate sufficient frequency data. In our system, cosegregation frequencies have proven to be invaluable in establishing a unique gene order all along the map. An example of relative cosegregation frequencies and their use in generating a linear map is given in Fig. 9. These data incorporate both Type II and Type III frequencies, but the same results are obtained with Type II frequencies alone. Type III cosegregations are too infrequent to be useful alone.

However, Type III segregation frequencies provide a different kind of information for mapping, namely, the distances between genes and a

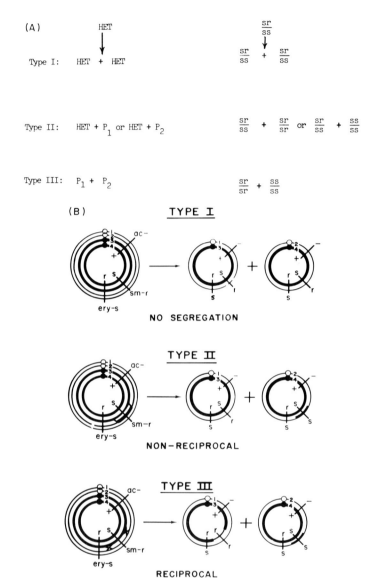

**Fig. 8.** (A) Segregation patterns seen with pairs of alleles. In this example, *sr* and *ss* are alleles of the sm2 gene. A typical zoospore from a cross of *sr* × *ss* is shown as a heterozygote (HET), sr/ss. At the first mitotic doubling, three alternative patterns of segregation are seen: Type I (no segregation); Type II (nonreciprocal recombination); and Type III (reciprocal recombination). (B) Segregation patterns in circular molecules. Each line represents a double-stranded DNA. The homologs from the two parents are distinguished by thick and thin lines, with their respective black and white attachment points. Exchanges occur after

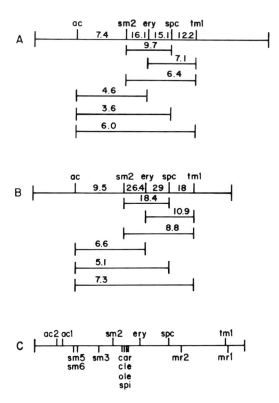

**Fig. 9.**  Maps of chloroplast genome based upon relative cosegregation frequencies. (A) Linear arrangement of five key markers from frequencies of Type II cosegregation. (B) Linear arrangement of five key markers from frequencies of Type II, Type III and zygote cosegregation. (C) Relative positions of 15 genes, based upon the data of Sager and Ramanis, 1970; 1976b. (From Sager and Ramanis, 1976b.)

centromere-like attachment point that regulates the distribution of chlDNA molecules to daughter cells at cell division. The rationale for this method is shown in Fig. 10. If reciprocal recombination events occur with equal frequency throughout the genome, the closer a gene is to the attachment point (*ap*), the less probable is an exchange between the gene and *ap*. In pedigree analysis, exchanges are seen as the recovery of the two parental alleles at cell division of a heterozygote.

---

replication but before cell division at the four-strand stage. Attachment points regulate distribution of sister strands to different daughter cells as in mitosis. If no exchanges occur, both daughters are heterozygous (Type I). Nonreciprocal (Type II) exchange is the result of a miscopying (conversion) event. Reciprocal (Type III) exchange requires two events, one on either side of the recombined region. (from Sager and Ramanis, 1976a)

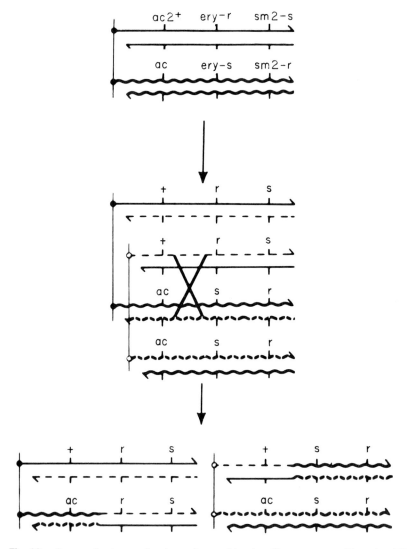

**Fig. 10.** Proposed scheme of reciprocal recombination. Heterozygous chloroplast DNA shown before replication, after replication, and after cell division. An exchange is shown, occurring at the four-strand stage (after replication) between the genes ac2 and ery. Daughter cells are heterozygous for ac2 and homozygous for genes ery and sm2 located beyond the exchange point. No molecular mechanism is implied in the figure. Solid lines: one parent; wavy lines: other parent; dotted lines: new replicated strands; •: old attachment points (*ap*); o: new attachment points. (From Sager and Ramanis, 1971.)

Statistically significant data have been generated by a simple method in which zygotes from multiply marked crosses are germinated synchronously, and the zoospores grown in liquid culture. Samples are taken at hourly intervals after germination, plated and scored for the frequencies of segregated pure types and of remaining heterozygotes for each marker. These data are plotted as heterozygote survival curves in which the rate of disappearance of heterozygotes is expressed as the slope of the survival curve (Fig. 11). Each locus has been found, in many different crosses, to give a constant characteristic slope, which is a function of the frequency of exchanges between it and *ap*. This method does not distinguish between one- and two-armed (one on each side of *ap*) arrangements, and the results must therefore be considered together with results of a different method, i.e., cosegregation frequencies in order to establish gene order along the map.

When the map based on gene–*ap* distances shown in Fig. 11 is compared with the map shown in Fig. 9, it is evident that they are related but different. Correlating the two maps has shown that the chloroplast map is genetically circular, as shown in Fig. 12.

In summary, pedigree analysis of the chloroplast genome has provided

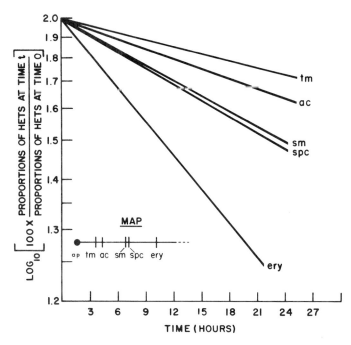

**Fig. 11.**    Segregation rates of principal genes. (From Singer *et al.*, 1976.)

RELATIVE MAP POSITIONS OF 15 CHLOROPLAST GENES

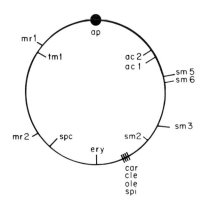

**Fig. 12.**   Relative map positions of 15 chloroplast genes. Markers shown inside of circle have been examined in 15 or more crosses; markers on outside of circle positioned from data of one or two crosses. (From Singer *et al.*, 1976.)

remarkable amounts of information about chloroplast DNA, some of it quite unexpected. In the zoospores and their vegetative progeny, chloroplast DNA molecules of biparental origin come together, pair and recombine at each cell doubling. The segregation and recombination patterns of linked markers show that these exchange events occur at a four-strand stage, formally identical with mitotic crossing-over, as described by Stern in *Drosophila* (1936) and by Pontecorvo in *Aspergillus* (1958). The chlDNA's are distributed in a regular manner to daughter cells, as shown by retention of all markers and by the ability to map distances unambiguously between each gene and a postulated attachment point.

### B. Mitochondrial Genome of Yeast

The cytoplasmic petite mutation in yeast was first described by Ephrussi *et al.* in a series of papers first published in 1949, and summarized by Ephrussi (1953). However, until 1968, genetic analysis of mitochondria remained at an impasse. All of the cytoplasmic petite mutations recovered conferred the same phenotype, loss of respiratory activity, and showed no complementation by the methods then available. Beginning in 1968, however, mutations conferring resistance to a series of antibiotics were selected (Thomas and Wilkie, 1968; Linnane *et al.*, 1968). The resulting mutants provided the means to examine single gene mutations in mitDNA with respect both to genetic analysis and to function. The genetic analysis proved to be more complex than initially anticipated, and only now is the overall pattern beginning to be understood. A short summary of the present status will be presented here (cf. Dujon *et al.*, 1974; Linnane *et al.*, 1976; Slonimski and Tzagaloff, 1976).

Two classes of mutants have been studied. One class is resistant to drugs, including chloramphenicol, erythromycin, spiramycin, and paromomycin, that inhibit mitochondrial but not cytoplasmic protein synthesis. The second class involves resistance to drugs, oligomycin and mikamycin, that block respiratory function.

For genetic analysis, haploid yeast strains of complementary mating types ($a$ and $\alpha$), carrying selective nuclear markers, are mated and the resulting diploid zygotes are allowed to form colonies. Samples of the progeny, i.e., cells from these zygote colonies, are classified for the mitochondrial alleles that are segregating in the cross. In most published data, samples from many zygote colonies have been pooled before analysis. In some studies cell samples from individual zygote colonies have been scored, and in a very few reports, pedigree studies have been performed by pulling off the first and subsequent buds from individual zygotes and scoring the colonies that grew from each.

The results of all methods are concurrent in showing certain major regularities:

1. Depending on the parental strains and on the mitochondrial markers in the cross, the progeny may show an excess of one parental type. The strain whose markers are in excess is called $w^+$ and the minority strain is called $w^-$. Most yeast strains can be classified as either $w^+$ or $w^-$, and this difference seems to be a property of the mitDNA.

2. Crosses of $w^+ \times w^-$ (called heterosexual or heteropolar) give ratios far from 1 : 1 (may be even $10^3 : 1$) for *some* genes but not for others. The genes showing deviations from 1 : 1 (i.e., polarity) confer resistance to chloramphenicol, erythromycin, and spiramycin and have been mapped in one region of the mitDNA map (see below). The other genes so far studied that confer resistance to oligomycin, mikamycin, and paromomycin do not show polarity in either heteropolar ($w^+ \times w^-$) or in homopolar ($w^+ \times w^+$; $w^- \times w^-$) crosses.

3. The relative deviation of alleles from 1 : 1 in the progeny provides a means for ordering the polar genes relative to one another. Thus far, there is agreement concerning the relative positions of three polar genes, called RIB1, RIB2, and RIB3. Alleles conferring chloramphenicol resistance are located at *RIB1*, and alleles conferring resistance to erythromycin and to spiramycin are located at both *RIB2* and at *RIB3*.

4. Recombination frequencies have not provided data for mapping, since all combination frequencies (except between RIB1, RIB2, and RIB3) have given 20–25% recombination.

5. Part of the difficulty in mapping results from the fact that early buds from the *ends* of the zygote usually contain only parental types, whereas recombination is high in early buds coming from the central or fusion region. For this reason pedigrees of buds from the central region should be especially informative for mapping but too few pedigrees of this sort have yet been described.

Recently a deletion map has been published by Molloy *et al.* (1975), using petite mutants in which the deletion was identified by loss of particular marker genes. Gene order was established by measuring the *coretention* of pairs of markers in strains converted to petites, starting with strains carrying four or more drug resistance markers. Several linear orders could be derived from the data, but they were circular permutations of each other, providing evidence that the mitochondrial genome is circular. Circularity was also demonstrated by Clark-Walker and Miklos (1974) using complementation among petites. These and other methods that combine physical and genetic methods are being used for mapping purposes, and it seems probable that a rough genetic map will soon be generally accepted.

What genetic questions remain unsolved in the mitochondrial system? Certainly the problem of prime importance is to saturate the map, i.e., to identify all the genes coding for proteins in mitDNA. As yet only eleven genes have been identified, and of these it is not clear whether the loci *RIB1, RIB2, RIB3* and *par* (paromomycin-resistance) code for proteins or represent regions in ribosomal RNA cistrons. More specifically with respect to recombination analysis, one would like to know how many copies of zygote DNA enter each bud in yeast, and whether recombination is restricted to the zygote or occurs also in buds. One would also like to know the molecular details of recombination, whether it is reciprocal or nonreciprocal, whether it occurs at a four-strand stage as in *Chlamydomonas*, and whether the same molecules undergo multiple rounds of recombination, as postulated by Dujon *et al.* (1974).

Mitochondrial mutants are now being recovered in organisms other than yeast, such as *Paramecium* (Beisson *et al.*, 1974) and mammalian cells in culture (Wallace *et al.*, 1975), and these systems may provide different mutants and insights from those of the yeast system. In particular, the interactions between mitochondrial and nuclear genomes may be quite different in obligate aerobic organisms than in a facultative anaerobe such as *Saccharomyces*.

## C. Mitochondrial Genomes of Other Organisms

The successful use of drugs that inhibit mitochondrial protein synthesis or respiration to select for mutations conferring drug resistance in yeast has led to application of this methodology to other organisms. In

*Paramecium*, mutants resistant to chloramphenicol and erythromycin have been described. In some strains the mutation to drug resistance is coupled with increased temperature sensitivity. As yet, recombination has not been found. In a cross of chloramphenicol-resistant × erythromycin-resistant paramecia, double resistance is seen initially, but as the animals multiply, the two kinds of resistance segregate independently and no permanently double resistant clones have been obtained. However, the mutant strains have provided good material for studies of selection, and differential multiplication rates of genetically different mitochondrial DNA's monitored by the marker mutations they carry. Evidence has been presented of different growth rates of genetically different mitochondrial DNA's within the same organism (Beisson *et al.*, 1974).

Cytoplasmically inherited mutations conferring cold sensitivity (Waldron and Roberts, 1973) and oligomycin resistance (Rowlands and Turner, 1973) have been described in *Aspergillus*, and the first evidence of recombination between them has been reported (Rowlands and Turner, 1974a,b). Two chloramphenicol-resistant strains have been isolated by Gunatilleke *et al.* (1975). When heterokaryons were formed between resistant and sensitive strains, the cytoplasmic markers segregated rapidly to give sectors or pure parental type heterokaryons, as a consequence of somatic segregation. The chloramphenicol markers recombined readily with the oligomycin resistance and cold sensitivity conferring genes, providing the investigators with three markers for further genetic analysis.

A cytoplasmic gene conferring chloramphenicol resistance has also been described in mammalian cell cultures, both in mouse lines and in the human HeLa cell line (Wallace *et al.*, 1975). Identification of the mutation as cytoplasmic was made possible by cell fusion between a nucleated drug-sensitive cell and an enucleated drug-resistant cell. [Enucleation by treating cells with cytochalasin B provides a source of enucleated cells or cytoplasts, which can be fused with nucleated cells to give viable cell products, called cybrids (Ringertz and Savage)]. The mitochondrial location of the mutation has not been firmly established as yet, but the nonnuclear location is unequivocally established by the enucleation procedure.

A beginning has been made in the study of nuclear–mitochondrial genomic interactions, using somatic cell hybrids between different mammalian species. When human cells are fused with rodent cells, either mouse, rat, or hamster, chromosomes of the human cells are preferentially lost. In human × mouse cell hybrids, the mitochondrial DNA's of the two species can be distinguished by buoyant density differences in CsCl gradients (mouse mitDNA = 1.692; human mitDNA = 1.700). Human mitochondrial DNA is lost from these hybrids, coincident with chromosomal elimination (Clayton *et al.*, 1971; Attardi and Attardi, 1972). However, conditions have been found in which the polarity of chromosome elimination has been reversed, with mouse chromosomes lost and human chromosomes retained in some cell hybrids, and different proportions of human and mouse nuclear DNA's retained in other hybrid cell lines (Dawid *et al.*, 1974). When the mitochondrial DNA's were compared with the nuclear DNA's in a series of such hybrid clones, it was found that in general mitochondrial retention paralleled nuclear. Evidence of mitochondrial recombination, based on recovery of molecules with intermediate densities in CsCl, was also obtained with this material.

## V. CONCLUDING REMARKS

Much has been learned about organelle genetic systems in the past 10 years, and much more remains to be discovered. We know from genetic analysis that chloroplasts and mitochondria contain unique DNA molecules that are not copies of nuclear DNA's. The organelle DNA's are transcribed as unique ribosomal, transfer, and messenger RNA's of the organelles; the mRNA's are translated into proteins. All of the proteins of organelle origin so far identified, including the chloroplast RUdP carboxylase, mitochondrial $F_1$–ATPase complex, cytochrome oxidase and cytochrome *c* reductase, as well as the organelle ribosomes themselves, are parts of complex assemblies that contain proteins coded by nuclear as well as organelle DNA's.

We know further that the organelle DNA's have been carefully preserved during evolution to maintain a relatively constant molecular ratio of organelle to nuclear DNA's. Complex mechanisms, possibly enzymatic, have evolved that block biparental inheritance, thus maintaining the homogeneity of organelle genomes. Genetic homozygosity is further facilitated by high frequencies of gene conversion, seen in *Chlamydomonas* as Type II segregation (Section IV,A) and postulated in yeast (Birky, 1975). Rare instances of biparental transmission could provide occasional opportunities for recombination in species with strict maternal inheritance. Recombination and gene conversion both require an apparatus for pairing of organelle DNA molecules, and genetic studies with *Chlamydomonas* have demonstrated the presence of this apparatus in chloroplasts and have established some of its properties. The occurrence of recombination of mitochondrial genes suggests that pairing of mitDNA's also occurs, but the details of this mechanism remain to be elucidated.

The availability of this background of knowledge should make possible a concerted attack upon some of the unsolved problems. Why do mitochondria and chloroplasts of eukaryotes require unique genomes and protein synthesizing apparatuses? A part of the answer probably lies in the hydrophobic nature of the proteins synthesized in the organelles and the importance of building them into the membrane as they are formed. However, hydrophobic proteins are incorporated into other cellular membranes as well, presumably without the presence of the coding DNA in the immediate vicinity of the membranes. Another consideration is the key role of organelle membranes in ATP production. Although the replication of organelle DNA's is tied to the cell cycle, the growth of chloroplast lamellar membranes and mitochondrial inner membranes is apparently regulated in relation to cellular metabolism, especially the energy requirements of the cell. The independence of organelle protein synthesis from that occurring in the cytosol may reflect one cellular device by which the rate of formation of organelle membranes is regulated.

Thus, one avenue of communication between the organelles and the rest of the cell may lie in the regulation of organelle protein synthesis. Another avenue may lie in signals of organelle origin feeding back from the organelle to the nucleus to regulate synthesis of particular proteins, or even to modulate the cell cycle itself. Evidence supporting this possibility comes from studies with *Chlamydomonas*, showing that inhibition of organelle protein synthesis blocks nuclear DNA replication (Blamire *et al.* 1974). In mammalian cells, which maintain a healthy metabolism in the nondividing state, signals of mitochondrial origin may play a key role in maintaining the cells in a resting state (Sager and Chang, 1975).

Another general problem of great interest is the manner in which mutations occurring in organelle DNA's become established in the cell. In most organisms, mitochondrial DNA's (and in plants, chloroplast DNA's as well) are present in many copies per cell, a condition which should lead to virtual immutability (at the cellular level)—that is, mutations may arise with conventional frequencies, but would have essentially no chance to become established in the absence of very strong selection. The availability of selectable mutations, such as drug resistance, provides material for model studies to follow the development of newly induced mutations, and to study intracellular interactions between normal and mutated DNA molecules. One question that needs to be resolved is whether selection can occur at the intracellular level, and if so, what is the detailed mechanism by which it occurs?

In an evolutionary sense, one would expect little variability in the organelle genome from species to species, in view of the restrictions on recombination and probably on the establishment of mutations. Nonetheless, the first studies of polymorphism of chloroplast proteins have shown the presence of small differences in proteins with the same function, e.g., RUdP carboxylase, both in the polypeptides coded by nuclear genes and those coded by the chloroplast DNA. Similar polymorphisms are suggested by recent studies of mitochondrial proteins (Jeffreys and Craig, 1976).

## ACKNOWLEDGMENT

The work of this laboratory was supported by grants from the U.S. Public Health Service and the American Cancer Society.

## REFERENCES

Aigle, M., and Lacroute, F. (1975). Genetical aspects of (URE3), a nonmitochondrial, cytoplasmically inherited mutation in yeast. *Mol. Gen. Genet.* **136,** 327–335.

Attardi, B., and Attardi, G. (1972). Fate of mitochondrial DNA in human–mouse somatic cell hybrids. *Proc. Natl. Acad. Sci. U.S.A.* **69,** 129–133.

Attardi, G., Costantino, P., England, J., Lynch, D., Murphy, W., Ojala, D., Posakony, J., and Storrie, B. (1975). The biogenesis in mitochondria in HeLa cells: A molecular and cellular study. *In* "Genetics and Biogenesis of Mitochondria and Chloroplasts" (C. W. Birky, Jr., P. S. Perlman, and T. J. Byers, eds.), pp. 3–65. Ohio State Univ. Press, Columbus.

Baur, E. (1909). Das Wesen und die Erblichkeitsverhältnisse der "Varietates albomarginatae hort" von *Pelargonium* zonale. *Z. Vererbungsl.* **1,** 330–351.

Beisson, J., Sainsard, A., Adoutte, A., Beale, G. H., Knowles, J., and Tait, A. (1974). Genetic control of mitochondria in *Paramecium. Genetics* **78,** 403–413.

Bernardi, G., Faures, M., Piperno, G., and Slonimski, P. P. (1970). Mitochondrial DNA's from respiratory-sufficient and cytoplasmic respiratory-deficient mutant yeast. *J. Mol. Biol.* **48,** 23–42.

Bernardi, G., Prunell, A., and Kopecka, H. (1975). An analysis of the mitochondrial genome of yeast with restriction enzymes. *In* "Molecular Biology of Nucleocytoplasmic Relationships" (S. Puiseux-Dao, ed.), pp. 85–90. Elsevier, Amsterdam.

Birky, C. W., Jr. (1975). Mitochondrial genetics of fungi and ciliates. *In* "Genetics and Biogenesis of Mitochondria and Chloroplasts" (C. W. Birky, Jr., P. S. Perlman, and T. J. Byers, eds.), pp. 182–224. Ohio State Univ. Press, Columbus.

Birky, C. W., Jr., Perlman, P. S., and Byers, T. J., eds. (1975). "Genetics and Biogenesis of Mitochondria and Chloroplasts." Ohio State Univ. Press, Columbus.

Blamire, J., Flechtner, V. R., and Sager, R. (1974). Regulation of nuclear DNA replication by the chloroplast in *Chlamydomonas. Proc. Natl. Acad. Sci. U.S.A.* **71,** 2867–2871.

Borst, P. (1972). Mitochondrial nucleic acids. *Annu. Rev. Biochem.* **41,** 333–376.

Borst, P., and Kroon, A. M. (1969). Mitochondrial DNA: Physicochemical properties, replication and genetic function. *Int. Rev. Cytol.* **26,** 107–190.

Boyer, H. W. (1974). Restriction and modification of DNA: Enzymes and substrates. *Fed. Proc., Fed. Am. Soc. Exp. Biol.* **33,** 1125–1127.

Britten, F. J., and Kohne, D. E. (1968). Repeated sequences in DNA. *Science* **161,** 529–540.

Bücher, Th., Neupert, W., Sebald, W., and Werner, S., eds. (1976). "Genetics and Biogenesis of Chloroplasts and Mitochondria," Interdisciplinary Conference on The Genetics and Biogenesis of Chloroplasts and Mitochondria, Munich, Germany, August 2–7, 1976. North-Holland, Amsterdam.

Burton, W. G. (1972). Dehydrospectinomycin binding to chloroplast ribosomes from antibiotic-sensitive and–resistant strains of *Chlamydomonas reinhardtii*. *Biochim. Biophys. Acta* **272**, 305–311.

Burton, W. G., Roberts, R. J., Myers, P. A., and Sager, R. (1976). A eukaryotic endonuclease with nonrandom cleavage specificity. *Fed. Proc., Fed. Am. Soc. Exp. Biol.* **35**, 1172.

Burton, W. G., Roberts, R. J., Myers, P. A., and Sager, R. (1977). A site-specific single strand endonuclease from the eukaryote *Chlamydomonas*. *Proc. Natl. Acad. Sci. U.S.A.* (in press).

Casjens, S., and King, J. (1975). Virus assembly. *Annu. Rev. Biochem.* **44**, 555–611.

Chun, E. H. L., Vaughan, M. H., and Rich, A. (1963). The isolation and characterization of DNA associated with chloroplast preparations. *J. Mol. Biol.* **7**, 130–141.

Clark-Walker, G. D., and Miklos, G. L. G. (1974). Mitochondrial genetics, circular DNA and the mechanism of the *petite* mutation in yeast. *Genet. Res.* **24**, 43–57.

Clayton, D. A., Teplitz, R. L., Nabholz, M., Dovey, H., and Bodmer, W. (1971). Mitochondrial DNA of human–mouse cell hybrids. *Nature (London)* **234**, 560–562.

Correns, C. (1909). Zur Kenntnis der Rolle von Kern and Plasma bei der Vererbung. *Z. Vererbungsl.* **2**, 331–340.

Correns, C. (1937). "Nicht Mendelende Vererbung" (F. von Wettstein, ed.). Borntraeger, Berlin.

Dawid, I., Horak, I., and Coon, H. G. (1974). The use of hybrid somatic cells as an approach to mitochondrial genetics in animals. *Genetics* **78**, 459–471.

Dujon, B., Slonimski, P. P., and Weill, L. (1974). Mitochondrial genetics. IX. A model for recombination and segregation of mitochondrial genomes in *Saccharomyces cerevisiae*. *Genetics* **78**, 415–437.

Ephrussi, B. (1953). "Nucleo-cytoplasmic Relations in Micro-organisms." Oxford Univ. Press (Clarendon), London and New York.

Gillham, N. W. (1969). Uniparental inheritance in *Chlamydomonas reinhardi*. *Am. Nat.* **103**, 355–388.

Gillham, N. W. (1974). Genetic analysis of the chloroplast and mitochondrial genomes. *Annu. Rev. Genet.* **8**, 347–391.

Gillham, N. W., and Levine, R. P. (1962). Studies on the origin of streptomycin resistant mutants in *Chlamydomonas reinhardi*. *Genetics* **47**, 1463–1474.

Green, B. R., Padmanabhan, U., and Muir, B. L. (1975). The kinetic complexity of *Acetabularia* chloroplast DNA. *In* "Proceedings of the 12th Botanical Congress," Vol. 2, p. 403. USSR Academy of Science, Leningrad.

Griffiths, D. E. (1975). Utilization of mutations in the analysis of yeast mitochondrial oxidative phosphorylation. *In* "Genetics and Biogenesis of Mitochondria and Chloroplasts" (C. W. Birky, Jr., P. S. Perlman, and T. J. Byers, eds.), pp. 117–135. Ohio State Univ. Press, Columbus.

Griffiths, D. E., Lancashire, W. E., and Zanders, E. D. (1975). Evidence for an extrachromosomal element involved in mitochondrial function: A mitochondrial episome? *FEBS Lett.* **53**, 126–130.

Grivell, L. A., Netter, P., Borst, P., and Slonimski, P. P. (1973). Mitochondrial antibiotic resistance in yeast: Ribosomal mutants resistant to chloramphenicol, erythromycin and spiramycin. *Biochim. Biophys. Acta* **312**, 358–367.

Guérineau, M., Slonimski, P. P., and Aver, R. P. (1974). Yeast episome: Oligomycin resistance associated with a small covalently closed nonmitochondrial circular DNA. *Biochem. Biophys. Res. Commun.* **61**, 462–469.

Gunatilleke, I. A. U. N., Scazzocchio, C., and Arst, H. N., Jr. (1975). Cytoplasmic and nuclear mutations to chloramphenicol resistance in *Aspergillus nidulans*. *Mol. Gen. Genet.* **137**, 269–276.

Hutchison, C. A., III, Newbold, J. E., Potter, S. S., and Edgell, M. H. (1974). Maternal inheritance of mammalian mitochondrial DNA. *Nature (London)* **251**, 536–538.

Jeffreys, A. J., and Craig, J. W. (1976). Interspecific variation in products of animal mitochondrial protein synthesis. *Nature (London)* **259**, 690–692.

Kolodner, R., and Tewari, K. K. (1975). The molecular size and conformation of chloroplast DNA of higher plants. *Biochim. Biophys. Acta* **402**, 372–390.

Kroon, A. M., and Saccone, C., eds. (1974). "The Biogenesis of Mitochondria." Academic Press, New York.

Kung, S. D., Sakano, K., and Wildman, S. G. (1974). Multiple peptide composition of the large and small subunits of *Nicotiana tabacum* Fraction 1 protein ascertained by fingerprinting and electrofocusing. *Biochim. Biophys. Acta* **365**, 138.

Laughnan, J. R., and Gabay, S. J. (1975). An episomal basis for instability of S male sterility in maize and some implications for plant breeding. *In* "Genetics and Biogenesis of Mitochondria and Chloroplasts" (C. W. Birky, Jr., P. S. Perlman, and T. J. Byers, eds.), pp. 330–349. Ohio State Univ. Press, Columbus.

Levine, R. P., and Goodenough, U. W. (1970). The genetics of photosynthesis and of the chloroplast in *Chlamydomonas reinhardi*. *Annu. Rev. Genet.* **4**, 397–408.

Linnane, A. W., Saunders, G. W., Gingold, E. B., and Lukins, H. B. (1968). The biogenesis of mitochondria. V. Cytoplasmic inheritance of erythromycin resistance in *Saccharomyces cerevisiae*. *Proc. Natl. Acad. Sci. U.S.A.* **59**, 903–910.

Locker, J., Rabinowitz, M., and Getz, G. S. (1974). Electron microscopic and renaturation kinetic analysis of mitochondrial DNA of cytoplasmic petite mutants of *Saccharomyces cerevisiae*. *J. Mol. Biol.* **88**, 489–507.

Luck, D. J. L., and Reich, E. (1964). DNA in mitochondria of *Neurospora crassa*. *Proc. Natl. Acad. Sci. U.S.A.* **52**, 931.

Marmur, J., and Doty, P. (1962). Determination of the base composition of deoxyribonucleic acid from its thermal denaturation temperature. *J. Mol. Biol.* **5**, 109–118.

Mets, L. J., and Bogorad, L. (1971). Mendelian and uniparental alteration in erythromycin binding by plastid ribosomes. *Science* **174**, 707–709.

Mets, L., and Bogorad, L. (1972). Altered chloroplast ribosomal proteins associated with erythromycin-resistant mutants in two genetic systems of *Chlamydomonas reinhardi*. *Proc. Natl. Acad. Sci. U.S.A.* **69**, 3779–3783.

Molloy, P. L., Linnane, A. W., and Lukins, H. B. (1975). Biogenesis of mitochondria: Analysis of deletion of mitochondrial antibiotic resistance markers in petite mutants of *Saccharomyces cerevisiae*. *J. Bacteriol.* **122**, 7–18.

Nass, M. M. K. (1976). Mitochondrial DNA. *Handb. Genet.* **5**, 497–533.

Nass, M. M. K., and Nass, S. (1962). Fibrous structures within the matrix of developing chick embryo mitochondria. *Exp. Cell Res.* **26**, 424–437.

Nomura, M., and Held, W. A. (1974). Reconstitution of ribosomes: Studies of ribosome structure, function and assembly. *In* "Ribosomes" (M. Nomura, A. Tissieres, and P. Lengyel, eds.), pp. 193–223. Cold Spring Harbor Lab., Cold Spring Harbor, New York.

Ohta, N., Inouye, M., and Sager, R. (1975). Identification of a chloroplast ribosomal protein altered by a chloroplast mutation in *Chlamydomonas*. *J. Biol. Chem.* **250**, 3655–3659.

Pontecorvo, G. (1958). "Trends in Genetic Analysis." Columbia Univ. Press, New York.

Prunell, A., and Bernardi, G. (1974). The mitochondrial genome of wild-type yeast cells. IV. Genes and spacers. *J. Mol. Biol.* **86**, 825–841.

Ris, H., and Plaut, W. (1962). Ultrastructure of DNA-containing areas in the chloroplast of *Chlamydomonas. J. Cell Biol.* **13**, 383–391.

Rowlands, R. T., and Turner, G. (1973). Nuclear and extranuclear inheritance of oligomycin resistance in *Aspergillus nidulans. Mol. Gen. Genet.* **126**, 201–216.

Ringertz, N. R., and Savage, R. E. (1976). "Cell Hybrids." Academic Press, New York.

Rowlands, R. T., and Turner, G. (1974a). Physiological and biochemical studies of nuclear and extranuclear oligomycin-resistant mutants of *Aspergillus nidulans. Mol. Gen. Genet.* **132**, 73–88.

Rowlands, R. T., and Turner, G. (1974b). Recombination between the extranuclear genes conferring oligomycin resistance and cold sensitivity in *Aspergillus nidulans. Mol. Gen. Genet.* **133**, 151–161.

Saccone, C. and Kroon, A. M., eds. (1976). "The Genetic Function of Mitochondrial DNA," Proceedings of the 10th International Bari Conference on the Genetic Function of Mitochondrial DNA, Riva dei Tessali, Italy, May 25–29, 1976. North-Holland, Amsterdam.

Sager, R. (1962). Streptomycin as a mutagen for nonchromosomal genes. *Proc. Natl. Acad. Sci. U.S.A.* **48**, 2018–2026.

Sager, R. (1972). "Cytoplasmic Genes and Organelles." Academic Press, New York.

Sager, R. (1975). Patterns of inheritance of organelle genomes: Molecular basis and evolutionary significance. *In* "Genetics and Biogenesis of Mitochondria and Chloroplasts" (C. W. Birky, Jr., P. S. Perlman, and T. J. Byers, eds.), pp. 252–267. Ohio State Univ. Press, Columbus.

Sager, R. (1977). Genetic analysis of chloroplast DNA, in *Chlamydomonas.* Adv. Genet. **19**, 287–340.

Sager, R., and Chang, F. (1975). Effects of mitochondrial protein synthesis inhibitors on nuclear thymidine incorporation by CHO cells. *Fed. Proc., Fed. Am. Soc. Exp. Biol.* **34**, 621a.

Sager, R., and Ishida, M. R. (1963). Chloroplast DNA in *Chlamydomonas. Proc. Natl. Acad. Sci. U.S.A.* **50**, 725–2730.

Sager, R., and Kitchin, R. (1975). Selective silencing of eukaryotic DNA. *Science* **189**, 426–433.

Sager, R., and Lane, D. (1972). Molecular basis of maternal inheritance. *Proc. Natl. Acad. Sci. U.S.A.* **69**, 2410–2414.

Sager, R., and Ramanis, Z. (1967). Biparental inheritance of nonchromosomal genes induced by ultraviolet irradiation. *Proc. Natl. Acad. Sci. U.S.A.* **58**, 931–937.

Sager, R., and Ramanis, Z. (1968). The pattern of segregation of cytoplasmic genes in *Chlamydomonas. Proc. Natl. Acad. Sci. U.S.A.* **61**, 324–331.

Sager, R., and Ramanis, Z. (1970). A genetic map of nonMendelian genes in *Chlamydomonas. Proc. Natl. Acad. Sci. U.S.A.* **65**, 593–600.

Sager, R., and Ramanis, Z. (1971). Methods of genetic analysis of chloroplast DNA in *Chlamydomonas. In* "Autonomy and Biogenesis of Mitochondria and Chloroplasts" (N. K. Boardman, A. W. Linnane, and R. M. Smillie, eds.), pp. 250–259. North-Holland Publ., Amsterdam.

Sager, R., and Ramanis, Z. (1974). Mutations that alter the transmission of chloroplast genes in *Chlamydomonas. Proc. Natl. Acad. Sci. U.S.A.* **71**, 4698–4702.

Sager, R., and Ramanis, Z. (1976a). Chloroplast genetics of *Chlamydomonas.* I. Allelic segregation ratios. *Genetics* **83**, 303–321.

Sager, R., and Ramanis, Z. (1976b). Chloroplast genetics of *Chlamydomonas.* II. Mapping by consegregation frequency analysis. *Genetics* **93**, 323–340.

Sager, R., and Schlanger, G. (1976). Chloroplast DNA: physical and genetic studies. *Handb. Genet.*, **5**, 371–423.

Sakano, K., Kung, S. D., and Wildman, S. G. (1974). Identification of several chloroplast DNA genes which code for the large subunit of *Nicotiana* fraction I proteins. *Mol. Gen. Genet.* **130**, 91–97.

Sanders, J. P. M., Weijers, P. J., Groot, G. S. P., and Borst, P. (1974). Properties of mitochondrial DNA from *Kluyveromyces lactis*. *Biochim. Biophys. Acta* **374**, 136–144.

Schatz, G., and Mason, T. L. (1974). The biosynthesis of mitochondrial proteins. *Annu. Rev. Biochem.* **43**, 51–87.

Schildkraut, C. L., Marmur, J., and Doty, P. (1962). Determination of the base composition of deoxyribonucleic acid from its buoyant density in CsCl. *J. Mol. Biol.* **4**, 430–433.

Schlanger, G., and Sager, R. (1974). Localization of five antibiotic resistances at the subunit level in chloroplast ribosomes of *Chlamydomonas*. *Proc. Natl. Acad. Sci. U.S.A.* **71**, 1715–1719.

Schlanger, G., Sager, R., and Ramanis, Z. (1972). Mutation of a cytoplasmic gene in *Chlamydomonas* alters chloroplast ribosome function. *Proc. Natl. Acad. Sci. U.S.A.* **69**, 3551–3555.

Schlanger, G., Lane, D., Blamire, J., and Sager, R. (1977). Regulation of the inheritance of chloroplast DNA in zygotes of *Chlamydomonas*. (In preparation.)

Schlanger, Gladys and Sager, R. (1974). Correlation of chloroplast DNA and cytoplasmic inheritance in *Chlamydomonas* zygotes. *J. Cell Biol.* **63**. 301A.

Schotz, F. (1970). Effects of the disharmony between genome and plastome on the differentiation of the thylakoid system in *Oenothera*. *Symp. Soc. Exp. Biol.* **24**, 39–54.

Singer, B., Sager, R., and Ramanis, Z. (1976). Chloroplast genetics of *Chlamydomonas*. III. Closing the circle. *Genetics* **83**, 341–354.

Slonimski, P., and Tzagoloff, A. (1976). Localization in yeast mitochondrial DNA of mutations expressed in a deficiency of cytochrome oxidase and/or coenzyme QH₂–cytochrome *c* reductase. *Eur. J. Biochem.* **61**, 27–41.

Stern, C. (1936). Somatic crossing over and segregation in *Drosophila melanogaster*. *Genetics* **21**, 625–730.

Stubbe, W. (1964). The role of the plastome in evolution of the genus *Oenothera*. *Genetica* **35**, 28–33.

Talen, J. L., Sanders, J. P. M., and Flavell, R. A. (1974). Genetic complexity of mitochondrial DNA from *Euglena gracilis*. *Biochim. Biophys. Acta* **374**, 129–135.

Thomas, D. Y., and Wilkie, D. (1968). Recombination of mitochondrial drug resistance factors in *Saccharomyces cerevisiae*. *Biochem. Biophys. Res. Commun.* **30**, 368–372.

Tilney-Bassett, R. A. E. (1975). Genetics of variegated plants. *In* "Genetics and Biogenesis of Mitochondria and Chloroplasts" (C. W. Birky, Jr., P. S. Perlman, and T. J. Byers, eds.), pp. 268–308. Ohio State Univ. Press, Columbus.

Tzagoloff, A., Rubin, M. S., and Sierra, M. F. (1973). Biosynthesis of mitochondrial enzymes. *Biochim. Biophys. Acta* **301**, 71–104.

von Wettstein, D., Henningsen, K. W., Boynton, J. E., Kannangara, G. C., and Nielsen, O. F. (1971). The genic control of chloroplast development in barley. *In* "Autonomy and Biogenesis of Mitochondria and Chloroplasts" (N. K. Boardman, A. W. Linnane, and R. M. Smillie, eds.), pp. 205–223. North-Holland, Amsterdam.

Waldron, C., and Roberts, C. F. (1973). Cytoplasmic inheritance of a cold–sensitive mutant in *Aspergillus nidulans*. *J. Gen. Microbiol.* **78**, 379–381.

Wallace, D. C., Bunn, C. L., and Eisenstadt, J. M. (1975). Cytoplasmic transfer of chloramphenicol resistance in human tissue culture cells. *J. Cell Biol.* **67**, 174–188.

Wells, R., and Sager, R. (1970). Denaturation and the renaturation kinetics of chloroplast DNA from *Chlamydomonas reinhardi*. *J. Mol. Biol.* **58**, 611–622.

Wildman, S. G., Lu-Liao, C., and Wong-Staal, F. (1973). Maternal inheritance, cytology and macromolecular composition of defective chloroplasts in a variegated mutant of *Nicotiana tabacum. Planta* **113,** 293–312.

Wildman, S. G., Chen, K., Gray, J. C., Kung, S. D., Kwanyuen, P., and Sakano, K. (1975). Evolution of ferredoxin and fraction I protein in the genus *Nicotiana. In* "Genetics and Biogenesis of Mitochondria and Chloroplasts" (W. C. Birky, Jr., P. S. Perlman, and T. J. Byers, eds.), pp. 309–329. Ohio State Univ. Press, Columbus.

# 8

# Inheritance of Infectious Elements

*Louise B. Preer and John R. Preer, Jr.*

## I. INTRODUCTION

Infectious heredity is due to elements that originally entered their present hosts by infection and now are transmitted by heredity. Although not generally regarded as integral parts of the cell, most lack the capacity for independent existence. Normally they are not harmful, and often they play major roles in the physiology and evolution of their hosts. In a few cases they have even become indispensable. A limited capacity for infectious transfer through the medium is retained by some, but hereditary transfer at cell division or at cell fusion is the favored means of spread. Some are simply free nucleic acid molecules such as the bacterial plasmids. Others are more complex, and include viruses, bacteria and even algae. Infectious elements of heredity occur with surprisingly high frequency. In this chapter space permits only a brief consideration of some of the better known cases.

Infectious elements of heredity have a handicap. Except for some bacterial plasmids and viruses, most are not bound to chromosomal or other specific cellular sites. Therefore, if they do not contain or are not linked to essential factors they often can be eliminated easily and without harmful consequences to the cell. Furthermore, most must rely for their transmission on the chance distribution of multiple copies at gametogenesis and cell division. High rate of loss is a well-known property of infectious elements of heredity.

It is less well known that freedom from the chromosome confers a major advantage on infectious elements of heredity. For a gene to increase in frequency in a population, it must either be favored by natural selection or depend upon inefficient processes such as mutation pressure or genetic drift; genes are always in competition with their alleles that occupy the same site. But an element of heredity that does not occupy a chromosomal site can often increase without the aid of natural selection, provided it has an efficient means of transmission. It may have to compete with other similar elements, but intracellular selection is quite different from natural selection on the whole organism. (For example, it is possible for one infectious element to replace another by intracellular selection even though it may reduce the fitness of the whole organism.) L'Héritier (1970) demonstrated mathematically in *Drosophila* in connection with the sigma virus, that an agent which is transmitted strictly maternally (always through the egg, never through the sperm) and which is occasionally lost and never replaced by infection from the medium, must be favored by natural selection (because of the advantage to the host) to be maintained. He also was able to prove the remarkable fact that it is possible for an element that is transmitted by both parents, even if only occasionally by males, to be maintained and spread without the aid of natural selection.

The results of L'Héritier can be verified and extended by considering the outcome, in the absence of selection, of random matings between given frequencies of infected and uninfected individuals and noting whether the frequencies of infected individuals rises in the next generation. It must also be recognized that, for this purpose, protoplasmic fusions of all types should be regarded as matings. Simple calculations made in this way show that numerous processes allow the establishment of elements without the aid of natural selection. Examples include not only biparental transmission in higher organisms, but also cytoplasmic transfer at conjugation in protozoa (W. G. Landis, personal communication), fusion of conidia and hyphae in fungi, and biparental transmission to zygotes from both mating types in isogamous organisms such as yeast and many algae. In such instances it is possible for infectious elements harmful to the host to be established. It is interesting that application of this line of reasoning to the B chromosomes of maize and other organisms shows that to be maintained they too may not have to be favored by selection, and in this way resemble infectious elements (Rhoades and Dempsey, 1972). Generally, hereditary modes of transmission are not efficient enough, i.e., are not transmitted to enough new individuals, to counterbalance a strong adverse selection. Therefore infectious elements of heredity are not usually found to be harmful. An exception results from heterokaryon formation in fungi. It is very efficient and, as predicted, often leads to the establishment of harmful agents. Infectious transfer through the medium is, of course, highly efficient. Often it results in serious harmful effects which constitute parasitism, with which we are not concerned in this chapter.

Finally, we should point out that the status of infectious heredity has recently undergone a change. For many years it was merely a growing list of heterogeneous curiosities. But now, because of work on the tumor viruses, we perceive that it may play a central role in the maintenance of the cancerous state in cells. Furthermore, it is apparent that resistance factors make infectious heredity a great impediment to our control of diseases with antibiotics. And finally, work on the recombinant plasmids and proviruses of vertebrates provides hope for establishing new genes in organisms through genetic engineering.

We have attempted to summarize in this chapter the main points of an extraordinarily large and varied number of lines of research. To conserve space, we have cited primarily only selected general papers and reviews with extensive bibliographies, supplementing them occasionally with significant or more recent references.

## II. PLASMIDS OF BACTERIA

A plasmid is an hereditary bacterial element that is able to exist physically separate from the chromosome and reproduce in a stable fashion autonomously. All known plasmids are DNA. They represent a small percentage of the total DNA in the bacterium, and are usually not essential for the survival of the cell. Many plasmids are self-transmissible and can also help transfer other, nontransmissible plasmids and chromosomes.

Plasmids are often carried by transducing phage. Their great mobility among the bacteria makes it likely that plasmids arose in organisms other than those in which they are now found.

Some workers include temperate bacteriophages in the category of plasmids. Temperate phages after infecting a bacterium may, like virulent phages, enter a lytic cycle; or they may establish a relationship with the host bacterium known as lysogeny. A typical lysogenic phage is $\lambda$. If the lytic cycle follows infection by $\lambda$, the DNA duplex of phage $\lambda$ circularizes and replicates repeatedly as an autonomous element. Lysis of the host bacterium and liberation of phage ensue. Alternatively, if lysogeny is established, the DNA of $\lambda$ becomes inserted into the host bacterial chromosome where it exists as a prophage, replicating under control of the host. Under certain conditions, a proportion of the prophage in a lysogenic culture is induced to leave the host chromosome to become autonomous phage and enter the lytic cycle. Thus, whenever existing in the autonomous condition, temperate phage DNA replicates uncontrollably, and eventually lysis of the host cell occurs. Plasmids, on the other hand, exist autonomously and replicate in a controlled fashion; lysis of the host bacterium does not occur. Except for this difference in control of replication when existing autonomously, temperate phages and plasmids are very similar. In fact, by a deletion in phage $\lambda$, a mutant, $\lambda dv$, is produced that is unable to integrate into the bacterial chromosome and is propagated autonomously in a controlled fashion in the cytoplasm. It now is indistinguishable from a plasmid. In this chapter the more restricted view of plasmids that excludes temperate phages is adopted, since phages will be considered elsewhere in this series.

Some plasmids are capable of integrating with the host chromosome. Plasmids capable of existing either in the integrated or autonomous state are known as *episomes*. While most temperate phages can integrate and may be considered episomes, most plasmids are incapable of integrating and are not episomes. F, the fertility or sex factor carried by *Escherichia coli*, is unusual in that it is a plasmid that integrates. F is autonomous in $F^+$ strains, integrated in Hfr strains and lacking in $F^-$ strains.

A wealth of information concerning plasmids is found in the following references: Novick (1969), Hayes (1970), Helinski and Clewell (1971), Clowes (1972), Willetts (1972), Meynell (1973), Richmond and Wiedeman (1974), and the collections edited by Wolstenholme and O'Connor (1969), and Schlessinger (1975).

## A. Chemical and Physical Properties

Every plasmid studied to date has been shown to be present during some of its existence in the cell as a small covalently closed circular (CCC)

DNA molecule. If one strand of the CCC duplex DNA becomes nicked, it is known as relaxed, nicked, or open circular (OC) DNA. One technique of isolation of plasmid DNA is based upon the fact that ethidium bromide intercalates into supercoiled CCC DNA to a lesser extent than it does into linear DNA, creating a difference in buoyant density. Molecular weights of plasmids range from $1.5 \times 10^6$ to $150 \times 10^6$ (Sherratt, 1974); the molecular weight of the *E. coli* chromosome is $2.5 \times 10^9$.

In some cases the plasmid DNA consists of more than one genome and is multimeric rather than monomeric. See Fig. 1 for a diagrammatic representation of configurations of plasmid DNA. If identical genomes are joined end to end (linearly or circularly), the structure is known as a concatenated oligomer or concatemer. When Col E1 is present in *E. coli*, it is a monomer; but when transferred to *Proteus mirabilis*, it is concatenated. A catenated oligomer or catenane consists of interlocking monomeric CCC or OC DNA. Errors in replication or circular monomers as well as reciprocal recombination have been proposed to account for the formation of concatenated genomes. Recombination is suggested for the formation of catenated genomes. The above terms are not used for unlike genomes linked together, as occurs in the case of recombinant plasmids.

Plasmids often combine to form a single structure bearing all or most of the information of two or more plasmids, i.e., become recombinant. Related plasmids cannot become established in the same cell, because of plasmid incompatibility (discussed in Section II,C) unless they become recombinant.

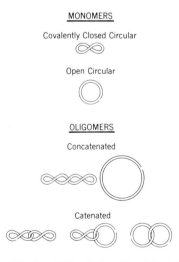

**Fig. 1.** Configurations of duplex DNA of plasmids. Adapted from Cohen *et al.* (1972) with permission of Springer-Verlag.

## B. Replication and Segregation

Since plasmids often exist in low numbers (for example, there is about one copy of F per bacterial chromosome), plasmid and chromosomal replication and segregation must be coordinated if plasmids are to be maintained stably through cell division. Jacob et al. (1963) proposed a "unit of segregation" in the bacterial cell made up of a replicon and its attachment site on the cell membrane. At cell division both components divide, one replicon with its attachment site segregating to each daughter cell. Since the attachment sites lie on a common cell membrane, chromosome and plasmid replication and segregation are coordinated. On the other hand, Kline and Miller (1975) have presented evidence that it is RNA, rather than membrane, that links the bacterial chromosome and F.

The original Cairns model for circular DNA replication postulated a fixed swivel (later thought to be a single-stranded nick) to allow unwinding and a single growing point. See Dressler (1975) for a discussion of models of DNA replication. Subsequent modifications have assumed transitory nicks and bidirectional growth. This last form of the hypothesis predicts CCC supercoiled intermediates which, when nicked by endonucleases in vitro, produce $\theta$-shaped structures visible with the electron microscope, as in the original model. Both $\theta$-shaped structures and evidence for CCC supercoiled intermediates have been found in studies on plasmids, but details of the process of replication are often obscure and appear to vary. Thus Col E1 growth is unidirectional from a single point of origin, according to Helinski et al. (1975). In the R factor NR1, replication is said to be either bidirectional or unidirectional and two points of origin of replication appear to be present, according to Rownd et al. (1975). The second well-known model of DNA replication, the rolling circle model, has also been demonstrated for plasmids and probably accounts for transfer of DNA during bacterial conjugation.

Hardy (1975) notes that protein synthesis is necessary for initiation of replication of some of the large, but not small plasmids. RNA is important in priming plasmid DNA synthesis in the small plasmid, Col E1 (Sakakibara and Tomizawa, 1974). Also, Helinski et al. (1975) showed that when protein synthesis in E. coli bearing Col E1 is inhibited by chloramphenicol, the RNA segment in the plasmid DNA, normally removed after replication, remains and replication continues. Between 1000 and 2000 plasmids accumulate, all bearing RNA.

In a number of plasmids investigated by Helinski's group, a high proportion of the plasmid molecules is present in the form of CCC supercoiled DNA complexed with protein. Such complexes are called "relaxation complexes," because one of the proteins of the complexes is an endonuclease which can, under certain conditions, nick one of the CCC DNA strands, converting the molecule to the OC relaxed form. The site of the nick in the plasmid Col E1 has been shown by Helinski and coworkers to occur always at a specific site in a specific strand. Relaxation complexes play a role in DNA replication, for Helinski's group has also demonstrated that in Col E1, replication is normally initiated at the site of the nick.

Usually one copy of the plasmid is present in the bacterial cell, and replication of the plasmid is said to be under stringent control. If multiple copies (usually between 5 and 25) per cell are produced, the plasmid is under relaxed control. Usually, large plasmids are under stringent, and small plasmids are under relaxed control. Whether relaxed or stringent, replication is controlled and does not end in the death of the cell carrying the plasmid (Arai and Clowes, 1975). Large penicillinase plasmids under stringent control and small plasmids under relaxed control were all found to segregate by a regular mechanism rather than randomly (Novick et al., 1975).

Plasmids may disappear spontaneously or as a result of the action of agents such as acridine, rifampin, etc. which are said to "cure" the hosts. These agents are thought to interfere with the replication of the plasmid.

## C. Sex Factors

Sex factors are plasmids that can effect their own transfer at conjugation from one bacterium to another and are known as transmissible, self-transmissible, infectious, or conjugative. They are able to establish a donor state in the host bacterium, conferring upon it the ability to form hair-like sex pili and undergo conjugation. They are usually rather large plasmids. Nonconjugative plasmids cannot establish the donor state and are normally transferred to other bacteria only if they are carried by a transducing phage or a conjugative plasmid. Nonconjugative plasmids are small. A common way of classifying plasmids is according to their function: as sex factors, colicinogenic (Col) factors, and resistance (R) factors. Col factors, carried by coliform bacteria, produce toxins known as colicins that affect related species of bacteria. Resistance factors, present in many strains of bacteria, specify resistance to drugs and other agents. Many colicinogenic and resistance factors, in addition to their distinctive functions, are also conjugative and could be considered sex factors. Sex factors are discussed in the references cited in Section II, p. 322. (See also Curtiss, 1969; Achtman, 1973; Curtiss and Fenwick, 1975.)

Sex factors have the remarkable ability to bring about conjugation. When conjugation occurs, the conjugative plasmid is transferred by means of the sex pilus from the donor to a recipient lacking the plasmid and sex pili. Upon entering the recipient, the conjugative plasmid converts it to a donor cell, able to form sex pili and transmit the plasmid. Thus the plasmid spreads throughout a bacterial population. The term sex factor, originally applied to F, is synonomous with conjugative plasmid according to some authors; others consider it appropriate only for a conjugative plasmid capable of transmitting other genetic material such as the bacterial chromosome or a nonconjugative plasmid in addition to its own genome. It is likely that all conjugative plasmids can transmit genes other than their own and are thus sex factors.

Achtman (1973) describes cistrons of F which control transfer of DNA, surface exclusion, replication and segregation, incompatibility, and female-specific phage restriction. Achtman and Helmuth (1975) give a map of F with fourteen cistrons; twelve of these are concerned with transfer of F, and most of these are involved in determining the sex pilus. The molecular weight of F ($64 \times 10^6$) suggests a much larger number of genes, and it is obvious that the function of large regions of F are unknown.

Two separate phenomena prevent identical or closely related conjugative plasmids from coexisting in the same cell. Surface or entry exclusion refers to the poor recipient ability of plasmid-bearing bacteria when mixed with donor cells carrying the same or a related sex factor. It has been shown that the presence of sex pili in the recipient cannot alone explain exclusion and a more generalized surface phenomenon is probably responsible. Exclusion can be avoided by transducing the similar plasmid with phage or by inducing the bacteria bearing autonomous F ($F^+$ bacteria) through starvation to change phenotype to become $F^-$ phenocopies. These $F^-$ phenocopies carrying F no longer exclude; but the incoming F may fail to become established because of the second phenomenon, plasmid incompatibility, also called superinfection immunity. To explain plasmid incompatibility it has been postulated that related plasmids compete for the single maintenance site necessary for replication. More recently it has been postulated that inhibitors of initiation of replication are involved.

In most naturally occurring conjugative plasmids, genes involved with sex functions are regulated by a typical repressor system, first demonstrated by Stocker *et al.* (1963) in Col I, a sex factor with a colicinogenic determinant. Sex factors such as Col I are normally transferred by only a few donor cells, demonstrating low frequency of transfer (LFT) of the plasmid. If Col I is newly transferred to a recipient Col$^-$ strain, it rapidly spreads to every Col$^-$ cell in the culture, showing high frequency of transfer (HFT); after several generations, however, it reverts to LFT. The explanation is that immediately after transfer to a Col$^-$ strain, the expression of genes in the operon for sex functions occurs before the Col I repressor gene is able to produce sufficient repressor to combine with the operator to shut off the operon, and an HFT culture results. When enough repressor is synthesized to inactivate the operator, an LFT culture results. Many R factors with sex functions are, like Col I, repressed and are transferred at conjugation very infrequently; sex factor F is derepressed, and is transferred by every F⁺ bacterial cell.

Sex factors have been classified according to whether they specify F-like or Col I-like sex pili. F-like pili adsorb specific spherical RNA and filamentous single-stranded DNA phages. I-like pili adsorb specific filamentous single-stranded DNA phages. Other kinds of pili, susceptible to other specific phages characterize additional groups (Datta, 1975).

Conjugative plasmids carry determinants to repress their own sex functions and also, in some cases, the sex functions of other plasmids. A bacterium carrying a conjugative R factor as well as F was found to repress F. The conjugative R factor was denoted fi⁺ (fertility inhibitor). fi⁺ R factors are like F, though not identical to it. Conjugative plasmids that do not repress F are denoted fi⁻. Some fi⁻ plasmids repress Col I and are classed as I-like; others do not, and fall into other groups.

The sexual process in bacteria is wholly dependent upon sex factor plasmids. At conjugation, according to the rolling circle model, thought to apply here, one of the strands of the CCC F plasmid DNA is nicked. The nicked strand migrates, 5′-end leading, to the recipient cell. A new strand of DNA is replicated alongside of the circular strand left in the donor. Replication of the single F strand in the recipient to form the DNA duplex is followed by circularization of the F plasmid. In the donor the two ends of the replicated strand join to form a circle. Exconjugant cells are now both F′.

On rare occasions F integrates linearly into the circular bacterial chromosome to form what is known as a high frequency of recombination (Hfr) clone. On the basis of genetic evidence, integration of the circular F into the circular bacterial chromosome is thought to occur by a single reciprocal crossover, according to the well-known Campbell model. Unlike some plasmids that integrate at one or a few sites, F is known to integrate in numerous sites on the *E. coli* chromosome, and in either clockwise or counterclockwise direction. Although integration of F is mediated by the generalized recombination (Rec) system of the *E. coli* host, it is also possible that F encodes for recombination enzymes (Davidson *et al.*, 1975). The integrated F plasmid, under replicative control of the bacterial chromosome is transmitted with the chromosome to the daughter cells at cell division. At conjugation between an Hfr donor and an F⁻ recipient, the bacterial chromosome is transferred by F. A nick in one strand of the Hfr duplex occurs at some point in the integrated F. The nicked single strand, 5′-end of the plasmid leading, enters the recipient cell first, followed by the adjoining bacterial chromosome, and at the very end, the remainder of the F plasmid. Usually the conjugating cells separate before the whole continuum of F segment chromosome–F segment passes to the recipient cell, which then remains F⁻, since it received only part of the F. If the entire plasmid and chromosome pass to the recipient, the cell can become an Hfr individual. By nonreciprocal interchanges, it is thought, segments of the donated chromosome become part of the recipient chromosome, and segments of the recipient chromosome are excised and lost at subsequent divisions. By this means the cell becomes haploid again.

F may revert from the integrated state in Hfr strains to become autonomous. Imprecise excisions sometimes occur, such that some of the adjoining chromosomal genes of the bacterium remain with F to form F prime (F′) factors (F*gal*, etc.), or some of the genes of F may remain in the bacterial chromosome, creating a sex factor affinity (*sfa*) region, where autonomous F subsequently integrates. The occurrence of imprecise excisions has made possible the determination of gene sequences in various F and F′ factors by electron microscopic heteroduplex analysis of DNA (see Davidson *et al.*, 1975). In this procedure, duplex F and F′ DNA's are mixed and denatured to produce dissociated strands of DNA. Heteroduplex DNA made up of complementary strands of F and F′ are formed during subsequent renaturation. When such heteroduplex DNA is viewed with the electron microscope, homologous sequences are observed as double-stranded regions, and nonhomologous sequences are single-stranded.

### D. Colicinogenic Factors

Colicinogenic bacteria harbor colicinogenic (Col) factors that mediate the synthesis of toxins known as colicins. Some Col factors are conjugative, others are not. Normally the sex functions are repressed but Col V is derepressed. Colicinogenic factors and colicins are discussed in the references cited in Section II, p. 322. (See also Reeves, 1972; Hager, 1973; Luria, 1973; Davies and Reeves, 1975; Hardy, 1975.)

Colicins, first found in *E. coli*, and later in *Shigella* and *Salmonella*, belong to a class of substances, bacteriocins, that inhibit growth and cause death, but not lysis of strains of the same or related species. In many cases the genetic determinants of bacteriocins have been shown to reside in plasmids. Although some purified bacteriocins are chemically complex, in all cases a protein component is responsible for the toxic activity.

Colicins were originally classified by Frédéricq (1957), who did much of the early work in this field. Classification is based on the specificity of adsorption of the colicins to receptors on the bacterial cell wall, with subgroups distinguished on the basis of patterns of immunity. Each col-icinogenic strain is immune to the action of the colicin it produces, but sensitive to certain other colicins.

Similarities between many colicins and temperate bacteriophages in-clude structure, control of synthesis, and mode of action. Some colicins, such as colicin of *E. coli* 15, appear to be structurally similar to phage heads or tails. However, many workers now exclude those forms struc-turally like phages from classification as colicins.

Recently an attempt has been made to classify Col factors on the basis of physiological criteria. Hardy *et al.* (1973) found on this basis that all eleven of the Col factors they studied fell into two distinct groups. Group I, comprised of low molecular weight ($5 \times 10^6$), nonconjugative Col factors D, E1, E2, E3, and K, that replicate under relaxed control and produce mostly free colicin; and Group II, made up of high molecular weight ($62$–$94 \times 10^6$) conjugative Col factors B, I, and V, that are under stringent replicative control and produce mostly cell-bound colicin. The two groups of Col factors also differ in their dependence on host functions for their maintenance. Two groups of colicins have been delineated by Davies and Reeves (1975), who note that these groupings correspond to the two groups of Col factors studied in Hardy's laboratory.

One or more Col factors may exist in the cell autonomously, or com-bined with other sex, Col, R factors, or phage, or infrequently with the bacterial chromosome. It has been possible to obtain in the laboratory very complex recombinant plasmids. A nonconjugative Col factor coexist-ing in a cell with a conjugative Col factor can be transferred at conjuga-tion without becoming united to the conjugative plasmid. Although it is rare among Col factors, stable high frequency of recombination clones (analogous to Hfr clones where F is integrated into the chromosome) have been reported by Frédéricq (1969) for Col B and by Kahn (1968) for Col V, where integration presumably is mediated by the host Rec system.

Ordinarily only a small proportion of cells bearing a col factor produces colicin at any given time. Synthesis of colicins is repressed in most cells carrying Col factors and spontaneously derepressed to produce the colicin and death of the cell which produced the colicin (Hardy, 1975). Some colicins may be produced constitutively (Meynell, 1973), possibly in small amounts. Colicin synthesis is induced in many colicinogenic bacteria by treatment with mitomycin C and other agents. Whether induction of col-icin synthesis is accompanied by a corresponding increase in Col factor DNA is under debate (see Durkacz *et al.*, 1974; Hardy, 1975).

Colicins adsorb to receptors in the outer membrane of the cell wall of sensitive *E. coli*. Following their adsorption, colicins are believed to be transported to make contact with the inner cytoplasmic membrane. Receptors for colicin E3 in the outer cell membrane of *E. coli* have been isolated by Sabet and Schnaitman (1973), and found to be protein. The change brought about by some colicins on the receptor site is reversible by treatment with trypsin. Bacterial mutants that have lost their sensitivity to a colicin may have lost the receptors on the cell membrane (Davies and Reeves, 1975).

Colicins may be grouped on the basis of their specific activity (Konisky, 1973) as follows. Colicins Ia, Ib, K and E1 inhibit all macromolecular synthesis. In addition, colicins E1 and K interfere with electron transport. Colicin E2 specifically acts to inhibit DNA synthesis or degrade DNA and inhibit cell division. Colicin E3 interferes with protein synthesis by acting on ribosomes. It acts directly *in vitro* in a membrane-free system, cleaving the 30 S ribosome at the same point as *in vivo*, and releasing a terminal fragment of the RNA. In addition to producing colicin E3, these colicinogenic cells also synthesize a small protein, absent in cells lacking Col E3, which reacts with colicin E3 to nullify its effect. Hardy (1975) notes that the immunity to a colicin given by a Col factor to its host can be overcome by high concentrations of the colicin.

## E. Resistance Factors

Resistance of bacteria to several antibacterial agents is usually by conjugally transmitted resistance (R) factors. Resistance factors are discussed in the references cited in Section II, p. 322. [See also Mitsuhashi (1971), Richmond (1975), Falkow (1975), and collections of papers edited by Dulaney and Laskin (1971), and Krčméry *et al.* (1972)].

R factors were first demonstrated by the transfer of multiple resistance in mixed cultures of *Shigella* and *E. coli*. Strains of *Shigella* resistant to tetracycline, streptomycin, sulfanilamide, and chloramphenicol made their appearance following the use of these antibiotics in Japan in the early nineteen fifties; within 12 years, over 70% of all clinical isolates of *Shigella* carried resistance to several antibiotics. Selective pressure of an antibiotic-rich environment has favored the striking increase of strains of bacteria bearing multiple R factors.

Two different molecular structures of R factors have been described, the plasmid cointegrate and the plasmid aggregate forms (Clowes, 1972). R factors having the cointegrate structure have been studied in *E. coli* and *P. mirabilis* in the laboratories of Rownd, Clowes, Falkow, Cohen, Punch, and others. There is agreement that in *E. coli*, the R factor exists as a cointegrate that is a composite circular molecule (one or a few copies per cell), made up of a large sex factor, referred to as the resistance transfer factor, RTF, and a smaller r-determinant(s) region carrying, for example, resistance genes C, S, and Su. A controversy exists, however, concerning the form of the R factor when the cointegrate is transferred to *P. mirabilis*. See Fig. 2 for a diagram incorporating the essential elements of the different views under discussion. According to many workers (see Falkow, 1975), when *P. mirabilis* is cultured in drug-containing medium, the R factor retains its composite form during exponential growth of the bacteria; but as the cells enter the stationary phase of the growth cycle, some of the R factors dissociate into their two component replicons. Replication of the r-determinant portion then proceeds at an increased rate. The dissociated RTF and r-determinant replicons are covalently closed circular DNA. During exponential growth there are about five copies of the R factor per cell; during the stationary growth phase, the RTF molecule and the composite molecule are present in about five copies per cell, whereas the r-determinant molecule increases to about fifty molecules per cell. The identity of the different classes of

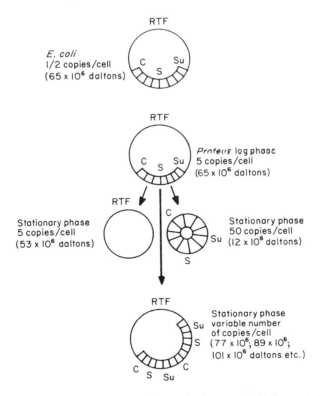

**Fig. 2.**   Diagrammatic representation of the molecular nature of R factors in *E. coli* and in *P. mirabilis*. After Falkow (1975) with permission of Pion, Limited, London.

molecules has been made on the basis of their differing buoyant densities in cesium chloride gradients and molecular weights estimated from contour measurements of DNA with the electron microscope. The different R factors vary, but the molecular weight of a composite molecule is equivalent to the sum of the molecular weights of its component RTF and r-determinant molecules. For example, R factor R1*drd* is reported to have a buoyant density of 1.711 and a molecular weight of $65 \times 10^6$. Upon dissociation of the R factor, the values for the RTF molecule are 1.709 and $53 \times 10^6$ and for the r-determinant molecule, 1.716 and $12 \times 10^6$.

On the other hand, Rownd states that the composite form of the R factor is maintained when *P. mirabilis* is cultured in drug-free medium (Rownd *et al.*, 1975); and only when *P. mirabilis* is cultured in drugs to which the R factor confers resistance does the R factor dissociate. He presents evidence that in *P. mirabilis* cultured in the presence of drugs, the dissociated r-determinants increase to become poly-r-determinants. They become covalently joined to composite R factors to produce very large circular molecules that comprise the major species of R factor. Generally during isolation of DNA, such molecules break down into linear fragments. Rownd believes that the circular molecules isolated by other workers represent a minor portion of the total R factor DNA. This is an active field, and undoubtedly differences among the participants will be resolved as more data accumulate.

A possible mechanism to account for the reversible dissociation between parts of the composite R factor has been suggested by Hu *et al.* (1975a,b) and Ptashne and Cohen (1975). They have discovered insertion sequences of DNA that are duplicates at the two sites on the circular R factor where the RTF and r-determinant portions are joined. They propose that recombination between the insertion sequences at the two sites could bring about dissociation of the components of the R factor.

The second kind of molecular structure of R factors, the plasmid aggregate form, is that in which two or more plasmids normally exist separately in a cell. Examples of this form are the $\Delta$A and $\Delta$S plasmid aggregates that exist stably in *Salmonella typhimurium*, studied by Anderson (1968) and others. The larger molecule, $\Delta$, is the transfer factor and the smaller molecule(s) bear the resistance gene(s). Segregant strains bearing $\Delta$, A, and S, singly, can be isolated. Each shows the appropriate phenotype, and the isolated DNA is unimolecular. Values for molecular weights of monomeric R factors are reported to range between 26 and $78 \times 10^6$.

Recombination between R factors to form a single structure occurs frequently and is dependent upon the Rec system of the host. Recombination between two related R factors makes it possible for them to coexist in a cell, otherwise impossible because of superinfection immunity. It is thought that deletion of the sex factor portion of one of the R factors may occur first, thus avoiding superinfection immunity, followed by recombination of the two R factors. R factors can promote a rather high frequency of oriented chromosomal transfer, indicative of integration with the chromosome, but in no case has a stable Hfr form been reported (Meynell, 1973).

Regulation of sex functions in conjugative R factors is by repressive control. Drug resistance brought about by antibiotic inactivating enzymes may be produced by derepression of genes, while erythromycin resistance, for example, is thought to be produced constitutively (Weisblum, 1975).

R factors specify resistance to an ever-lengthening list of agents. They include antibiotics of various classes, heavy metals, colicins, X-rays, etc. (see Benveniste and Davies, 1973; Weisblum, 1975; Dowding and Davies, 1975). Resistance due to R factors brought about by enzymatic inactivation of an antibacterial agent was demonstrated in studies by Pollack (1962) of inactivation of penicillin. A second mechanism of resistance involves a barrier in the cell to action of the antibiotic, such as a block to permeability. A third mechanism is chemical reduction which occurs in resistance to heavy metals. Thus, by different mechanisms, R factors nullify the effects of many antibacterial agents. The design and effective use of new antibiotics seems invariably to be followed by the appearance of bacteria resistant to them.

The classification of conjugative plasmids is based primarily on incompatibility (see Datta, 1975). Closely related plasmids cannot coexist in a cell and are included in the same compatibility group; unlike plasmids are able to coexist and are placed in different compatibility groups. DNA–DNA hybridization studies show extensive hybridization between plasmids of the same compatibility group, but not of different groups. Twenty-seven groups have now been distinguished among the plasmids of enteric bacteria on the basis of incompatibility, according to de Vries *et al.* (1975).

## F. Restriction and Modification of DNA

Some 10–20% of all R factors specify systems which restrict and modify infectious DNA (Roulland-Dussoix *et al.*, 1975). They thus afford the host bacterium protection from invading phage or plasmid DNA by producing endonucleases called restricting enzymes that cleave infectious DNA (Nathan and Smith, 1975). One class of restricting enzymes attacks the incoming DNA at specific sites. In order to prevent damage to the cell's own DNA, specific restricting enzymes are accompanied by specific modifying enzymes. Modifying enzymes generally have been shown to be methylases which add methyl groups from $S$-adenosyl-L-methionine to DNA at the same sites that are otherwise recognized by the restricting enzymes, thereby rendering their own DNA resistant to the restricting enzyme. Each specific restricting enzyme is thus paired with a matching modifying enzyme to form a restriction modification system. If phage $\lambda$ particles invade a strain containing such a system, they are often unsuccessful and are said to have a low efficiency of plating (EOP). Occasionally an invading $\lambda$ does multiply, and its DNA is found to have been modified (methylated). When these $\lambda$ particles are now replated on the same host they are no longer susceptible to the restricting enzyme, i.e., they have a high EOP and grow well. Such modifications, of course, are not hereditary and always reflect the particular restriction and modification system of the specific host in which the $\lambda$ has been grown. To define a given system, one cultures a phage such as $\lambda$ on cells containing the system. When the resulting phages are plated on a series of bacterial strains bearing a variety of systems, the resulting pattern of EOP's defines the system.

Many fi⁻ R factors determine the *E. coli* R factor II system (*Eco*RII), while fi⁺ R factors specify the *Eco*RI system. The *Eco*RI and *Eco*RII endonucleases both cleave one strand of the DNA duplex near one end of a given specific base sequence and the other strand near the opposite end of the sequence. This process generates fragments of duplex DNA with complementary single strands at either end of the fragment. Two important uses for these restriction enzymes that produce such highly specific cleavages and fragments are the characterization and sequencing of DNA's, and the production of hybrid molecules. The use of such DNA fragments with cohesive ends to join with other fragments provides a major new technique in plasmid research.

A review of the construction of hybrid plasmids and their use in bacterial transformation of *E. coli* is given by Cohen and Chang (1975). Thus, a nonconjugative plasmid pSC101 (MW = $5.8 \times 10^6$), formed by mechanical shearing of an R factor and subsequent circularization, is cleaved at only one site when attacked by restriction endonuclease *Eco*RI. The genes for replication and tetracycline resistance remain intact on this small plasmid, which becomes a linear structure after cleavage with *Eco*RI. When cleaved pSC101 can hybridize with a similarly produced cleavage fragment of DNA from any source, and hydrogen bonding of the complementary base pairs occurs. The 3'- and 5'-ends are joined by ligase, forming a circular DNA hybrid structure. Col E1 is also capable of being cleaved in only one site by a restricting endonuclease. Col E1 has the experimental advantage of being under relaxed control of replication, and as noted earlier, chloramphenicol treatment of *E. coli* cells bearing Col E1 results in the production of over a thousand copies of the plasmid per cell. Hybrid plasmids can be constructed *in vitro* and can be made to enter competent cells by the process of bacterial transformation. The constructed molecule can replicate in *E. coli*, even though portions of the molecule may have originated from very different sources, such as prokaryote *E. coli* plasmid DNA and eukaryote *Xenopus laevis* ribosomal DNA.

Experiments of this type, of course, appear to open the way for genetic engineering. The possible dangers, however, of creating new pathogens have also been recognized. An international committee of scientists met at Asilomar, California, to establish safeguards for such research (Wade, 1975). Research sponsored by the United States government at present follows the National Institutes of Health guidelines for research involving recombinant DNA molecules (1976).

## III. ENDOSYMBIONTS OF PROTOZOA

As originally used by deBary in 1879, the term symbiosis referred to the living together of unlike species in an intimate and constant relationship, and included harmful as well as beneficial associations (Henry, 1966). It is used in this sense here. Lengthy compilations of organisms living in or on protozoa have been made by Kirby (1941), Buchner (1965), and Ball (1969). Although there have been many cytological demonstrations in protozoa of endosymbionts, the evidence to establish the nature of these intracellular inclusions, their heritability, and whether they are infectious is in many cases not available.

Ball lists algae, viruses, and bacteria living in various protozoans. Cyanellae (blue-green algae) are found in shelled amoebae, cryptomonads, flagellates, and rhizopods. Zoochlorellae (green algae), usually *Chlorella* or *Pleurococcus*, are present in *Condylostoma*, *Strombidium*, *Paramecium*, and *Stentor*. *Chlorella* and *Paramecium bursaria* constitute a stably inherited system which has been well studied in recent years (see Karakashian, 1975). Zooxanthellae (dinoflagellates) are found in marine protozoa. Structures which resemble viruses in the electron microscope occur in *Plasmodium*, *Amoeba*, and *Entamoeba*. Fungal symbionts and associations of protozoa with other protozoa are generally parasitic. Bacterial endosymbionts are especially common. They have been reported in flagellates in the gut of termites, amphibia, and mammals. In *Amoeba* Jeon and Jeon (1976) have shown that certain of the bacterial forms which became established within the cytoplasm are essential for the life of their host. Bacterial endosymbionts have been studied in a number of ciliates such as *Euplotes* and *Paramecium*. The xenosomes of several strains of marine ciliates described by Soldo *et al.* (1974) appear to be bacterial in nature. Three cases of bacterial endosymbiosis have been selected for presentation in some detail here.

## A. Bacteria in *Crithidia* and *Blastocrithidia*

A recent publication by Tuan and Chang (1975) provides references to the literature. *Crithidia oncopelti*, a flagellate found in the intestine of the milk-weed bug, carries in each cell two bacterial endosymbionts known as "bipolar bodies." The factors regulating the number and distribution of the two bacteria at cell division are unknown. Symbiont-free strains require the addition of hemin, lysine, and other substances to promote growth, while symbiont-bearers do not. Tuan and Chang have isolated the DNA of the bacterial endosymbionts of *C. oncopelti* and of another flagellate *Blastocrithidia culculis*. They found the density of the DNA of the symbiont to be different from that of the host in *C. oncopelti*, but similar to that of the kinetoplast DNA in *B. culculis*. The densities of the DNA of the two symbionts are similar. Studies on renaturation kinetics and amount of DNA per cell show that the genome of the endosymbiont of *B. culculis* has a multiplicity of ten. Tuan and Chang note that in this respect it is similar to that of lambda, an endosymbiont of the *Paramecium aurelia* complex (Sonneborn, 1975) studied by Soldo and Godoy (1973).

## B. Bacteria in *Euplotes*

In *Euplotes* two functionally different kinds of endosymbionts have been found. One is concerned with a killer trait, the other is involved with normal cell function. Heckmann *et al.* (1967) note that, with the exception of *Paramecium bursaria* and *P. polycaryum*, all killing or mate-killing strains of protozoa are characterized by the possession of particles in the cytoplasm that are not present in sensitive strains. These strains include killers of the *P. aurelia* complex, *Euplotes patella*, and *E. minuta*. The killers of *E. minuta* have bacteria-like particles in their cytoplasm, designated epsilon. *Euplotes minuta* killers liberate toxic particles into the medium that are precipitated by centrifugation at relatively low speeds.

Observations by Fauré-Fremiet (1952) on *E. patella* and *E. eurystomas* led him to conclude that the symbionts found in these organisms were essential, since removal of them by treatment with low doses of penicillin resulted in death of the host. The possibility that death was due to sensitivity to penicillin rather than to loss of essential symbionts was not excluded, however.

More recently, work with *Euplotes aediculatus* (Heckmann, 1975) has revealed that the presence of the endosymbiont in this species is indeed essential for the host. He found gram-negative rods in the cytoplasm resembling bacteria and numbering about one thousand. The endosymbiont, called omikron (Fig. 3), was removed by treatment with penicillin, whereupon the host *Euplotes* failed to divide. Reinfection of a small percentage of such cells with omikron was achieved, and in these animals normal growth was resumed. No killing activity is associated with omikron. It has not been determined whether omikron is the same as the endosymbiont found in *E. patella* and *E. eurystomas* by Fauré-Fremiet.

## C. Bacteria in the *Paramecium aurelia* Complex

Some of the best studied examples of endosymbionts in protozoans are found in *Paramecium*. Some strains of paramecia liberate a toxin into the medium that kills other sensitive strains. This killing phenomenon was first observed in paramecia carrying kappa. Kappa is a gram-negative bacterial endosymbiont present in the cytoplasm of killers; it renders the bearer immune to the toxin it produces, but not to the toxins of other kinds of killers. Kappa in turn has *its* endosymbionts—phage-like elements found in a small percentage of the kappas that are specifically implicated in producing the killer trait.

Earlier work on kappa is detailed in a review by Sonneborn (1959). Recent reviews of the bacterial endosymbionts of the *P. aurelia* complex are provided by Soldo (1974), Preer *et al.* (1974), and Gibson (1974). See Sonneborn (1975) for current species designations in this complex. Only a few of the interesting features of the earlier work on kappa will be referred to here. Sonneborn (1943) demonstrated by genetic studies that the killer trait he discovered was cytoplasmically inherited; later Preer (1950) determined the size of the killing particles in the cytoplasm and demonstrated them cytologically. Historically, kappa is interesting because it was considered for many years a classic case of cytoplasmic inheritance, although suggestions were made soon after its discovery that kappa might be a virus or a relative of the algal symbiont of *P. bursaria*. Kappas are of two kinds: "brights," so-called because they possess a refractile (R) body when seen with bright phase microscopy, and "nonbrights," which lack it (see Figs. 4 and 5). Nonbrights are infective and do not kill; brights are responsible for killing. Nonbrights give rise to brights, which after a time lyse. Variant kappas have arisen spontaneously; in all cases the mutation has occurred in kappa and not the paramecium genome.

Immediately following is a discussion of the different endosymbionts of this group, their distinguishing characteristics, general maintenance, infec-

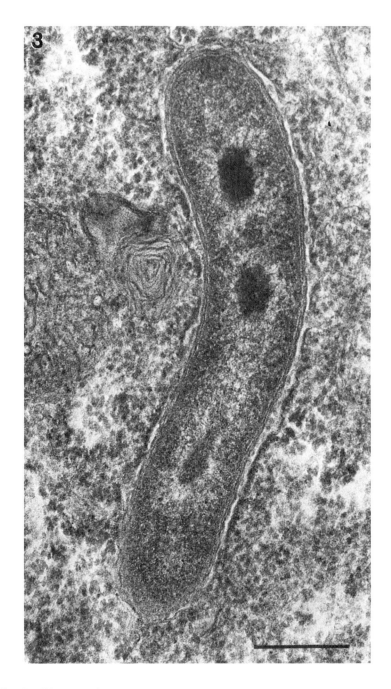

**Fig. 3.** Electron micrograph of section of *Euplotes aediculatus* bearing omicron, an essential bacterial endosymbiont. × 100,000, scale 0.25 μm. After Heckmann (1975) with permission of *J. Protozool.*

tivity, effect on the host, and nature of their DNA. Finally consideration is given to the phages and their role in the killing phenomenon.

Endosymbionts called mu (Fig. 6), which are similar to kappa, were found by Siegel (1954) in paramecia known as mate-killers. Mate-killers kill only by cell-to-cell contact at conjugation. Subsequently, other kinds of killing, dependent upon different endosymbionts, were discovered in the various killer stocks. There are similarities among the endosymbionts as well as differences. They are present in large numbers, in the hundreds or even thousands per cell. Most produce toxic effects on sensitives, are approximately 0.5 $\mu$m wide and 1–2 $\mu$m long, and with a single exception, are found in the cytoplasm. The distinctive features of the different endosymbionts follow.

Kappa is found only in stocks of *P. biaurelia* and *P. tetraurelia*; a characteristic R body is found in a low percentage of the paramecia bearing kappa. Pi lacks an R body and does not kill. It was originally thought to be a mutant of kappa, and early DNA–DNA hybridizations by Behme (see Preer *et al.*, 1974) indicated that kappa and pi were homologous. Recently however, it has been shown that kappa and pi differ in the size and buoyant density of their DNA (J. A. Dilts, 1977), and show little homology in DNA–DNA hybridization studies (R. L. Quackenbush, 1977). Mu is found only in *P. primaurelia*, *P. octaurelia*, and in one stock of *P. biaurelia*; cell-to-cell contact (but not nuclear exchange) is necessary for killing. Tau, delta (Fig. 7), and nu produce no toxic effects, although some may have done so when originally collected; they are nondescript except that delta has a thick cell wall and occasionally a few flagella. Gamma (Fig. 8) is found in only two stocks of *P. octaurelia*, it is small and possesses an extra set of outer membranes. Lambda (Fig. 9) is found only in *P. tetraurelia* and *P. octaurelia*; it is large, covered with peritrichous flagella, and kills by rapid lysis. Sigma is found only in one stock of *P. biaurelia*; it is large, curved or spiral, and resembles lambda in its peritrichous flagellation, rapid lysis killing, and range of sensitives killed. Alpha (Fig. 10) has been found only in one stock of *P. biaurelia*, although it can be made to infect a few stocks of this species. It does not kill. It is crescent or spiral-shaped and large. It is present in large numbers only in the macronucleus, is found rarely in the cytoplasm, and never in the micronucleus. It appears to be similar to omega, a micronuclear symbiont of *P. caudatum* described by Ossipov and Ivakhnyuk (1972).

Numerous genes affecting the maintenance of the various endosymbionts have been described. For example, kappa of stock 51 is present in the cytoplasm only if the nuclear gene K is present. If killers are made homozygous for gene k from stock 32, kappa fails to grow. It is not known whether K supports kappa or k prevents kappa from being maintained. In

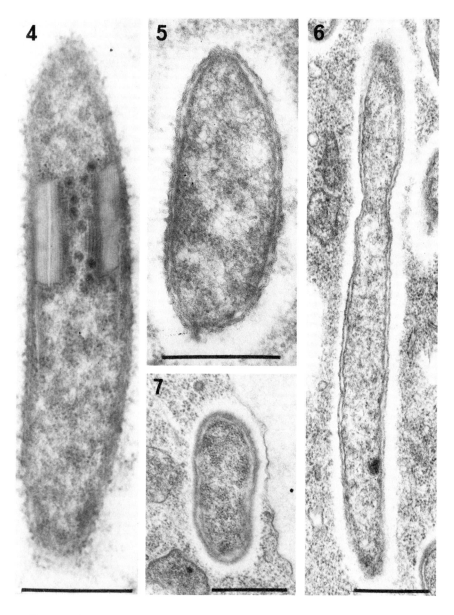

**Fig. 4.**   Electron micrograph of section of stock 7 *Paramecium biaurelia* bright kappa. Note dark-staining spherical phages inside coiled refractile body. × 60,000, scale 0.5 μm. After Preer and Jurand (1968) with permission of *Genet Res*.

    **Fig. 5.**   Electron micrograph of section of stock 1039 *Paramecium biaurelia* nonbright kappa. × 62,700, scale 0.5 μm (courtesy of Dr. A. Jurand).

stock 29 two different loci, $S_1$ and $S_2$, increase the probability of loss of kappa. There have been several studies on the kinetics of loss of kappa when the genotype changes at autogamy from Kk to kk. Similar studies have been made on mu. Although the time of loss appears to be very variable in these experiments, it is clear that when the loss begins, all of the endosymbionts disappear rather quickly. Nongenic factors also influence the maintenance of endosymbionts. For example, loss of kappa occurs under a variety of circumstances: at extreme temperatures, treatment with X-rays or with too rapid feeding, etc. Ordinarily only one kind of endosymbiont is present in an animal, although exceptions are known. Delta and mu coexist in stock 131, and possibly in a number of other stocks. Cytoplasmic exchange at conjugation between animals carrying sigma and kappa produce offspring carrying both kinds of endosymbionts. Gibson (1973) indicates that microinjection offers a means of bringing together different endosymbionts in a single cell, although there are conflicting reports concerning what transfers are possible.

Bacteria-free culture of paramecia is now possible, and some strains bearing endosymbionts are regularly maintained in high concentrations in axenic medium. Soldo (1974) has cultured stocks bearing mu, pi, or lambda without loss of the endosymbionts in axenic medium for the past several years. Using a chemically defined medium, Soldo and Godoy (1973a) showed that paramecia carrying lambda produce enough folic acid to enable the animals to grow without an external source of the vitamin. Lambda-free paramecia require folic acid and fail to grow in medium lacking it.

*In vitro* growth of stock 299 lambda in a complex medium for 24 weeks was reported by van Wagtendonk *et al.* (1963). More recently Williams (1971) reported limited success in culturing stock 138 mu and stock 299 lambda *in vitro*. Culture of endosymbionts on a routine basis has clearly not been achieved, and contradictory results have been obtained in different laboratories.

Several workers have shown that under special laboratory conditions kappa and lambda are infective. As noted above, nonbright kappas are infective, brights are not. Alpha can be transmitted from alpha-bearing to alpha-free animals of a few stocks in *P. biaurelia*. Gibson (1973) reports numerous successful transfers of kappa, mu, lambda, and alpha into stocks of different species of the *P. aurelia* complex by microinjection,

**Figs. 6 and 7.**    Electron micrograph of stock 131 *Paramecium biaurelia*, bearer of two kinds of endosymbionts.
**Fig. 6.**    Section of mu.
**Fig. 7.**    Section of delta. Note electron-dense material surrounding endosymbiont. × 39,900, scale 0.5 μm. Courtesy of Dr. A. Jurand.

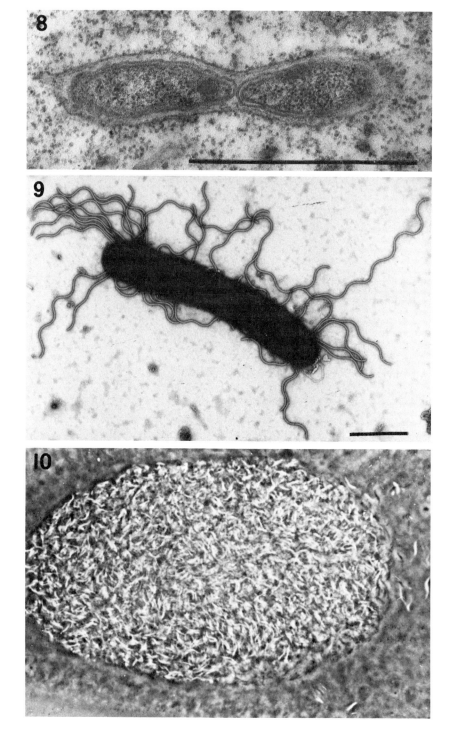

while S. Koizumi (personal communication) finds transmission by this means rare.

The presence of endosymbionts appears to have little effect on the host paramecium other than to confer on it the killing character and to provide it with resistance to its own kind of killing. Whether the killing trait has any selective advantage in nature is not known. If infection is rare in nature, as seems likely on the basis of laboratory experience, selective pressure favoring the presence of endosymbionts in paramecia may exist, since symbiont-bearers are common in natural collections. The beneficial effect of lambda in providing the host with folic acid is unique among the endosymbionts of paramecium, though whether this confers any benefit in nature might be doubted. Under conditions of stable maintenance of the endosymbionts, the numbers and growth rate of the endosymbiont and the host paramecium are balanced. Substantial increases in the symbiont population in paramecia occasionally have been observed; the death of the host paramecia soon followed.

Determinations of the buoyant density of the DNA of the endosymbionts have been made by a number of workers. The values reported for alpha, gamma, kappa, pi, mu and nu fall within a range of 1.694–1.702 gm/cm$^3$, compared to the density of the nuclear DNA of stocks of paramecium which has been reported to be in the range of 1.685–1.693. A wide discrepancy exists in the determinations in different laboratories for lambda, which might indicate that a difference in strains had arisen. DNA densities of the defective phages associated with kappa have been determined in a few stocks and will be discussed presently. Their values are close to but consistently slightly less than the density of the DNA of the kappa with which they are associated.

Soldo and Godoy (1973b) from studies of renaturation kinetics have determined the size of the lambda genome to be $0.39 \times 10^9$ daltons. The amount of DNA per lambda was found to be equivalent to a molecular size of $7.5 \times 10^9$ daltons, indicating that lambda DNA exists as a multiple genome, perhaps as many as ten to twenty copies per lambda. Similar studies of the pi and mu genome indicate that they contain multicopy

---

**Fig. 8.** Electron micrograph of section of stock 565 *Paramecium octaurelia* showing gamma, apparently dividing. Note presence of extra outer membrane. × 60,000, scale 1 μm. After Beale *et al.* (1969) with permission of *J. Cell Sci.*

**Fig. 9.** Electron micrograph of purified stock 327 lambda of *Paramecium octaurelia*. Note peritrichous flagella; negatively stained with phosphotungstic acid. × 15,000, scale 1 μm. After Preer *et al.* (1974) with permission of *Bacteriol. Rev.*

**Fig. 10.** Osmium-lactoorcein preparation of stock 562 *Paramecium biaurelia* showing vegetative macronucleus containing alpha, seen as numerous bright spiral rods; bright phase contrast. × 1,000. After Preer (1969) with permission of *J. Protozool.*

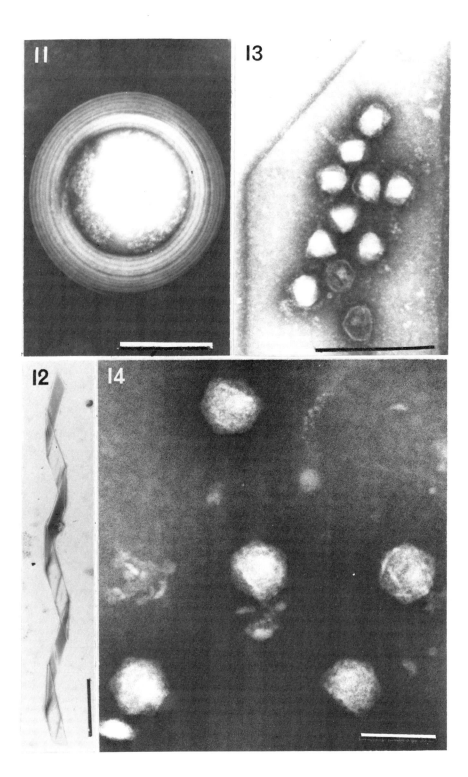

genomes with five to six copies each. Multiple genomes have not been reported for free-living bacteria, but have for some mitochondria and chloroplasts.

Studies by Preer et al. (1974) using the electron microscope have revealed interesting structures within kappa. The R body, seen earlier by others, is the most striking feature. R bodies were found to consist of a thin sheet of protein about 0.4 $\mu$m wide and 12 $\mu$m long, wound around itself into a tight coil (Fig. 11). Under certain conditions, the R body can unroll completely to form a wavy ribbon or a long tube (Fig. 12). On the inside end of the R body, structures thought to be defective phages or parts of phages are observed (Fig. 13). In some cases these look like phage heads and are icosahedral in form (Fig. 14). They contain about equal amounts of DNA and protein in stock 562. The buoyant density of the phage DNA as determined by density gradient centrifugation in cesium chloride is estimated at 1.700 gm/cm$^3$, while that of 562 kappa DNA is 1.702 gm/cm$^3$. In other strains, the phage heads are empty of DNA. In still others, such as stock 51, the R bodies unroll from the inside (Fig. 15). Helical structures (Fig. 16) that may be unassembled capsid protein are seen on the inner end of the R body. Since infection with these phages has never been accomplished, they are regarded as defective. On the basis of kind of killing produced, how the R body unrolls, shape of the outer end of the R body, and the kind of phage-like structures associated with the R body, three major classes of kappa were distinguished among the kappas of sixteen different killing stocks. The three classes are typified by the kappa found in stocks 7, 51, and 562. Almost without exception the R body is found only when the phage like structures are present. For this reason and others to be discussed presently, the R body is believed to be a product of the phages. It has been proposed that in nonbright kappas latent prophage is present. Spontaneous induction occurs, giving rise to mature phage with an increase in phage DNA and the protein it specifies,

**Fig. 11.** Electron micrograph of intact coiled refractile body of kappa of stock 511 *Paramecium biaurelia*; phosphotungstic acid. × 100,000, scale 0.25 $\mu$m. After Preer *et al.* (1974) with permission of *Bacteriol. Rev.*

**Fig. 12.** Electron micrograph of unrolled refractile body of kappa of stock 1039 *Paramecium biaurelia*; phosphotungstic acid. × 15,000, scale 1 $\mu$m. After Preer *et al.* (1974) with permission of *Bacteriol. Rev.*

**Fig. 13.** Electron micrograph of unrolled refractile body of kappa of stock 249 *Paramecium biaurelia* with adhering spherical phages on tip; phosphotungstic acid. × 130,000, scale 0.25 $\mu$m. After Preer *et al.* (1972) with permission of *J. Cell Sci.*

**Fig. 14.** Electron micrograph of stock 562 *Paramecium biaurelia* purified spherical phages prepared by centrifugation in cesium chloride density gradient; phosphotungstic acid. × 200,000, scale 0.1 $\mu$m. After Preer *et al.* (1972) with permission of *J. Cell Sci.*

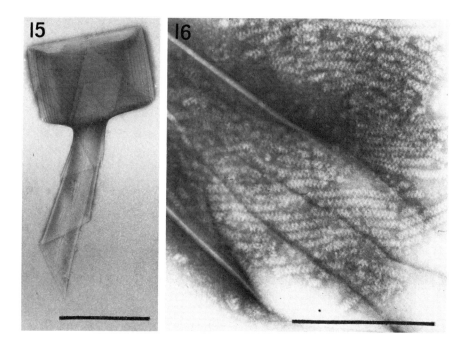

**Fig. 15.** Electron micrograph of stock 51 *Paramecium tetraurelia* refractile body of kappa unrolling from inside. × 48,000, scale 0.5 μm. After Preer *et al.* (1972) with permission of *J. Cell Sci.*

**Fig. 16.** Electron micrograph of stock 298 *Paramecium tetraurelia* unrolled refractile body of kappa with associated helical structures. × 150,000, scale 0.25 μm. After Preer *et al.* (1972) with permission of *J. Cell Sci.*

including the R body. Upon induction, the nonbright kappa is transformed to a bright kappa, which is toxic. Subsequently the bright kappa dies.

How killing of sensitives is accomplished is being investigated. In no case can anything less than an intact, rolled-up R body kill. It has been demonstrated that the site of entry of the toxin is the food vacuole (Preer, 1975), since particles that interfere with the uptake of kappa into the food vacuole also inhibit killing. After being taken into a sensitive, the R body is unrolled. Studies with the electron microscope show that the food vacuole membrane is ruptured and the R body and the phage enter the cytoplasm when killing occurs. Killing is thought to be the effect of a toxic protein on the sensitive, rather than a result of phage multiplication, since the action of the toxin can be modified by chymotrypsin.

Singler (1974) has shown that phages and R bodies of stock 562 have antigenic sites in common, since both are agglutinated by antiserum against the phage. She suggests that the R body protein and the phage capsid protein are related, that both are coded by the phage DNA, and that perhaps the R body represents an incomplete assembly of phage coat molecules. She found that phage antiserum inhibited the killing effect of isolated kappa, thus showing that phage is implicated in killing. She concludes that it is not known whether the toxin is the R body, the phage capsid, or some other substance, but that it is likely that they are all related to each other and specified by the phage genome.

Covalently closed circular (CCC) DNA has been demonstrated by Dilts (1975) in kappa isolated from stocks 51 (Fig. 17), 298, and 6g2 of *P. tetraurelia*, which have the 51 class of kappa. Isolation of CCC DNA was accomplished by centrifugation in ethidium bromide–cesium chloride density gradients. The CCC DNA was found to have a density of 1.698 gm/cm³ in cesium chloride alone, while the linear DNA had a density of 1.700–1.701 gm/cm³. When whole kappa DNA was centrifuged in cesium

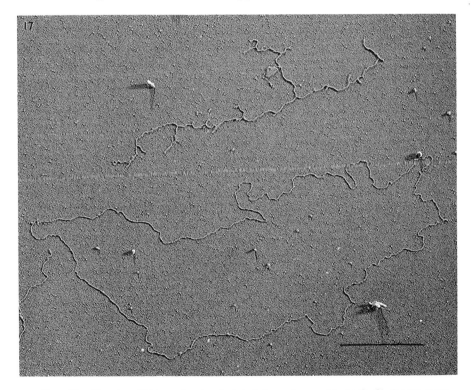

**Fig. 17.**  Circular DNA molecules isolated from kappa of stock 51 *Paramecium primaurelia*: covalently closed circular DNA (CCC), top; open circular DNA (OC), bottom. Scale 1 μm. Electron micrograph courtesy of Dr. D. Lang; DNA preparation of Dr. J. Dilts.

chloride, two peaks were found at these same densities, one due to kappa chromosomal DNA, and the other due to the extrachromosomal DNA. The contour length of the isolated CCC DNA is 13.75 ± 0.04 $\mu$m.

Helical phage-like structures are present in R bodies of bright kappas of stock 51. Dilts (1975) obtained preparations containing high and low percentages of bright kappas by zonal centrifugation in a sucrose gradient. Isopycnic centrifugation in cesium chloride of preparations containing a high percentage of bright kappas were found to have more extra-chromosomal DNA than preparations with a low percentage of brights. This correlated increase in extrachromosomal DNA and bright kappas suggests that the extrachromosomal DNA is the determinant for the helical phage-like structures and the R body.

The evidence that led finally to the conclusion that the endosymbionts present in the *P. aurelia* complex are indeed bacteria has been summarized by Preer (1975). In addition to size and shape, the endosymbionts have other visible features of bacteria: they are surrounded by two double membranes, which look much like those of gram-negative bacteria; typical peritrichous bacterial flagella are present on some; a nuclear membrane and mitochondria are lacking in all of them. The chemical composition of the endosymbionts is bacteria-like, with DNA, RNA, protein and lipids in the usual bacterial proportions. Mu contains ribosomal RNA's which hybridize with the DNA of *E. coli*, but not that of paramecia, and the ribosomal RNA's of mu and alpha have sedimentation coefficients like those of bacteria, not paramecia. They have DNA with a base ratio distinctly different from that of the nuclear or mitochondrial DNA of the paramecium. They produce enzymes and other complex substances involved in cellular metabolic processes. For example, oxidative enzymes are present in kappa; diaminopimelic acid and possibly muramic acid, constituents of bacterial cell walls, are found in mu. Mu and lambda have been reported to be cultured extracellularly, although results are conflicting, see Preer *et al.* (1974). Phage-like structures associated with a refractile protein structure, the R body, have been demonstrated. Induction of R bodies occurs upon treatment of kappa with ultraviolet light, suggesting a situation similar to induction of prophage of lysogenic bacteria or induction of colicins in colicinogenic bacteria. Moreover, as in the case of bacteria carrying phage or plasmids, covalently closed circular DNA associated with kappa bearing phage-like structures has been characterized.

The endosymbionts of the *P. aurelia* complex have recently been classified (Preer *et al.*, 1974) on the basis of their morphology, physiology, distribution, killing properties, and DNA densities. Alpha is described as a new species of the gliding bacteria of the genus *Cytophaga*, called *C. caryophila*. One new bacterial family is described, the *Caedobacteriaceae*, comprising three new genera: *Caedobacter*, *Lyticum*, *Tectobacter*. The species described in these genera, with their common names in parentheses, are: *C. taeniospiralis* (kappa), *C. conjugatus* (mu), *C. minutus* (gamma), *C. falsus* (nu), *L. flagellatum* (lambda), *L. sinuosum* (sigma), *T. vulgaris* (delta).

## IV. VIRUSES IN FUNGI

Many small, polyhedral, double-stranded RNA viruses are found in fungi (Lemke and Nash, 1974). Transmission through the medium is rare; transmission at sexual and asexual reproduction and transmission at heterokaryon formation (fusion involving hyphae and asexual spores) are the main means of spread. Although spread by heterokaryon formation is generally regarded as a form of infection, it surely must also be regarded as hereditary transfer. So effective is spread by heterokaryon formation that elements may be easily maintained in the face of adverse selection against infected individuals. Thus, many fungal viruses are harmful to their hosts. Others, although harmful in some strains, show little effect in others.

It is possible that many of the reported cases of cytoplasmic inheritance in fungi are due to viruses. This possibility is especially likely in cases where harmful effects have been found, such as, for example, a case of lethal cytoplasms in aging clones of *Aspergillus* studied by Jinks (1959). Genetic techniques are completely ineffective in resolving the basis for such characters. Only biochemical or microscopical studies suffice.

Two cases, the killer systems of *Ustilago maydis* (the fungus which causes corn smut disease) and the yeast *Saccharomyces cerevisiae* deserve special comment. Killer strains of both organisms liberate into the medium in which they live toxic proteins that kill sensitive strains of their own species. In both cases the genetic basis has been shown to involve cytoplasmic determinants.

Puhalla (1968) showed that resistant killers, resistant nonkillers, or sensitives may be present among strains of *Ustilago*. Resistant killers carry the killing cytoplasmic factor (I) and the resistance factor (S), irrespective of what nuclear genes are present. Resistance is conferred upon strains by the presence of either the (S) factor or the $s^+$ allele. Only strains which lack both cytoplasmic elements and have the $s$ allele are sensitive. Although it is generally assumed (Day and Anagnostakis, 1973) that (I) confers killing ability and not resistance, the possibility that both (I) and (S) confer resistance (similar to the situation in killer paramecia) is not ruled out. Strains carrying only the (I) element have never been isolated. It has

been shown that all strains bearing (I) and (S) or only (S) contain a spherical dsRNA (double-stranded RNA) virus. Moreover, all strains lacking
the cytoplasmic elements are free of the virus. It is thus likely that (S) is
the virus that is responsible for the toxin. It is also, of course, quite
possible that (I) and (S) represent variants of the same element.

It is reported that killer strains of *Saccharomyces cerevisiae*, first discovered by Bevan and Makower (1963), are quite common. Woods and
Bevan (1968) showed that they liberate a proteinaceous toxin into the
medium that can kill sensitive strains of *S. cerevisiae*. Somers and Bevan
(1969) showed that a cytoplasmic element is responsible for the killer trait.
It confers on its host not only the ability to produce toxin, but also resistance to the toxin.

In addition to the usual killers, there are cytoplasmic mutants yielding
temperature-sensitive killers that only kill below 30°C, and superkillers
that kill very strongly. Since killers produce their toxin under certain
conditions, not others, it has also been possible to demonstrate cytoplasmic mutants that kill themselves when placed under conditions that induce
toxin production (Vodkin *et al.*, 1974). A gene mutation, *rex 1*, which has
this same ability to kill itself has also been described by Wickner (1974).
Neutral strains that are resistant nonkillers are also determined by a similar cytoplasmic element. Cytoplasmic mutations to nonkiller sensitives
may occur naturally, or can be induced by cycloheximide or by
fluorouracil. Other nonkiller cytoplasmic sensitives are suppressive and
when crossed to killers or neutrals convert them to sensitives (Somers,
1973). Nuclear genes necessary for maintenance of the elements have also
been described (reviewed in Wickner, 1974). Other genes are necessary
for expression of the killer trait.

The killer and neutral phenotypes are perfectly correlated with the presence of a medium sized double-stranded (ds) RNA that is absent in sensitives (Berry and Bevan, 1972; Bevan *et al.*, 1973; Vodkin *et al.*, 1974). In
addition, still smaller dsRNA molecules that vary from strain to strain are
found in suppressives and the killers that kill themselves, but not in sensitives. These correlations of the characteristic dsRNA's with the killer
phenotypes are maintained in the progeny of various kinds of crosses, and
leave little doubt that they form the cytoplasmic basis for the inheritance
of the traits. The picture is complicated, however, by the fact that all
strains of killers and most sensitives too, carry a dsRNA that is larger than
all of these. Its constancy and lack of correlation with variations in the
killer phenotype indicate that it bears no killer or resistant determinants.

Herring and Bevan (1974) have demonstrated virus capsids in both
killers and sensitives that enclose these dsRNA's (see Fig. 18). They were
also able to show that the large and medium dsRNA molecules are sepa-

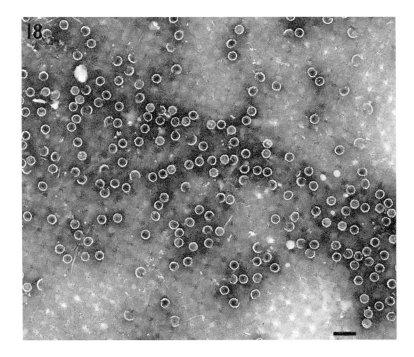

**Fig. 18.**   Virus-like particles from a strain of killer yeast. × 60,000, scale 0.1 μm. After Herring and Bevan (1974) with permission of *J. Gen. Virol.*

rately encapsidated in the viruses. Furthermore, the capsids bearing the two species of RNA appear to be constructed of identical proteins (Bevan, 1976). It is concluded that most *S. cerevisiae* bear dsRNA viruses which are closely related. Certain ones of these, identified by their characteristic dsRNAs are the basis for the killer trait.

## V. ELEMENTS IN INSECTS

Many bacterial and viral endosymbionts have been found in insects and many are transmitted through the gametes. Three especially well-studied examples have been chosen to illustrate the different cases. First are "sex ratio" spiroplasmas, bacterial endosymbionts of *Drosophila willistoni* and related species. Second, is the thoroughly studied sigma virus which causes $CO_2$ sensitivity in *D. melanogaster*. Third is delta, a well-studied, but nevertheless poorly understood element in *D. melanogaster*. The reader is referred to Preer (1971) for references and a discussion of a number of other cases as follows: Mycoplasmas of *D. paulistorum*, according to Ehrman, play a role in cytoplasmic sterility which she believes is important in the evolution of that species. Virus-like elements determine sex ratio in *D. bifasciata* and *D. robusta*. An episome-like, maternally inherited agent is responsible for chromosome breakage in *D. robusta*, according to Levitan (Levitan and Williams, 1965). Virus-like elements appear to be involved in the production of melanotic tumors in *D. melanogaster*. Cytoplasmic sterility in interspecific crosses in the common mosquito, *Culex*, has been studied for many years by Laven (1967) and others, but its basis has been unknown. More recently it has been shown by Yen and Barr (1971, 1973, 1974) that it is due to bacterial rickettsia-like endosymbionts classified as *Wolbachia pipientis*. They, along with sterility, may be eliminated by tetracycline. The numerous strains of endosymbionts are a major factor in speciation in *Culex* and probably in other mosquitos as well.

## A. Sex Ratio Spiroplasmas

Sex ratio variants in *Drosophila* show an excess of females, and the phenomenon is called SR. Males may be missing altogether, and lines can be maintained only by outcrossing. SR in the *obscura* group of *Drosophila* is chromosomally determined and appears to be an example of meiotic drive in which X and Y gametes are not produced in equal frequency (Policansky, 1974). Although many cases of SR have been shown to have a chromosomal basis, in some it is transmitted only by the mother. Such maternal sex ratio, as noted above, in *D. bifasciata* and *D. robusta* is due to virus-like elements. However, in certain species of *Drosophila* of the *willistoni* group, maternal SR has been shown to be due to infections of *Spiroplasma*. Eggs fertilized by Y-bearing sperm die as zygotes or early in development (see review by Poulson, 1963). For many years it was thought that the infectious elements were spirochaetes. However, they lack the cell wall and axial filament of spirochaetes (Poulson and Oishi, 1973). More recently, Williamson and Whitcomb (1974, 1975) have shown that they are mycoplasmas of the genus *Spiroplasma*.

The spiroplasmas are 5–6 $\mu$m long, 0.1–0.2 $\mu$m wide and have a spiral form (see Fig. 19). They are found in various tissues and especially in the hemolymph. Although they are normally transmitted only through the egg they may be transferred by injection in the laboratory. They have not been cultured, but related spiroplasmas from plants have.

Not only is it possible to infect spiroplasma-free strains of the *willistoni* group, but the spiroplasmas have been transferred by injection into *D. melanogaster*, *D. bifasciata*, *D. robusta*, and *D. pseudoobscura*. As a result

**Fig. 19.**    Sex ratio spiroplasma from *Drosophila*, after Oishi and Poulson (1970).

SR generally appears in the new hosts, but in *D. bifasciata*, although the spiroplasmas grow, they have little or no effect on the sex ratio. Transfer back to other species results in restoration of SR (Ikeda, 1965). All transfers are not successful, however, for although the spiroplasma from *D. equinoxiales* can be established in *D. pseudoobscura*, it does not persist in *D. melanogaster* (Sagaguchi and Tuchiyama, 1974).

Spiroplasmas have been eliminated from some, but not all strains, by culturing the *Drosophila* at high temperatures. Hemolymph may be rendered inactive in injections by treating it with X-rays or penicillin. Although penicillin treatment of whole *Drosophila* reduces the population of spiroplasmas, it does not bring about permanent cure (Poulson, 1969). Some strains of *Drosophila*, called "disrupter" strains, carry polygenes which do not permit maintenance of the spiroplasmas.

When SR spiroplasmas from different sources are mixed, either *in vitro* or *in vivo* by injection, the spiroplasmas clump and later lyse. An agent separable from the spiroplasmas is responsible for these phenomena. The agent has been isolated and found to be a DNA bacteriophage 500–600 Å in diameter (Oishi and Poulson, 1970). The spiroplasmas from different species cause clumping and lysis of each other, but usually spiroplasmas from different strains of the same species of *Drosophila* do not interact in this way. Spiroplasmas (with their phages) from one species may be injected into another SR species of *Drosophila*, thereby curing the *Drosophila* of both spiroplasmas.

Studies on the spiroplasma strain WSR revealed an unusual kind of phage (Oishi, 1971). Normally, if a strain of spiroplasmas such as NSR, which bears the normal kind of phage, is homogenized, the phage can be recovered, as judged by its ability to clump and lyse other spiroplasmas (such as WSR). The phage is thus said to be in a "manifest" state. On the other hand, if the aberrant WSR is homogenized, no phage is recovered. Nevertheless WSR does carry a phage, for if WSR-bearing flies are exposed to the phage from NSR, the phages carried within WSR in a "latent" state multiply (along with the newly introduced phages from NSR) and can be demonstrated by their ability to lyse NSR and other strains of spiroplasmas. It may be that the phages within WSR are defective but can be repaired by combination with the infective phage from NSR. In any event, both phages produce the SR phenotype. In fact, all spiroplasmas in *Drosophila* produce SR, and all spiroplasmas have phages that cause clumping and lysis.

The androcidal effect on male zygotes appears to be generated by the spiroplasmas, possibly the phage or a product of the phage, called "androcidin." When spiroplasmas are injected into *D. robusta*, some injected flies become infected, exhibit SR, and transmit SR to future generations. Other injected flies exhibit SR, yet fail to transmit either spiroplasmas or SR, suggesting that SR and the spiroplasmas may be separated (Williamson, 1966). Injection of phage from NSR spiroplasmas into *Drosophila* bearing WSR lyses the WSR, the androcidin is lost and an immediate and permanent cure of the SR condition occurs. Injection of active phage from induced WSR spiroplasmas into *Drosophila* bearing NSR spiroplasmas lyses the spiroplasmas, but the SR trait does not disappear until the next generation (Oishi, 1971). It is assumed that in this case the androcidin produced by NSR is either in higher titer or else is more stable than the androcidin of WSR. How the androcidin manages to kill male, but not female, zygotes is entirely unknown. [For a discussion of the role of intracellular symbionts in sex determination of certain coccids see Chapter 4 (p. 139). Eds.]

## B. Sigma Virus

Most of the work in this section has been reviewed by Seecof (1968). Normally, when insects are exposed to high concentrations of $CO_2$ they are anesthetized, but then quickly and completely recover when the $CO_2$ is removed. L'Héritier and Tessier (1937) discovered a sensitive strain of *D. melanogaster* in which flies do not recover from $CO_2$. The trait is widespread, having been found in many wild strains of *D. melanogaster* and in other species of *Drosophila* as well.

Crosses between normal and $CO_2$-sensitive strains show that transmission of the determinant of sensitivity, sigma, is usually through the female, rarely through the male. Additional crosses between resistants and sensitives reveal a remarkable phenomenon. The sensitive progeny produced by crossing sensitive females to resistant males are entirely normal in their breeding behavior, transmitting sensitivity in virtually 100% of the cases from females and incompletely from males, as already described. However, in sensitive progeny produced by the reciprocal cross of sensitive males to resistant females, the females transmit only incompletely (to less than 100% of their progeny), and the males never transmit the sensitivity. Thus two kinds of sensitive flies may be distinguished on the basis of their performance in breeding tests. The females from the original sensitive strain that always transmit the trait and the males that sometimes do are said to be stabilized. Sensitive males and females from a cross of a sensitive male with a resistant female that transmit sensitivity with a low frequency are said to be nonstabilized.

Important clues concerning the basis of the stabilized and nonstabilized states come from the fact that sigma may be transferred by injecting an extract of sensitive flies into the abdomen of resistant larvae, pupae, or adults. Flies often develop resistance, and are transformed into nonstabilized individuals that behave exactly like the nonstabilized flies described above. Although, as already described, males of nonstabilized flies never transmit the trait, females often do. The sensitive progeny of nonstabilized females, however produced, consist of some newly stabilized and some nonstabilized flies. It is thought that in stabilized flies, sigma has invaded the germ line, and in nonstabilized flies it has not. The transformation from nonstabilized to stabilized lines is therefore called "passage to the germ line."

Passage to the germ line, however, seems to involve more than just the invasion of a new tissue. Sigma particles may be titered by injecting appropriate dilutions of extracts. Measurement of the number of infecting sigma particles by titration shows that the total amount of infective sigma in nonstabilized flies (which transmit poorly to the next generation in crosses) is much greater than the amount in stabilized flies (which transmit much better). This finding indicates that the relationship between sigma and its host is different in stabilized and nonstabilized flies. It has been suggested by several workers that in stabilized flies, most of the sigma exists in a noninfective provirus form and is transmitted by cellular heredity, but in nonstabilized flies it is maintained in the infective form. The biochemical characteristics of sigma in the two states is of great interest, and the problem is clearly not solved. It may be that the two states have more to do with repression and derepression of genes, not only virus genes, but even genes of *Drosophila* (L'Héritier, 1970).

The maintenance of sigma is affected by several genes in *Drosophila* as well as genes of sigma. Other genes of sigma affect the stabilized and nonstabilized states differentially. The sigma mutants *ultrarho* and *rho* are impaired in their ability to multiply in the nonstabilized state. Extracts of *ultrarho* contain no infectious virus, and *ultrarho* flies are resistant to $CO_2$. *Ultrarho* can only be detected by its resistance to superinfection by infectious sigma. (Superinfection immunity is a general feature of all sigma and is readily demonstrated by utilizing mutant strains of sigma.) *Rho* is much like *ultrarho*, but appears to produce a small number of wild-type sigma by reverse mutation. Other sigma mutants are normal in their nonstabilized state, but are modified in their behavior in stabilized flies. $v^-$ cannot exist in the male germ line and is never transmitted by males. $g^-$ mutants can exist in the germ line of males and females and can be transferred by injection to resistant flies. However, they are unable to undergo "passage to the germ line." Thus injected $g^-$ causes $CO_2$ sensitivity, but it is never transmitted to the next generation.

Infections with mixed mutants have shown that it is possible to get recombination between multiply marked mutants, but recombinants have proved to be unstable.

Berkaloff and co-workers (1965) have found sigma with the electron microscope. Work with inhibitors suggests that it is an RNA virus (Herforth, 1973). Although it has not been isolated chemically, it can be maintained in transferred imaginal discs and in cells and tissue *in vitro*. Its highly characteristic structure suggests that it is a rhabdovirus related to vesicular stomatitis virus (VSV) and rabies virus.

VSV is normally found in horses and cattle, where it is pathogenic. The morphological similarities between sigma and VSV have led to attempts to establish VSV in *Drosophila*. The work has been reviewed by Printz (1973). One of the several known serologically different strains of VSV, the Indiana strain, was found to grow in *D. melanogaster* after injection. After several passages it adapted to its new host and even produced $CO_2$ sensitivity. However, it was never transmitted by heredity, and could be transferred only by injection. Obviously VSV never became as well adapted to *Drosophila* as sigma.

## C. Delta

Delta is an element which is rather different from most of the others discussed in this chapter. It is found in certain strains of *Drosophila melanogaster* and often kills its host. The work on delta b and delta r has been presented in a number of papers by Minamori and co-workers beginning in 1969. Two of the most recent papers that give references to all of the earlier papers are Minamori and Sugimoto (1973) and Ito (1974). Perhaps the easiest way to understand the complex facts about to be presented is initially to regard the deltas simply as gene products with somewhat unusual properties. The nature and significance of these properties will be considered later. Certain alleles at a locus on the second chromosome (the *Da* locus) can carry delta; i.e., under the proper conditions these alleles can initiate and allow the deltas to persist in the cytoplasm. Although it is conceivable that a chromosome might be found that cannot initiate delta yet can allow its existence, in fact, none has been found. The ability to carry delta is inherited as a dominant. Another property, sensitivity to the killing and mutagenic action of delta, is strongly (but not absolutely) correlated with the ability to carry delta. Sensitivity is determined by the same second chromosomal locus. It is unknown why sensitivity to delta and the ability to carry delta are related. Five kinds of second chromosomes, $S^b$, $S^r$, $S^c$, $ID^b$, $ID^c$, may be identified in respect to sensitivity and the ability to carry delta b and delta r. Sensitivity is designated by S, insensitivity by ID or I, the ability to carry a given delta by the superscripts b for delta b, r for delta r and c for neither. The combinations $S^b/S^b$, $S^b/S^r$, $S^b/S^c$ are sensitive to delta B. Only $S^r/S^r$ is sensitive to delta r.

Perhaps the most striking and unusual property of the deltas is exhibited most clearly by delta b. While delta b can be transmitted in high concentration by the egg, only small quantities can be transmitted by the sperm. If delta b is introduced by means of a spermatozoan carrying an $S^b$ chromosome into a cytoplasm previously lacking delta b, the concentration of delta b is initially too low to kill, but after many generations it gradually accumulates to high levels in female lines of descent. These facts mean that the rate of synthesis of delta b is high when delta b is in high concentration but low when delta b is in low concentration. If it were not, equilibrium should be reached very quickly after a small amount of delta b is introduced into a delta-free line. This autocatalytic property is one of the main reasons for viewing delta as a virus-like agent. Obviously, there are other interpretations possible. We return to this point later.

Crosses illustrating delta killing and the differences in paternal and maternal transmission of low and high levels of delta b are as follows. In the cross $bw^D/bw^D$ ♀ × $Cy/Pm$ ♂ ($bw^D$, $Cy$ and $Pm$ are second chromosome marker genes), if there were no delta, one would expect a 1 : 1 ratio of $Cy$ to $Pm$ in the $F_1$. However, the $bw^D/bw^D$ parental females ($S^b/S^b$, i.e., the $bw^D$-bearing chromosome is $S^b$) carry high concentrations of delta b; the parental $Cy/Pm$ males ($ID^c/S^c$, i.e., the $Cy$ chromosome is $ID^c$ and the $Pm$ chromosome is $S^c$) carry none. As a result, $Cy$ offspring are about twice as frequent as $Pm$ because (since sensitivity is recessive) the $S^b/ID^c$ ($Cy$) $F_1$ progeny are insensitive to delta b, while many of the $S^b/S^c$ ($Pm$) $F_1$ progeny are killed by delta b. In the reciprocal cross of $Cy/Pm$ ♀ × $bw^D/bw^D$ ♂ a normal 1 : 1 ratio of $Cy$ : $Pm$ is found. This result is obtained because the $Cy/Pm$ line carries no delta, and although the $bw^D/bw^D$ males transmit $S^b$ chromosomally (transmission is dominant), only very small amounts are transmitted through the male, too low in fact to produce killing.

Minamori showed that the killing of the $S^b/S^c$ individuals in crosses such as that described above occurred in zygotes or early embryos and that more females than males were killed. In some cases, however, it has been shown that killing can completely eliminate certain classes of gametes.

Not only does delta kill, it also induces mutations. Minamori and Ito (1971) showed that delta b induces lethals at 3–4 times the spontaneous rate in various second chromosomes ($S^b$, $ID^b$, $ID^c$). The lethal mutations do not occur at random but tend to be clustered in the three regions 0–15, 55–65, 70–85 crossover units.

All the effects of delta, however, are not harmful. By growth at appropriate temperature or by means of suitable crosses it is possible to obtain $Cy/ID45$ ($ID^c/ID^b$) flies with high or low levels of delta b. If delta b is low, the flies are partially sterile. Thus in flies of this genotype, delta b is necessary for gametogenesis in both males and females. Indeed, Minamori and Ito suggest that delta may have an essential function in gametogenesis, the killing effect appearing only when it occurs in an abnormal genotype.

Other crosses illustrate paternal transmission of delta b by chromosome 2 and the ensuing slow cytoplasmic accumulation of high levels of delta b. First delta b was established in high concentration in the line $Cy/S5$ ($ID^c/ID^b$). ($Cy$ and S5 are both lethal when homozygous and differ by complex inversions as well: they form a balanced lethal stock.) Males from this line were now crossed to females from a line of $Cy/Pm$ ($ID^c/S^c$), flies free of delta b and incapable of maintaining it. Therefore, $Cy/S5$ males from this cross, if they received delta b, could have obtained it only from their fathers. They were backcrossed to females from the delta-free $Cy/Pm$ stock. $Cy/S5$ males were again selected for backcrossing and the process was continued repeatedly. Any cytoplasmic delta should be eliminated in this way. After 20 backcross generations $Cy/S5$ males were used to establish a new line of $Cy/S5$ flies like the original by mating them with their $Cy/S5$ sisters. When appropriate crosses to test for ratio distortion were now made, no evidence for delta b could be found, indicating that delta had indeed been eliminated. The new $Cy/S5$ lines were maintained by inbreeding to allow delta b to reappear if it could. The appropriate crosses to test for ratio distortion were made at every generation of inbreeding. No distortion was found for some time, but as successive generations of inbreeding passed, distortion slowly appeared and became stronger. Thirty-five generations of inbreeding were required before the lines acquired their full level of cytoplasmic delta, as judged by the degree of ratio distortion. Minamori concludes that although the capacity for production of delta b is an inseparable property of the $S^b$ chromosome, the long time required for distortion to appear indicates that delta b has self-replicating properties. Therefore, he suggests that it cannot be thought of as a simple gene product which accumulates to high levels after a lag in gene expression.

Minamori and Sugimoto found that the ability of the $S^b$ chromosome to retain delta b was due to a gene, $Da$, at 24.9 on the second chromosome. They also found that they were unable by recombination to separate sensitivity to delta b from the capacity of the $S^b$ chromosome to carry delta b. No explanation of why maintenance of delta b and sensitivity are associated is apparent. It should be recalled, however, that these two properties are separate in the naturally occurring chromosomes $S^c$ (sensitive but does not transmit) and $ID^b$ (transmits but is insensitive).

Delta r is similar to delta b, yet differs in some important properties. Only the $S^r$ chromosome carries delta r and only individuals with two $S^r$ chromosomes are sensitive to delta r. Delta r is transmitted biparentally with no evidence of more being transmitted by females than males. Transmission and apparently sensitivity to delta r maps on the $S^r$ chromosome very close to $Da$, and both are presumed to be determined by an allele at the same locus. Like delta b, delta r does not reach lethal levels in the offspring of gametes which are produced by flies cultured at 18°C, but does in 25°C lines. For an unknown reason, flies homozygous for S20 ($S^r/S^r$) have only low levels of delta r. Surprisingly, after introduction of an insensitive $Cy$ chromosome to yield S20/$Cy$ ($S^r/ID^c$), delta r showed a gradual increase over nine generations to its normal high level. The parallels between the activities of delta b and delta r suggest that the two are closely related. It is noted that the $S^r$ chromosome found in strain SD 20 bears the Segregation–Distorter gene on its second chromosome. SD causes SD/SD$^+$ males to produce a reduced number of functional SD$^+$ bearing spermatozoa, an effect somewhat similar to the effects of delta b and delta r. Nevertheless, Minamori has found no other similarities between the two phenomena and it is assumed that they are unrelated.

Minamori and Sugimoto (1973) exposed $Cy/I521$ ($ID^c/ID^c$) lines to ethyl methane sulfonate (EMS) and the I521 chromosomes were screened for sensitivity; 5 out of 1970 mutant chromosomes were found. In four cases $ID^c$ mutated to $S^b$, and delta b was produced; in one case $ID^c$ mutated to $S^r$, producing delta r. 2131 controls showed no mutation. Other data indicated that similar mutations could be induced by delta b and delta r. In one case delta r appeared to induce a delta b. The data thus suggest that noncarrier resistant second chromosomes of *D. melanogaster* contain the information for delta at 24.9 crossover units in an unexpressed form. Such chromosomes can mutate to yield a *Da* allele which now both carries delta and bestows sensitivity to delta.

What is delta? Is it a simple gene product? It may be a gene product, but it is certainly not a simple one, for its production is autocatalytic. Although this fact suggests self-replicating elements, differing states of gene activity may also be responsible for such effects. (See, for example, the discussion of the inheritance of mating type, serotype and other characters in *Paramecium* by Preer, 1969; see also Chapter 9.) Minamori and Ito (1971) cite parallels between delta and a number of other elements; thus, certain viruses when introduced into *Drosophila* induce mutations and kill cells as well. Sigma virus of *D. melanogaster* can also produce mutations in its host. Injected free DNA from various sources has been shown to induce heritable changes in *Drosophila*. The mutagenic effects of delta are similar to the action of the well-known controlling elements in various organisms, such as *Drosophila* and maize. Delta also resembles chromosome breaker in *D. robusta* (see above), and the agents found in certain mutator strains of *Drosophila*. If delta is a virus-like agent, then the data suggest that it is well-established in *Drosophila* and is possibly akin to chromosomally borne oncogenic viruses.

## VI. ONCOGENIC VIRUSES

Most of our information concerning infectious heredity of viruses in vertebrates comes from work on the oncogenic or tumor-inducing viruses. The very large body of work on these forms is treated only briefly here with emphasis on their role as agents of infectious heredity. A recent book documenting most of the information given here is Tooze (1973); the reader is also referred to the *Symposia in Quantitative Biology*, Cold Spring Harbor, 1974, volume 29 for recent references.

The usual mode of transmission of viruses is laterally by infection from one individual to another in the population. In some cases, however, vertical transmission from parent to offspring occurs. If infective viruses are transferred vertically, the process is termed congenital infection. On the other hand, if noninfective proviruses are transmitted vertically, they do so while integrated into the chromosomes of their hosts and are said to show true genetic transfer.

Congenital infection might be expected to include, and, in fact, does include a number of diverse modes of transmission. A classic case is the maternally transmitted mouse mammary tumor virus of Bittner (MTV-S), which passes from mother to offspring through the mother's milk after birth. If the embryos from infected mothers are removed by cesarean section and allowed to suckle on mothers from a low mammary tumor strain, the young do not acquire the virus. In the case of maternal transmission of RIF (Rous interfering factor), an RNA virus of chickens, it appears likely that infective virus is transmitted intracellularly through the egg (Rubin *et al.*, 1961) but not the sperm. Experimentally it is not easy to distinguish those cases of congenital infection which arise from invasion of the embryo by extracellular virus from the mother (true infection), from those cases due to intracellular extrachromosomal transmission through the germ line to the gametes (true extrachromosomal inheritance) as with RIF. Besides the cases of RIF virus in chickens and the MTV-S virus in mice, a number of cases of maternal transmission involving other viruses in mice have been described (Gross, 1961; Haas, 1941; Smith, 1959). No case of paternal transmission of infectious virus particles has been reported. In any event it should be clear that genetic transfer of proviruses, as well as all cases of congenital infection that involve intracellular transfer, constitute infectious heredity. Even transfer from parental to daughter cells at mitosis, as seen in tissue cultures, constitutes infectious heredity at the cellular level.

The chief oncogenic viruses are all spherical. The papovaviruses, the adenoviruses and the herpesviruses all contain DNA. The smallest and best known of the DNA viruses belong to the papova group (polyoma, SV40, and papilloma). See Dulbecco (1969) for a succinct account of much of the basic pioneering work on these viruses. They are distinguished by having covalently closed circular DNA of $3-5 \times 10^6$ daltons. Polyomavirus induces connective tissue tumors (carcinomas) of various kinds in mice. Polyomavirus grows in permissive mouse cells (such as 3T3 fibroblasts) in tissue culture in which it usually behaves as a lytic virus. Hamster cells (such as BHK fibroblasts) in tissue culture, however, are nonpermissive: upon exposure to the virus they are transformed and become neoplastic, a number of other properties becoming stably modified. Permissive 3T3 cells may also be transformed, but much less frequently. Transformed cells are neoplastic and form tumors when transplanted into mice. Transformed cells do not contain infective virus.

SV40 (originally isolated from a monkey) is very similar to polyomavirus. Various lines of cells may be permissive or nonpermissive for it, just as they are for polyomavirus. If cells of an uninfected line permissive for SV40 are fused with virus-free cells which have been transformed by SV40, infectious SV40 is released. So, transformed cells contain the complete SV40 genome in proviral form. A different technique has been used to demonstrate that complete viruses can be recovered from cells transformed by a temperature sensitive line of polyomavirus.

When polyoma and SV40 virus DNA's are used as templates *in vitro* to synthesize complementary RNA, the RNA can hybridize with DNA from transformed (but not untransformed) cells. The results indicate that at least 5 to 60 copies of virus genomes are integrated per cell. Further experiments show that the viral DNA is covalently linked to the DNA of the chromosomes. Hybridization between viral DNA and cellular DNA yields similar conclusions. Studies on hybridization between viral and cellular RNA from transformed cells show that much of the viral DNA is transcribed while integrated. Furthermore a number of antigens coded for by the viral genotype has been found in transformed cells. Reversion of transformed cells may also occur, but the process is poorly understood. The role of host and viral genes and site of integration in establishing transformation is being actively investigated.

No satisfactory *in vitro* system for studying transformation has been found for the other oncogenic DNA-containing viruses. Nevertheless all cause cellular transformations, and transformed cells appear to retain the viral genome as a provirus integrated into the cellular DNA. The papillomaviruses cause warts in most mammals. They are closely related to polyoma and SV40, the other papovaviruses. The adenoviruses are much larger. They contain linear DNA of about $23 \times 10^6$ daltons. Human adenoviruses induce tumors in hamsters. Finally, the herpesviruses are large, common viruses found in frogs, birds and mammals. Many are oncogenic. The Epstein–Barr virus, a herpesvirus, is associated with Burkitt's lymphoma and is the cause of infectious mononucleosis in humans.

Although the DNA viruses induce tumors when injected into animals, they do not appear to be very important in inducing tumors in natural populations. The RNA viruses, however, are much more widespread and are thought to be of much greater importance. The RNA tumor viruses are classified in relation to their morphology, the kind of tumors they produce, and the vertebrates in which they are found. Morphologically they fall into two groups, B-type and C-type. (The A-type was found to be an earlier developmental stage of the other two.) An internal region containing the RNA is called the nucleoid. An outer unit membrane which is formed by budding from the nuclear membrane of the cell is the envelope. B-type particles may be distinguished from C-type because the B-type particles have an eccentrically (rather than centrally) located nucleoid that is placed within an additional membrane (lying between the nucleoid and the envelope). Also the surface of the envelope is more prominently spiked in the B-type viruses (see Fig. 20).

B-type viruses are synonymous with the agents associated with mammary tumors (carcinomas, malignant tumors derived from epithelial tissue). The C-type viruses are associated with two kinds of tumors. First are the sarcomas (tumors of connective tissue) and second are various neoplasms of the hemopoietic and reticuloendothelial systems (collectively called leukoses in fowl and mice). These latter include leukemias, in which the blood contains large numbers of circulating tumor cells, as well as solid tumors such as blastomas.

B-type viruses are common as agents of mammary tumors in mice, and recently a similar virus has been isolated from a human mammary tumor (Furmanski *et al.*, 1974). C-type viruses are extremely common in birds, reptiles, and mammals, and a C-type virus has recently been isolated from leukemic cells in man (Gallagher and Gallo, 1975). Most of the work on C-type viruses has been done on the avian sarcoma viruses (such as the Rous sarcoma virus, RSV), the avian leukosis viruses (ALV), the murine sarcoma viruses (MSV), and the murine leukemic viruses (MLV).

Results on RSV are typical for the C-type viruses. RSV was the first well-studied tumor virus, and the techniques for studying it are generally better than for the other C-type viruses. It can be grown in tissue culture, and assays for transformed cells can be carried out in culture. The Bryan strain of RSV is defective; it can, however, infect in the presence of another virus called Rous helper factor (RHF), an ALV. The helper factor can supply a membrane component needed for infection. After infection the Bryan strain of RSV can still transform its host cells.

B       C

**Fig. 20.**   B- and C-type RNA tumor viruses.

Temin and others have shown that upon entering a cell, RSV produces a DNA provirus copy of its RNA using reverse transcriptase. The DNA provirus acts as a replicative intermediate. Reverse transcriptase is found in all RNA tumor viruses, and evidence supports the view that it is specified by the viral genome. The existence of the provirus is demonstrated by the action of inhibitors of DNA synthesis in preventing replication of the virus, by DNA–RNA hybridization experiments, and finally, by the demonstration that DNA recovered from infected cells can infect other cells and result in transformation and normal progeny viruses. Furthermore the results on hybridization show the the proviral DNA is attached to molecules larger than the viral RNA, suggesting that it is integrated into the nuclear DNA of the host cells. Transformation is apparently a result of the integration of provirus.

The role of genes on the transmission of the RNA tumor viruses has been extensively studied for ALV, MLV, and the murine mammary tumor viruses (MTV). Unfortunately, although not generally recognized, one cannot obtain from crosses decisive genetic data that bear on the question of whether the viral genome is transmitted via the chromosomes. To illustrate, MTV-S, as indicated above, is transmitted maternally through the mother's milk. However, in another strain of mice it has been found that the presence of MTV-P (see Bentvelzen, 1972) depends upon a Mendelian gene. There is no way of knowing from the genetic data whether the gene for the presence of MTV-P contains the information for MTV-P, or is a regulatory gene that merely allows the expression of the viral information present in all lines. Nevertheless, experiments on nucleic acid hybridization and viral antigen expression indicate that the information for RNA tumor viruses is integrated into the chromosomes and regularly transmitted by them. Bentvelzen (1972) has produced a model for the inheritance of MTV in various strains that assumes that the strains differ in regulatory genes. Similar roles for the genes have been hypothesized for ALV and MLV.

Finally, mention must be made of two recent theories of the C-type viruses. Huebner and Todara (1969) have suggested that all animals have the information for C-type viruses which has become integrated into chromosomes as virogenes. Furthermore, oncogenes are present among the virogenes. Both are normally repressed. Expression of virogenes yields active virus; expression of oncogenes yields transformation. Expression may occur in either or both at once. Expression is induced by carcinogens or even spontaneously with age to produce neoplasia, viruses, or both. Temin (1971) has suggested that the proviral oncogenes are actual protoviruses which do not carry the information for the whole viral genome, but instead carry similar base sequences which are transcribed into RNA, then are reverse transcribed back to DNA, then reintegrated in modified ways that eventually lead to complete viral or oncogenic sequences. It is clear that cancer workers are likely to be preoccupied for a long time with the finding that viral information exists within the eukaryotic chromosome.

## VII. VIRUSES IN HIGHER PLANTS

While intracellular bacteria are virtually unknown in higher plants (Lange, 1966), viruses are extremely common (Matthews, 1970). Furthermore, viruses are readily transmitted intracellularly by asexual means. Thus vegetative propagation by bulbs, cuttings, runners, grafts, etc. in horticulture insures the continued maintenance of viruses. In some varieties of potatoes all individuals tested have potato virus X.

Although seeds and pollen often carry viruses and transmit them to their offspring, most viruses are not transmitted in this way, and when they are, the frequency of infected offspring is low in all but a few cases. Transmission by seeds is generally assumed to be intracellular, but the actual presence of viruses within cells of the embryo has only occasionally been demonstrated, as in the case of false stripe of barley virus (Gold *et al.*, 1954). In fact, the resistance of seeds to infection has given rise to considerable speculation as to its cause. Although the mechanism of resistance is unknown, resistance is obviously adaptive, for plant viruses are generally harmful to the host.

It has been suggested that maternally transmitted pollen sterility may in some instances be due to viruses or viruslike elements. Grafts have been shown to induce sterility in their hosts in a number of cases, and certain viruses are known to cause sterility (see Preer, 1971, for references). Although many cases of cytoplasmically inherited male sterility may be due to incompatibility between genes and chloroplasts (see, for example, Göpel, 1970) or possibly genes and mitochondria, other cases may very well be due to viruses or viruslike elements.

*Agrobacterium tumefaciens* induces crown gall disease or tumors in numerous species of higher plants. See Lippincott and Lippincott (1975) for a recent review. The bacterium enters a wound, remains extracellular, and induces the cells to transform to a non-self-limiting pattern of growth. The transformed state is heritable in cell lines both in the plant and in tissue culture and is perpetuated in the absence of the bacteria. There is evidence that the bacteria produce a tumor-inducing principle (TIP) that can transform cells at a distance from the bacterium. Although the nature of TIP is unknown, a number of recent studies suggest that it is RNA

(Beljanski *et al.*, 1974). Of particular interest is the recent finding that all virulent strains of bacteria that have been tested contain a large covalently closed circular DNA plasmid and all avirulent strains tested lack it (Zaenen *et al.*, 1974). Furthermore, when virulent strains of bacteria are made avirulent by culture at high temperature, the plasmid is lost (Chilton *et al.*, 1975). The possibility that the bacterial plasmid becomes established in the plant cells, thereby inducing transformation is not supported by recent nucleic acid hybridization studies. In fact, there is no evidence that either bacterial or plasmid DNA is contained within the transformed plant cells (Chilton *et al.*, 1974, 1975; Hanson and Chilton, 1975). Although the plasmid contains genes essential for transformation, transformation appears to be a complex process requiring bacterial genes as well.

## VIII. CONCLUSIONS

A wide assortment of elements with genetic continuity, including bare nucleic acid, viruses, bacteria and even occasionally lower eukaryotes have become established intracellularly in numerous organisms. Generally there is hereditary transfer from one generation to the next, asexually during cell division or sexually during processes involving protoplasmic fusion. Transfer by infection through the medium is rare or absent. Only a few of the elements are capable of multiplication outside their host. Sometimes though not always, the host clearly benefits from the presence of the infectious elements. In some cases the host is harmed. Harmful extrachromosomal elements are not as readily eliminated by natural selection as harmful genes, for the population genetics of extrachromosomal elements is such that efficient means of transfer can counterbalance harm to the host. Adaptations of host and element often involve numerous major changes in both, and suggest that the relationships, while delicately balanced, nevertheless are ancient, and play fundamental roles in the life and evolution of the host.

Examples of the importance of infectious heredity to man include its role in bacterial resistance to antibiotics, its probable role in cancer, and the techniques it offers for manipulating the genomes of various organisms.

## ACKNOWLEDGEMENT

The authors are indebeted to Public Health Service Grant GM20038 for financial support of their research. Contribution Number 1028 from the Department of Zoology, Indiana University.

## REFERENCES

Achtman, M. (1973). Genetics of the F sex factor in *Enterobacteriaceae. Curr. Top. Microbiol. Immunol.* **60,** 79–123.

Achtman, M., and Helmuth, R. (1975). The F factor carries an operon of more than 15 × 10⁶ daltons coding for deoxyribonucleic acid transfer and surface exclusion. *In* "Microbiology—1974" (D. Schlessinger, ed.), pp. 95–103. Am. Soc. Microbiol., Washington, D.C.

Anderson, E. S. (1968). The ecology of transferable drug resistance in enterobacteria. *Annu. Rev. Microbiol.* **22,** 131–180.

Arai, T., and Clowes, R. (1975). Replication of stringent and relaxed plasmids. *In* "Microbiology—1974" (D. Schlessinger, ed.), pp. 141–155. Am. Soc. Microbiol., Washington, D.C.

Ball, G. H. (1969). Organisms living on and in Protozoa. *Res. Protozool.* **3,** 567–718.

Beale, G. H., Jurand, A., and Preer, J. R., Jr. (1969). The classes of endosymbiont of *Paramecium aurelia. J. Cell Sci.* **5,** 65–91.

Beljanski, M., Aaron-da Cunha, M. I., Beljanski, M. S., Manigault, P., and Bourgarel, P. (1974). Isolation of the tumor-inducing RNA from oncogenic and nononcogenic *Agrobacterium tumefasciens. Proc. Natl. Acad. Sci. U.S.A.* **71,** 1585–1589.

Bentvelzen, P. (1972). Hereditary infections with mammary tumor viruses in mice. *In* "RNA Viruses and Host Genome in Oncogenesis" (P. Emmelot and P. Bentvelzen, eds.), pp. 309–337. North-Holland Publ., Amsterdam.

Benveniste, R., and Davies, J. (1973). Mechanisms of antibiotic resistance in bacteria. *Annu. Rev. Biochem.* **42,** 471–506.

Berkaloff, A., Bregliano, J. C., Ohanessian, A. (1965). Mise en évidence de virions dans des Drosophilies infectées par le virus héréditaire *sigma*, C. R. Acad. Sci. **260,** 5956–5959.

Berry, E. A., and Bevan, E. A. (1972). A new species of double-stranded RNA from yeast. *Nature (London)* **239,** 279–280.

Bevan, E. A. (1976). Virus-host cell relationships in killer yeast. *Heredity* **37,** 150.

Bevan, E. A., and Makower, M. (1963). The physiological basis of the killer character in yeast. *Proc. Intern. Congr. of Genet., 11th* **1,** 203.

Bevan, E. A., Herring, A. J., and Mitchell, D. J. (1973). Preliminary characterization of two species of dsRNA in yeast and their relationship to killer character. *Nature* **245,** 81–86.

Buchner, P. (1965). "Endosymbiosis of Animals with Plant Microorganisms." Wiley (Interscience), New York.

Chilton, M. D., Currier, T. C., Farrand, S. K., Bendich, A. J., Gordon, M. P., and Nester, E. W. (1974). *Agrobacterium tumefaciens* DNA and PS8 bacteriophage DNA not detected in crown gall tumors. *Proc. Natl. Acad. Sci. U.S.A.* **76,** 3672–3676.

Chilton, M. D., Farrand, S. K., Eden, F., Currier, T., Bendich, A. J., Gordon, M. P., and Nester, E. W. (1975). Is there foreign DNA in crown gall tumor DNA? *In* "Second Annual John Innes Symposium" (R. Markham *et al.*, eds.), pp. 297–311. Am. Elsevier, New York.

Clowes, R. C. (1972). Molecular structure of bacterial plasmids. *Bacteriol. Rev.* **36,** 361–405.

Cohen, S. N., Sharp, P. A., and Davidson, N. (1972). Investigations of the molecular structure of R factors. *In* "Bacterial Plasmids and Antibiotic Resistance" (V. Krčméry, L. Rosival, and T. Watanabe, eds.), pp. 269–282. Springer-Verlag, New York.

Cohen, S. N., and Chang, A. C. Y. (1975). Transformation of *Escherichia coli* by plasmid chimeras constructed *in vitro*: A review. *In* "Microbiology—1974" (D. Schlessinger, ed.), pp. 66–75. Am. Soc. Microbiol., Washington, D.C.

Curtiss, R., III. (1969). Bacterial conjugation. *Annu. Rev. Microbiol.* **23**, 69–136.
Curtiss, R., III, Fenwick, R. G., Jr. (1975). Mechanism of conjugal plasmid transfer. *In* "Microbiology—1974" (D. Schlessinger, ed.), pp. 156–165. Am. Soc. Microbiol., Washington, D.C.
Datta, N. (1975). Epidemiology and classification of plasmids. *In* "Microbiology—1974" (D. Schlessinger, ed.), pp. 9–15. Am. Soc. Microbiol., Washington, D.C.
Davidson, N., Deonier, R. C., Hu, S., and Ohtsubo, E. (1975). Electron microscope heteroduplex studies of sequence relations among plasmids of *Escherichia coli*. *In* "Microbiology—1974" (D. Schlessinger, ed.), pp. 56–65. Am. Soc. Microbiol., Washington, D.C.
Davies, J. K., and Reeves, P. (1975). Genetics of resistance to colicins in *Escherichia coli* K-12: Cross resistance among colicins of group A. *J. Bacteriol.* **123**, 102–117.
Day, P. R., and Anagnostakis, S. L. (1973). The killer system in *Ustilago maydis*: Heterokaryon transfer and loss of determinants. *Phytopathology* **63**, 1017–1018.
de Vries, J. K., Pfister, A., Haenni, C., Palchaudhuri, S., and Maas, W. (1975). F incompatibility. *In* "Microbiology—1974" (D. Schlessinger, ed.), pp. 166–170. Am. Soc. Microbiol., Washington, D.C.
Dilts, J. A. (1976). Covalently closed, circular DNA in kappa endosymbionts of *Paramecium*. *Genet. Res.* **27**, 161–170.
Dilts, J. A. (1977). Chromosomal and extrachromosomal deoxyribonucleic acid from four bacterial endosymbionts derived from stock 51 of *Paramecium tetraurelia*. *J. Bacteriol.* **129**, 888–894.
Dowding, J., and Davies, J. (1975). Mechanisms and origins of plasmid–determined antibiotic resistance. *In* "Microbiology—1974" (D. Schlessinger, ed.), pp. 179–186. Am. Soc. Microbiol., Washington, D.C.
Dressler, D. 1975. DNA replication: Portrait of a field in mid-passage. *Symp. Soc. Gen. Microbiol.* **25**, 51–76.
Dulaney, E. L., and Laskin, A. I. (1971). The problems of drug-resistant pathogenic bacteria. *Ann. N.Y. Acad. Sci.* **182**, 5–415.
Dulbecco, R. (1969). Cell transformation by viruses. *Science* **166**, 962–968.
Durkacz, B. W., Kennedy, C. K., and Sherratt, D. J. (1974). Plasmid replication and the induced synthesis of colicins E1 and E2 in *Escherichia coli*. *J. Bacteriol.* **117**, 940–946.
Falkow, S. (1975). "Infectious Multiple Drug Resistance." Pion, Ltd., London.
Fauré-Fremiet, E. (1952). Symbiontes bactériens des ciliés du genre *Euplotes*. *C. R. Acad. Sci.* **235**, 402–403.
Frédéricq, P. (1957). Colicins. *Annu. Rev. Microbiol.* **11**, 7–22.
Frédéricq, P. (1969). The recombination of colicinogenic factors with other episomes and plasmids. *In* "Bacterial Episomes and Plasmids" (G. E. W. Wolstenholme and M. O'Connor, eds.), pp. 163–174. Churchill, London.
Furmanski, P., Longley, C., Fouchey, D., Rich, R., and Rich, M. A. (1974). Normal human mammary cells in culture: Evidence for oncornavirus–like particles. *J. Natl. Cancer Inst.* **52**, 975–977.
Gallagher, R. E., and Gallo, R. C. (1975). Type C RNA tumor virus isolated from cultured human acute myelogenous leukemia cells. *Science* **187**, 350–353.
Gibson, I. (1973). Transplantation of killer endosymbionts of *Paramecium*. *Nature* **241**, 127–129.
Gibson, I. (1974). The endosymbionts of *Paramecium*. *Crit. Rev. Microbiol.* **3**, 243–273.
Gold, A. H., Suneson, C. A., Houston, B. R., and Oswald, J. W. (1954). Electron micros-

copy and seed and pollen transmission of rod-shaped particles associated with the false-stripe disease of barley. *Phytopathology* **44**, 115–117.

Göpel, G. (1970). Plastomabhängige Pollensterilität bei *Oenothera*. *Theor. Appl. Genet.* **40**, 111–116.

Gross, L. (1961). "Vertical" transmission of passage of a leukemic virus from inoculated C3H mice to their untreated offspring. *Proc. Soc. Exp. Biol. Med.* **107**, 90–93.

Haas, V. H. (1941). Studies on the natural history of the virus of lymphocytic choriomeningitis in mice. *Public Health Rep.* **56**, 285–292.

Hager, L. P., ed. (1973). "Chemistry and Functions of Colicins." Academic Press, New York.

Hanson, R. S., and Chilton, M.-D. (1975). On the question of integration of *Agrobacterium tumefaciens* deoxyribonucleic acid by tomato plants. *J. Bacteriol.* **124**, 1220–1226.

Hardy, K. G. (1975). Colicinogeny and related phenomena. *Bacteriol. Rev.* **39**, 464–515.

Hardy, K. G., Meynell, G. G., Dowman, J. E., and Spratt, B. G. (1973). Two major groups of colicin factors: their evolutionary significance. *Molec. Gen. Genet.* **125**, 217–230.

Hayes, W. (1970). "The Genetics of Bacteria and Their Viruses." Wiley, New York.

Heckmann, K. (1975). Omikron, ein essentieller Endosymbiont von *Euplotes aediculatis*. *J. Protozool.* **22**, 97–104.

Heckmann, K., Preer, J. R., Jr., and Straetling, W. H. (1967). Cytoplasmic particles in the killers of *Euplotes minuta* and their relationship to the killer substance. *J. Protozool.* **14**, 360–363.

Helinski, D. R., and Clewell, D. B. (1971). Circular DNA. *Annu. Rev. Biochem.* **40**, 899–942.

Helinski, D. R., Lovett, M. A., Williams, P. H., Katz, L., Kupersztoch–Portnoy, Y. M., Guiney, D. G., and Blair, D. G. (1975). Plasmid deoxyribonucleic acid replication. *In* "Microbiology—1974" (D. Schlessinger, ed.), pp. 104–114. Am. Soc. Microbiol., Washington, D.C.

Henry, S. M., ed. (1966). "Symbiosis," Vol. 1. Academic Press, New York.

Herforth, R. S. (1973). Effect of actinomycin D and mitomycin C on sigma virus multiplication in *Drosophila melanogaster*. *Virology* **51**, 47–55.

Herring, A. J., and Bevan, E. A. (1974). Virus-like particles associated with the double-stranded RNA species found in killer and sensitive strains of the yeast *Saccharomyces cerevisiae*. *J. Gen. Virol.* **22**, 387–394.

Herring, A. J., and Bevan, E. A. (1976). Purification, characterisation and comparison of virus-like particles from killer and sensitive strains of the yeast *Saccharomyces cerevisiae*. (In press.)

Hu, S., Ohtsubo, E., and Davidson, N. (1975a). Electron microscope heteroduplex studies of sequence relations among plasmids of *Escherichia coli*: Structure of F13 and related F-primes. *J. Bacteriol.* **122**, 749–763.

Hu, S., Ohtsubo, E., Davidson, N., and Saedler, H. (1975b). Electron microscope heteroduplex studies of sequence relations among bacterial plasmids: Identification and mapping of the insertion sequences IS1 and IS2 in F and R plasmids. *J. Bacteriol.* **122**, 764–781.

Huebner, R. J., and Todaro, G. J. (1969). Oncogenes of RNA tumor viruses as determinants of cancer. *Proc. Natl. Acad. Sci. U.S.A.* **64**, 1087–1094.

Ikeda, H. (1965). Interspecific transfer in the "sex ratio" agent of *Drosophila willistoni* in *D. bifasciata* and *D. melanogaster*. *Science* **145**, 1147.

Ito, K. (1974). Mutagenicity of an extrachromosomal element delta for X chromosomes in *Drosophila melanogaster*. *Jpn. J. Genet.* **49**, 25–31.

Jacob, F., Brenner, S., and Cuzin, F. (1963). Regulation of DNA replication. *Cold Spring Harb. Symp. Quant. Biol.* **28**, 329–348.

Jeon, K. W., and Jeon, M. S. (1976). Endosymbiosis in amoebae: Recently established endosymbionts have become required cytoplasmic components. *J. Cell. Physiol.* **89**, 337–344.

Jinks, J. L. (1959). Lethal suppressive cytoplasms in aged clones of *Aspergillus glaucus. J. Gen. Microbiol.* **21**, 397–409.

Kahn, P. L. (1968). Isolation of high-frequency recombining strains from *Escherichia coli* containing the V colicinogenic factor. *J. Bacteriol.* **96**, 205–214.

Karakashian, M. (1975). Symbiosis in *Paramecium bursaria. Symp. Soc. Exp. Biol.* **29**, 145–173.

Kirby, H., Jr. (1941). Organisms living on and in Protozoa. *In* "Protozoa in Biological Research" (G. N. Calkins and F. M. Summers, eds.), pp. 1009–1113. Columbia Univ. Press, New York.

Kline, B. C., and Miller, J. R. (1975). Detection of nonintegrated plasmid deoxyribonucleic acid in the folded chromosome of *Escherichia coli*: Physicochemical approach to studying the unit of segregation. *J. Bacteriol.* **121**, 165–172.

Konisky, J. (1973). Chemistry of colicins. *In* "Chemistry and Functions of Colicins" (L. P. Hager, ed.), pp. 41–58. Academic Press, New York.

Krčméry, V., Rosival, L., and Watanabe, T. (1972). "Bacterial Plasmids and Antibiotic Resistance." Avicenum Czech. Med. Press, Prague.

Lange, R. T. (1966). Bacterial symbiosis with plants. *In* "Symbiosis" (S. M. Henry, ed.), Vol. 1, pp. 99–170. Academic Press, New York.

Laven, H. (1967). Speciation and evolution in *Culex pipiens. In* "Genetics of Insect Vectors of Disease" (J. W. Wright and R. Pal, eds.), pp. 251–275. Elsevier Publ., Amsterdam.

Lemke, P. A., and Nash, C. H. (1974). Fungal viruses. *Bacteriol. Rev.* **38**, 29–56.

Levitan, M. and Williamson, D. L. (1965). Evidence for the cytoplasmic and possibly episomal nature of a chromosome breaker. *Genetics* **52**, 456.

L'Héritier, P. (1970). *Drosophila* viruses and their role as evolutionary factors. *Evol. Biol.* **4**, 185–209.

L'Héritier, P., and Tessier, G. (1937). Une anomalie physiologique héréditaire chez la Drosophile. *C. R. Acad. Sci.* **205**, 1099–1101.

Lippincott, J. A., and Lippincott, B. B. (1975). The genus *Agrobacterium* and plant tumorigenesis. *Annu. Rev. Microbiol.* **29**, 337–405.

Luria, S. (1973). Colicins. *In* "Bacterial Membranes and Walls" (L. Lieve, ed.), pp. 293–320. Dekker, New York.

Matthews, R. E. F. (1970). "Principles of Plant Virology." Academic Press, New York.

Meynell, G. G. (1973). "Bacterial Plasmids." MIT Press, Cambridge, Massachusetts.

Minamori, S., and Ito, K. (1971). Extrachromosomal element delta in *Drosophila melanogaster.* VI. Induction of recurrent lethal mutations in definite regions of second chromosomes. *Mutat. Res.* **13**, 361–369.

Minamori, S., and Sugimoto, K. (1973). Extrachromosomal element delta in *Drosophila melanogaster.* IX. Induction of delta-retaining chromosome lines by mutation and gene mapping. *Genetics* **74**, 477–487.

Mitsuhashi, S. (1971). "Transferable Drug Resistance Factor R." Univ. Park Press, Baltimore, Maryland.

Nathan, D., and Smith, H. O. (1975). Restriction endonucleases in the analysis and restructuring of DNA molecules. *Annu. Rev. Biochem.* **44**, 273–293.

Novick, R. P. (1969). Extrachromosomal inheritance in bacteria. *Bacteriol. Rev.* **33**, 210–235.

Novick, R. P., Wyman, L., Bouanchaud, D., and Murphy, E. (1975). Plasmid life cycles in
    *Staphylococcus aureus*. *In* "Microbiology—1974" (D. Schelssinger, ed.), pp. 115–129.
    Am. Soc. Microbiol., Washington, D.C.
Oishi, K. (1971). Spirochaete-mediated abnormal sex-ratio (SR) condition in *Drosophila*: A
    second virus associated with spirochaetes and its use in the study of the SR condition.
    *Genet. Res.* **18**, 45–56.
Oishi, K., and Poulson, D. F. (1970). A virus associated with SR–spirochaetes of *Drosophila
    nebulosa*. *Proc. Natl. Acad. Sci. U.S.A.* **67**, 1565–1572.
Ossipov, D. V., and Ivakhnyuk, I. S. (1972). Omega particles—micronuclear symbiotic
    bacteria of *Paramecium caudatum* clone M1–48. *Cytologia* **14**, 1414–1419.
Policansky, D. (1974). "Sex ratio," meiotic drive, and group selection in *Drosophila
    pseudoobscura*. *Am. Nat.* **108**, 75–90.
Pollack, M. R. (1962). Penicillinase. *In* "Resistance of Bacteria to the Penicillins"
    (A. V. S. de Rueck and M. P. Cameron, eds.), pp. 56–75. Little, Brown, Boston,
    Massachusetts.
Poulson, D. F. (1963). Cytoplasmic inheritance and heredity infections in *Drosophila*. *In*
    "Methodology in Basic Genetics" (W. J. Burdette, ed.), pp. 404–424. Holden-Day,
    San Francisco, California.
Poulson, D. F. (1969). Nature, stability, and expression of hereditary SR infections in
    *Drosophila*. *Proc. Int. Congr. Genet., 12th, 1968* Vol. 2, pp. 91–92.
Poulson, D. F., and Oishi, K. (1973). New aspects of hereditary androcidal infectious SR in
    neotropical species of *Drosophila*. *Genetics* **74**, No. 2, Part 2, s216.
Preer, J. R., Jr. (1950). Microscopically visible bodies in the cytoplasm of the "killer" strains
    of *P. aurelia*. *Genetics* **35**, 344–362.
Preer, J. R., Jr. (1969). Genetics of Protozoa. *Res. Protozool.* **3**, 133–278.
Preer, J. R., Jr. (1971). Extrachromosomal inheritance: Hereditary symbionts, mitochondria,
    chloroplasts. *Annu. Rev. Genet.* **5**, 361–406.
Preer, J. R., Jr. (1975). The hereditary symbionts of *Paramecium aurelia*. *Symp. Soc. Exp.
    Biol.* **29**, 125–144.
Preer, J. R., Jr., and Jurand, A. (1968). The relation between virus-like particles and R
    bodies of *Paramecium aurelia*. *Genet. Res.* **12**, 331–340.
Preer, J. R., Jr., Preer, L, B., and Jurand, A. (1974). Kappa and other endosymbionts in
    *Paramecium aurelia*. *Bacteriol. Rev.* **38**, 113–163.
Preer, L. B., Jurand, A., Preer, J. R., Jr., and Rudman, B. M. (1972). The classes of kappa in
    *Paramecium aurelia*. *J. Cell Sci.* **11**, 581–600.
Printz, P. (1973). Relationship of sigma virus to vesicular stomatitis virus. *Adv. Virus Res.*
    **18**, 143–157.
Ptashne, K., and Cohen, S. N. (1975). Occurrence of insertion sequence (IS) regions on
    plasmid deoxyribonucleic acid as direct and inverted nucleotide sequence duplications.
    *J. Bacteriol.* **122**, 776–781.
Puhalla, J. E. (1968). Compatibility reactions on solid medium and interstrain inhibition in
    *Ustilago maydis*. *Genetics* **60**, 461–474.
Quackenbush, R. L. (1977). Phylogenetic relationships of bacterial endosymbionts of
    *Paramecium aurelia*: polynucleotide sequence relationships of 51 kappa and its mu-
    tants. *J. Bacteriol.* **129**, 895–900.
Reeves, P. (1972). "The Bacteriocins." Chapman & Hall, London.
Rhoades, M. M., and Dempsey, E. (1972). On the mechanism of chromatin loss induced by
    the B chromosome of maize. *Genetics* **71**, 73–96.
Richmond, M. H. (1975). R factors in man and his environment. *In* "Microbiology—1974"
    (D. Schlessinger, ed.), pp. 27–35. Am. Soc. Microbiol., Washington, D.C.

Richmond, M. H., and Wiedeman, B. (1974). Plasmids and bacterial evolution. *Symp. Soc. Gen. Microbiol.* **24**, 59–85.

Roulland-Dussoix, D., Yoshimori, R., Greene, P., Betlach, M., Goodman, H. M., and Boyer, H. W. (1975). R factor-controlled restriction and modification of deoxyribonucleic acid. *In* "Microbiology—1974" (D. Schlessinger, ed.), pp. 187–198. Am. Soc. Microbiol., Washington, D.C.

Rownd, R. H., Perlman, D., and Goto, N. (1975). Structure and replication of R-factor deoxyribonucleic acid in *Proteus mirabilis*. *In* "Microbiology—1974" (D. Schelssinger, ed.), pp. 76–94. Am. Soc. Microbiol., Washington, D.C.

Rubin, H., Cornelius, A., and Fanshier, L. (1961). The pattern of congenital transmission of an avian leukosis virus. *Proc. Natl. Acad. Sci. U.S.A.* **47**, 1058–1069.

Sabet, S. F., and Schnaitman, C. A. (1973). Chemistry of the colicin E receptor. *In* "Chemistry and Functions of Colicins" (L. P. Hager, ed.), pp. 59–86. Academic Press, New York.

Sakaguchi, B., and Tuchiyama, S. (1974). Characteristics of transmission of SR spirochaete of *Drosophila equinoxiales*. *Jpn. J. Genet.* **49**, 320.

Sakakibara, Y., and Tomizawa, J. (1974). Replication of colicin E1 in cell extracts. II. Selective synthesis of early replicative intermediates. *Proc. Natl. Acad. Sci. U.S.A.* **71**, 1403–1407.

Schlessinger, D., ed. (1975). "Microbiology—1974." Am. Soc. Microbiol., Washington, D.C.

Seecof, R. (1968). The sigma virus infection of *Drosophila melanogaster*. *Curr. Top. Microbiol. Immunol.* **42**, 60–93.

Sherratt, D. (1974). Bacterial plasmids. *Cell* **3**, 189–195.

Siegel, R. W. (1954). Mate-killing in *Paramecium aurelia*, variety 8. *Physiol. Zool.* **27**, 89–100.

Singler, M. J. (1974). Antigenic studies on the relationship of the viral capsid, R body, and toxin of kappa in stock 562, *Paramecium aurelia*. *Genetics* **77**, Suppl. s60–61.

Smith, M. G. (1959). The salivary gland viruses of man and animals (cytomegalic inclusion disease). *Prog. Med. Virol.* **2**, 171–202.

Soldo, A. T. (1974). Intracellular particles in *Paramecium aurelia*. *In* "Paramecium: A Current Survey" (W. van Wagtendonk, ed.), pp. 375–442. Elsevier, Amsterdam.

Soldo, A. T., and Godoy, G. A. (1973a). Observations on the production of folic acid by symbiont lambda particles of *Paramecium aurelia* stock 299. *J. Protozool.* **20**, 502.

Soldo, A. T., and Godoy, G. A. (1973b). The molecular complexity of *Paramecium* symbiont lambda DNA: evidence for the presence of a multicopy genome. *J. Mol. Biol.* **73**, 93–108.

Soldo, A. T., Godoy, G. A., and Brickson, S. (1974). Infectious particles in a marine ciliate. *Nature (London)* **249**, 284–286.

Somers, J. M. (1973). Isolation of suppressive sensitive mutants from killer and neutral strains of *Saccharomyces cerevisiae*. *Genetics* **74**, 571–579.

Somers, J. M. and Bevan, E. A. (1969). Inheritance of the killer characteristic in yeast. *Genet. Res.* **13**, 71–83.

Sonneborn, T. M. (1943). Gene and cytoplasm. I. The determination and inheritance of the killer character in variety 4 of *P. aurelia*. II. The bearing of determination and inheritance of characters in *P. aurelia* on problems of cytoplasmic inheritance, pneumococcus transformations, mutations, and development. *Proc. Nat. Acad. Sci. U.S.* **29**, 329–343.

Sonneborn, T. M. (1959). Kappa and related particles in *Paramecium*. *Adv. Virus Res.* **6**, 229–356.

Sonneborn, T. M. (1975). The *Paramecium aurelia* complex of fourteen sibling species. *Trans. Am. Microsc. Soc.* **94,** 155–178.

Stocker, B. A. D., Smith, S. M., and Ozeki, H. (1963). High infectivity of *Salmonella typhimurium* newly infected by the Col I factor. *J. Gen. Microbiol.* **30,** 201–221.

Temin, H. M. (1971). The protovirus hypothesis: Speculations on the significance of RNA-directed DNA synthesis for normal development and for carcinogenesis. *J. Natl. Cancer Inst.* **46,** III–VII.

Tooze, J. (1973). "The Molecular Biology of Tumor Viruses." Cold Spring Harbor Lab., Cold Spring Harbor, New York.

Tuan, R. S., and Chang, K. P. (1975). Isolation of intracellular symbiotes by immune lysis of flagellate Protozoa and characterization of their DNA. *J. Cell Biol.* **65,** 309–323.

U.S. Public Health Service. 1976. National Institutes of Health Guidelines for Research Involving Recombinant DNA Molecules. Supt. Doc., U.S. Govt. Print. Off., Washington, D.C.

van Wagtendonk, W. J., Clark, J. A. D., and Godoy, G. A. (1963). The biological status of lambda and related particles in *Paramecium aurelia. Proc. Nat. Acad. Sci., U.S.* **50,** 835–838.

Vodkin, M., Katterman, F., and Fink, G. R. (1974). Yeast killer mutants with altered double-stranded ribonucleic acid. *J. Bacteriol.* **117,** 681–688.

Wade, N. (1975). Genetics: Conference sets strict controls to replace moratorium. *Science* **187,** 931–935.

Weisblum, B. (1975). Altered methylation of ribosomal ribonucleic acid in erythromycin-resistant *Staphylococcus aureus. In* "Microbiology—1974" (D. Schlessinger, ed.), pp. 199–206. Am. Soc. Microbiol., Washington, D.C.

Wickner, R. B. (1974). Chromosomal and nonchromosomal mutations affecting the "killer character" of *Saccharomyces cerevisiae. Genetics* **76,** 423–432.

Willetts, N. (1972). The genetics of transmissible plasmids. *Annu. Rev. Genet.* **6,** 257–268.

Williams, J. (1971). The growth *in vitro* of killer particles from *Paramecium aurelia* and the axenic culture of this protozoan. *J. Gen. Microbiol.* **68,** 253–262.

Williamson, D. L. (1966). Atypical transovarial transmission of sex-ratio spirochaetes by *Drosophila robusta* Sturtevant. *J. Exp. Zool.* **161,** 425–430.

Williamson, D. L., and Whitcomb, R. F. (1974). Helical, wall-free prokaryotes in *Drosophila*, leaf hoppers, and plants. *In* "Mycoplasmas of Man, Animals, Plants, and Insects" (J. M. Bové and J. F. Duplan, eds.), pp. 283–290. Inst. Nat. Santé Rech. Méd., Paris.

Williamson, D. L., and Whitcomb, R. F. (1975). Plant mycoplasmas: A cultivable *Spiroplasma* causes corn stunt disease. *Science* **188,** 1018–1020.

Wolstenholme, G. E. W., and O'Connor, M., eds. (1969). "Bacterial Episomes and Plasmids." Churchill, London.

Woods, D. R. and Bevan, E. A. (1968). Studies on the nature of the killer factor produced by *Saccharomyces cerevisiae. J. Gen. Microbiol.* **51,** 115–126.

Yen, J. H., and Barr, A. R. (1971). New hypothesis of the cause of cytoplasmic incompatibility in *Culex pipiens* L. *Nature (London)* **232,** 657–658.

Yen, J. H., and Barr, A. R. (1973). The etiological agent of cytoplasmic incompatibility in *Culex pipiens. J. Invertebr. Pathol.* **22,** 242–250.

Yen, J. H., and Barr, A. R. (1974). Incompatibility in *Culex pipiens. In* "The Use of Genetics in Insect Control" (R. Pal and M. J. Whitten, eds.), pp. 97–118. Elsevier, Amsterdam.

Zaenen, I., von Larebeke, N., Teuchy, H., von Montagu, M., and Schell, J. (1974). Supercoiled circular DNA in crown-gall inducing *Agrobacterium* strains. *J. Mol. Biol.* **86,** 109–127.

# 9

# Non-nucleic Acid Inheritance and Epigenetic Phenomena

*Janine Beisson*

## I. INTRODUCTION

Our knowledge of the nature and functions of the genes and of the mechanisms ensuring their replication and hereditary transmission supports the conclusion that all the properties of a cell can be explained by the specificities of its genes. However, it is most likely that a genome isolated from its cell, even if provided with all the necessary synthesizing machinery, energy sources, and elementary molecules, would have only a very low probability of reproducing this cell, whereas within its normal organized environment, the genome is always successful in directing this reproduction. To make two new cells, generally identical to the mother cell in every detail of their anatomy, physiology, and biochemical specifities, an entire cell is needed, not only its genes. This is not a new concept, but our knowledge of genetics still does not fully explain why cell heredity requires cell continuity. The questions still to be answered are, What do the nongenic parts of the cell contribute to cell heredity? Do they simply provide an operational framework for gene expression or do they contain some kind of intrinsic information superimposed on the basic genetic—or nucleic—information? The tentative answers to these questions will be discussed in this chapter.

Various experimental systems are suitable for the analysis of cell heredity, in particular, unicellular organisms, "simple" animals or plants like *Hydra* or fungi, and cells or tissues isolated from higher organisms and grown *in vitro*. The meaningful phenomena whose study might answer the above questions comprise all the situations where hereditary differences are observed between cell lines of *identical genotype*. The existence of such differences has long been recognized and the first that were studied were examples of non-Mendelian or extranuclear inheritance in Protozoa, fungi, and *Drosophila*. Since, in most cases, no change in the basic genetic information was evident, these phenomena suggested the existence of self-perpetuating systems located in the nongenic parts of the cells and responsible, by direct or indirect action, for the observed hereditary differences.

To designate these unknown systems responsible for the maintenance of alternative phenotypes associated with an apparently invariant genome, the word *epigenetic* was coined by D. Nanney (1958), while a more pessimistic author, R. D. Hotchkiss, called them "UNclear" (as opposed to "NUclear"). Since then, however, the frontiers between clear and unclear mechanisms ensuring cell heredity have moved. A number of previously unclear phenomena have been clarified and have turned out to be neither epigenetic nor nuclear, but truly genetic and controlled by extranuclear (chloroplastic or mitochondrial) DNA or by viruses or episomes. These types of extranuclear inheritance are dealt with in other chapters of this treatise. Furthermore, the mechanisms responsible for nuclearly controlled heredity are clearly more complicated than previously thought: (1) a large part of the genome of eukaryotes consists of quantitatively variable "simple" repetitive sequences of unknown function, so that genetic information can no longer be simply equated to DNA sequences coding for RNA and polypeptidic chains, and (2) DNA molecules can undergo chemical modifications (methylations) or structural changes in their organization (tandem duplication, amplification, restriction) or in their tridimensional conformation (due in particular to the variable but intimate association with chromosomal proteins). For these reasons the limits of genome invariance have become difficult to define.

Therefore, clear-cut evidence for the existence of non-nucleic acid inheritance can presently be provided only by hereditary phenotypic differences independent of any change in gene expression. Whenever phenotypic differences are correlated with changes in gene expression, their nucleic versus non-nucleic acid basis cannot be definitively ascertained. Although it is likely that a given pattern of transcribed genes can result in different phenotypes (since there is evidence that cell properties can be controlled at various levels beyond transcription: transport, translation, membranes), it remains to be demonstrated that such peripheral mechanisms can actually account for stable, self-maintained differences. Conversely, although the various chemical or structural changes occurring in the DNA and the chromatin are clearly correlated with differential gene activity, it is generally not known to what extent such changes are self-reproducible and can explain stable changes in cellular phenotype. Nevertheless, a wide range of phenomena that cannot be explained by mutations in chloroplastic or mitochondrial DNA or by viruses or episomes show or suggest that different phenotypes can be inherited through multiplication of cells of identical or supposedly identical genotype.

This chapter is divided into two main parts. The first part (Phenomenology) is devoted to a description, on the basis of a few examples, of the phenomenology of phenotypic variations affecting either the structural or the functional state of a cell. These examples circumscribe the *mode* and *range* of phenotypic variations encountered in cell heredity and serve to outline the frame within which the unknown underlying mechanisms might be operating. These examples show or suggest that, aside from all the nuclear and cytoplasmic genetic information stored in nucleic acids, cell heredity involves more or less stable interactions, embodied in the network of structures and metabolic circuits and laid down stepwise in the course of each cell's history.

In the second part (Mechanisms) the general significance of the chosen examples and of some possible underlying mechanisms are discussed. Since most of these phenomena remain incompletely analyzed, their interpretation is highly speculative. It may well turn out that their real explanation lies, at least in part, in still unknown biological processes (for example, some functions of the repetitive DNA sequences of eukaryotes) or in the existence of new species of cytoplasmic DNA, such as the 2 $\mu$m circles associated with pleotropic drug resistance in yeast (Guérineau et al., 1974). The possible role of viruses and of stable and possibly self-replicating RNA molecules should not be disregarded either. Whatever the detailed mechanisms eventually prove to be, it seems reasonable to admit that cells do not contain any other information than that encoded in nucleic acids. However, it seems equally likely that the gene-coded specificity of cell components does not exclude a few degrees of freedom in molecular interactions, resulting in alternative self-maintained states that may short circuit the expression of the genome and provide the cell with both phenotypic inertia and a limited range of variations which would not therefore pertain *directly* to nucleic acid inheritance. Finally, it is tempting to speculate that cell membranes (because of their physical continuity and their flexible organization) might be the privileged place where the limited freedom of molecular interactions might be expressed.


## II. PHENOMENOLOGY

One class of heritable phenotypic variations that can safely be considered as involving some mechanism of non-nucleic acid inheritance includes a few examples of variations in the structural arrangement of normal gene products, in the absence of any change in gene expression. The term of inheritance of structural organization, or *structural inheritance*, will be used to designate this particular aspect of cell heredity.

In all of the other cases of heritable phenotypic variations that are not due to a modification in nucleic acid specificity, the variations involve the functional state of the cell, i.e., its pattern of gene expression. Regardless of their nucleic or non-nucleic acid basis, the underlying mechanisms reveal another aspect of cell heredity which can be called inheritance of functional state or *functional inheritance*. From our present state of knowledge, functional inheritance can be divided into two categories: cases that seem unlikely to be based on some mechanism of non-nucleic acid inheritance, and cases that seem a little more likely to be so based. Figure 1 illustrates operational definitions of these two categories which correspond, respectively, to cases of *nuclear* functional inheritance and to cases of *cellular* functional inheritance. To the first category (Fig. 1B) belong X-chromosome inactivation (Lyon, 1972), transposition of genetic controlling elements (reviewed by Fincham and Sastry, 1974), or translocations of chromosome segments into heterochromatic regions (Baker, 1968). Although in these cases, discussed in Chapter 3, the nature of the chromosomal change and/or the mechanisms ensuring its stability are not well understood, the basis of the phenotypic variation is unambiguously chromosomal. To the same class B belong other cases where the modification of the expression of the genome is restricted to a single

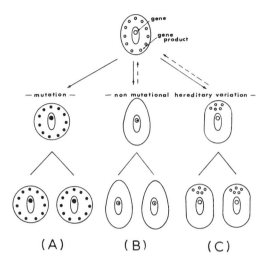

**Fig. 1.** Nucleic acid versus non-nucleic acid inheritance. The shape of the "cell" symbolizes its phenotype. The oval in the center of the cell represents all of the genetic information available in the nucleus and in DNA-containing cytoplasmic organelles; the open circles (○) represent a gene and its product. (A): Mutational change. The modified phenotype results from a mutated gene (●) and its *modified product*. (B), (C): Nonmutational changes. (B) The modified phenotype is due to a stable (for instance, inactive) gene configuration (⊙) that can be transmitted by the nucleus alone. (C) The modified phenotype does not involve any change in the nucleic acids but results from a stable change in the activity or localization of a normal gene product.

locus, as paramutation in maize (reviewed by Brink, 1965) or phase variation in *Salmonella* (reviewed by Iino, 1969). These modified "genic states" may be based on changes in DNA, for instance superstructural changes as proposed by Cook (1974). While the nature of such changes, as well as the mechanisms of their hereditary maintenance, remains speculative, it is a fact that they are transmitted through sexual or parasexual processes and are genetic.

In contrast to these nonmutational changes of the genome, there are other examples of functional inheritance (Fig. 1C) for which the existence of a heritable change in the genome is more questionable. These heritable phenotypic variations display two distinctive properties: (1) they are transmitted *only through cell continuity*, and (2) they appear to be *reversible* under the influence of changes in the cytoplasmic or nucleocytoplasmic milieu or even under particular environmental conditions. Their heredity does not reside only in genic, chromosomal, or nuclear continuity, but requires the continuity of the cell as a whole.

To this category of variations may belong various examples of "cytoplasmic" or maternal inheritance characterized by alternative phenotypes corresponding to the stable expression or nonexpression of a particular gene: for instance, in the ascomycete *Podospora anserina*, the inheritance of two alternative phenotypes [s] and [s$^s$] associated with gene *s* (Rizet, 1952; Beisson-Schecroun, 1962), or in yeast the inheritance of the non-mitochondrial cytoplasmic variation *URE$_3$* characterized by an abnormal regulation of glutamic dehydrogenases (Aigle and Lacroute, 1975).

To the same class of heritable phenotypic variations may belong also all the differentiated states in higher organisms. As generally assumed, differentiation does not involve loss or modification of genetic potentialities but the selective expression of different genes in different differentiated cells. The fact that differentiated states are maintained through cell continuity has been demonstrated for a number of cell types. The reversibility of a differentiated pattern of gene expression can be observed in particular situations such as lens differentiation from iris cells (Yamada and McDevitt, 1974), and is also shown by two types of classic experiments. (1) A nucleus from a differentiated cell, injected into an enucleated *Xenopus* egg, is able to control the development of a more or less normal organism (Gurdon, 1970; Gurdon *et al.*, 1975). (2) The completely inactive nucleus of a chick erythrocyte, fused with a HeLa cell or a mouse cell, resumes transcription and bird proteins eventually can be detected (Harris *et al.*, 1969; Cook, 1970). Since the phenotypic expression of a differentiated nucleus appears modulable by its cytoplasmic environment (as in the just mentioned cases of cytoplasmic inheritance in fungi), the maintenance of differentiated states seems not to be based *only* on some transmissible change in the genome but rather on a nucleocytoplasmic or a *cellular* condition, in which the involvement of mechanisms of non-nucleic acid inheritance can be looked for.

Concerning cellular functional inheritance, if reversibility is a common feature, its heredity itself might be questioned. Without arguing about how long a particular phenotype has to be perpetuated in order to be accepted as hereditary, and whether heredity concerns only those features which are transmitted through sexual reproduction, it seems reasonable to admit, as suggested to Preer (1959a), that a phenotype maintained for over 60 cell generations is hereditary. A phenotype maintained for that many generations must be controlled by a precise self-maintained mechanism and cannot be explained by the long persistence, in spite of dilution, of some previously synthesized gene product acting as a repressor, inducer, etc.

In the following sections, a few examples of structural inheritance and of cellular functional inheritance will be described. Because in most cases the "unclear" aspects of cell heredity remain poorly understood, it seems pointless to present a comprehensive review, and the few chosen examples will necessarily reflect the author's personal bias.

## A. Inheritance of Structural Organization

A prototype of structural inheritance is provided by inheritance through cell reproduction of changes in the cortical pattern of *Paramecium tetraurelia* (formerly *P. aurelia* syngen 4; new nomenclature of Sonneborn, 1975). The surface of this protozoan is covered by thousands of cilia whose basal bodies (bb) or kinetosomes are inserted at regular intervals in the plasma membrane and are arranged in a precise pattern. Each cilium, and its bb has accessory structures (a kinetodesmal fiber, kf, and a parasomal sac, ps) which endow it with a polarity with respect to the antero–posterior and left–right axes of the animal. In normal cells, all bb's have the same polarity (Fig. 2A). The maintenance of this pattern requires *a priori* at least two sets of conditions: (1) gene-controlled production of all necessary normal building blocks, and (2) physiochemical and/or structural conditions permitting assembly of bb, kf, etc, and their insertion at precise places in the cell cortex. Gene mutations affecting either directly the specificity of the building blocks or indirectly the micro-milieu for assembly and positioning of the organelle can be expected to alter the cortical pattern and indeed such mutations have been described (Whittle and Chen-Shan, 1972; Beisson *et al.*, 1976b). However, changes in the pattern can also be obtained by microsurgical intervention. A most striking modification of the pattern consists of the inversion of the polarity of all bb's along one or several kineties (longitudinal rows of cilia). This inversion is obtained by "grafting" on one animal a small piece of cortex taken from a sister cell of identical genotype and inserting it on the host *upside down* with respect to its polarity. The graft may comprise only a few bb's but develops, after a few fissions, into complete inverted kineties (Fig. 2B). Such reversed stripes have been maintained through cell continuity for over 800 cell generations without a reversion to the normal phenotype (Beisson and Sonneborn, 1965), and they were also shown to be cytoplasmically inherited in crosses with normal cells (Fig. 3).

Therefore, the inheritance of this modified pattern is clearly independent not only of changes in genetic information (as, e.g., following an exchange of nuclei during conjugation) but also of changes in gene expression, since the variation affects nothing but the polarity of otherwise normal basal bodies.

How the small grafted piece can yield, after a few divisions, complete kineties homogeneous from pole to pole with respect to the polarity of

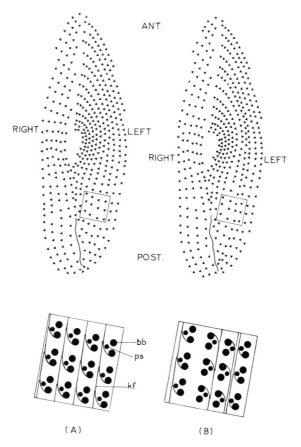

**Fig. 2.** Cortical pattern of *Paramecium* (ventral surface) in a normal cell (A) and in a variant (B) with two rows of inverted—rotated by 180°—basal bodies (bb). The actual number of ciliary rows and of basal bodies per row has been reduced, but this simplified scheme nevertheless conveys a correct view of the pattern. Each triangular dot represents a cortical unit comprising one or two basal bodies and their accessory structures (kinetosdesmal fiber, kf, and parasomal sac, ps), as shown in the enlarged details.

their component basal bodies, is understandable. During division, "new" bb's are intercalated between "old" ones, in a line along kineties which constitute longitudinal units of growth. Furthermore, thanks to Dippell (1968), enough is known of the development of new kinetosomes in *Paramecium* to clearly delineate the meaning of this hereditary variation. Each new kinetosome assembles close to an old one, anteriorly if the latter has a normal polarity, posteriorly if it has a reversed polarity. Therefore, *both the site where a new bb develops and its polarity are deter-*

*mined only by the polarity of the cortical unit around the old bb, and independently of the overall polarity of the animal.*

A few other examples of structural inheritance have been described for ciliates. The pioneer work in this field was done by Tartar (1956, 1961) who produced a variety of new (more or less stable) cell types by microsurgical translocation, inversion, addition, or deletion of pieces of cortex in *Stentor* and drew attention to the morphogenetic information stored in preexisting patterns. In *Paramecium*, complete "doublets" obtained by fusion of two conjugating animals or incomplete doublets which resulted from integration by one conjugant of a piece of cortex torn away from its partner were shown to maintain, over a large number of cell divisions, all of their extranumerary kineties and other surface organelles, gullet, cytopyge, contractile vacuole pores (Sonneborn and Dippell, 1960, 1961; Sonneborn, 1963). In *Tetrahymena*, Nanney (1966, 1968b) demonstrated the stability through cell multiplication of various "corticotypes" (differing mainly in the total number of ciliary rows per cell).

Altogether, these facts demonstrate that the inheritance of the cortical pattern involves at least *two* sources of information: first, the genes which specify the building blocks of basal bodies and accessory structures; second, the properties of the small territory where the new bb's are assembled. This territory provides twofold information which is not nucleic acid-based: it imprints its own *polarity* on that of the developing organelle, and it determines the *site* of its development, thus ensuring the "self-maintenance" of both the polarity of a row and the total number of rows. These facts do not mean, however, that the cortical pattern is genetically autonomous. For instance, a gene-controlled regulation of the number of ciliary rows has been demonstrated in the progeny of crosses between two species of *Euplotes* differing by their number of ciliary rows (Heckman and Frankel, 1968). The cited examples of structural inheritance do not demonstrate either that the type of non-nucleic acid information thus revealed, and for which the terms of "cytotaxis" (Sonneborn, 1963) or of "structural guidance" have been proposed (Frankel, 1974), is the *only* mechanism involved in pattern determination. As reviewed and discussed by Frankel (1974) positional information, referred to as "biochemical signaling" by Lynn and Tucker (1976), might play a role in the morphogenesis of ciliates, and it might also be significant in metazoan development.

In addition to these examples, which show unambiguously the existence on the surface of ciliates of a particular information superimposed on the nucleic acid information, there are a few other "unclear" phenomena for which a role of structural inheritance has been considered. In *Chlamydomonas reinhardi*, an abnormal hereditary transmission of

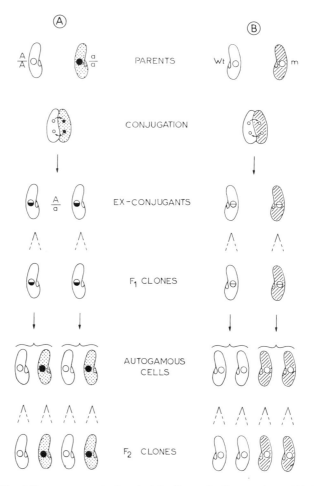

**Fig. 3.** Mendelian versus cytoplasmic inheritance in *Paramecium*. (A) Hereditary transmission of a nuclear single gene difference. The complex nuclear apparatus is simply represented by the open, black or "hybrid" circles which symbolize, respectively, the homozygous wild-type $A/A$, mutant $a/a$, and heterozygous $A/a$ genotypes. Dotted versus white cytoplasm symbolize the phenotypic difference between the two parental strains. (B) Hereditary transmission of a cytoplasmic difference between wild-type (Wt) and mutant (m) cells. Hatched versus white cytoplasm symbolize the phenotypic difference between the two parental strains.

During conjugation (see Fig. 5 for the details of nuclear processes) a reciprocal exchange of haploid nuclei takes place so that the two exconjugants acquire an identical heterozygous nuclear genotype. If the difference between the parental strains is nuclearly controlled (A), the two $F_1$ clones deriving from the two exconjugants display identical phenotypes. On the contrary, if the difference has a cytoplasmic basis (B), the two exconjugants and the two $F_1$ clones retain their phenotypic difference despite their nuclear identity.

In each $F_1$ clone, autogamy (see Fig. 5) can be induced. Each autogamous cell becomes cell-wall properties has been described. Some mutants characterized by a defective cell wall (in particular CW18, which usually segregates in a Mendelian fashion) show occasional abnormal segregations when crossed with wild-type cells (Davies and Plaskitt, 1971; Davies 1972). The results support the interpretation (Davies 1972; Davies and Lyall, 1973) that haploid cells of either genotype, CW or CW$^+$, can perpetuate either a normal or a mutant cell wall, as if once established during the first post-zygotic divisions, the complex pattern of organization of this cell wall (Hills *et al.*, 1973) could be maintained under either of the two alternative patterns. However, in this case the lack of knowledge of the "history" of the cell wall through the apparently decisive stage of zygote formation does not permit speculations as to the possible mechanisms involved, and the phenomenon might also be explained by a particular genetic instability of the loci involved.

Structural inheritance, at the level of mitochondrial membranes, has also been proposed (Linnane *et al.*, 1972; Beisson *et al.*, 1974) as the basis of abnormal hereditary transmission of mitochondrial properties in *Saccharomyces cerevisiae* and in *Paramecium*. As the properties of mitochondria are controlled by both nuclear and mitochondrial genes, their inheritance can be either Mendelian or cytoplasmic. However, in yeast certain mikamycin (Linnane *et al.*, 1972) or oligomycin (Avner and Griffith, 1973) resistant mutations behave as both nuclear and cytoplasmic; this is also the case for the mutation $cl_1$ described in *Paramecium* (Sainsard *et al.*, 1974). The nuclear recessive mutation $cl_1$, which segregates in a Mendelian fashion in crosses (Fig. 3A) modifies both the cellular phenotype (slow growth rate, thermosensitivity) and the mitochondrial phenotype (cytochrome oxidase deficiency and increased cyanide-insensitive respiration). However, in crosses $cl_1 \times$ wild-type some ab-

---

homozygous for all its genes. In (A) cells become $A/A$ or $a/a$ with a 1 : 1 probability. In (B) the same is true for any genic difference that may have existed between the parental strains beside their cytoplasmic difference. Therefore a 1 : 1 segregation of the two parental nuclear genotypes is observed among the $F_2$ clones deriving from autogamous cells. In (A) the two parental phenotypes are recovered in $F_2$ clones regardless of their cytoplasmic origin, while in (B) the parental mutant phenotype is maintained only in the $F_2$ clones of mutant cytoplasmic origin and among these clones, no segregation of the m phenotype occurs.

The situations depicted in the figure correspond to the most frequent case where no cytoplasmic exchange accompanies the nuclear exchange at conjugation. Some conjugating pairs may exchange cytoplasm through a cytoplasmic bridge developed at the end of conjugation. Occurrence of a cytoplasmic bridge will not change the picture in (A). But in (B), the cytoplasms of the two exconjugants become "mixed" and a variety of situations can emerge; in particular vegetative segregations of the mutant/wild-type cytoplasmic factors may be observed in the course of cell divisions, independently of nuclear processes at autogamy.

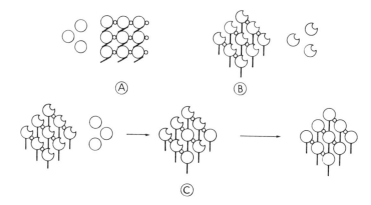

**Fig. 4.** Pattern differences in membrane structure. It is not specified whether the circles and dashes represent molecules or molecular complexes. The ordered arrangement does not necessarily extend over the entire membrane but can concern only a limited group of molecules or molecular complexes. (A) Pattern in wild-type cells, maintained as long as the wild-type building blocks (○) are available. (B) In a mutated cell, such as a $cl_1$ *Paramecium*, the modified arrangement depends on the properties of the mutated building blocks ( ◓ ). (C) When the mutated products are no longer available in a cell of mutant origin receiving a wild-type nucleus, if the wild-type building blocks can fit the ''mutant'' pattern, the latter will be maintained. The two alternative patterns may or may not be functionally identical.

normal features* of the transmission of the $cl_1$ characters show that the $cl_1$ phenotype depends upon both the nuclear gene $cl_1$ and a mitochondrial "factor." As there is yet no clear-cut evidence that the DNA of $cl_1$ mitochondria carries a mutation, a mechanism of structural inheritance might be involved as shown in Fig. 4. This diagram represents a situation which is formally analogous to that of the observed inheritance of pattern for ciliates cortical organelles, where the site and polarity of newly formed organelles is determined by the preexisting ones. However, while this is indeed the case for the cortical organelles, it remains to be demonstrated that membranes contain patterns of molecular organization endowed with this capacity for self-maintenance. Whatever the case may

* The abnormalities observed in crosses $cl_1$ × wild-type are the following: (1) In the $F_2$ clones which inherit a wild-type cytoplasm and become homozygous $cl_1/cl_1$, the wild-type mitochondria display an incompatibility with this gene: they are severely disorganized resulting in an extremely slow growth rate. This crisis persists for over 30 cell generations, then nearly all cells return to the parental $cl_1$ phenotype. (2) In the $F_2$ clones which inherit $cl_1$ cytoplasm and become homozygous $cl_1^+/cl_1^+$, some properties of the $cl_1$ strain are retained indefinitely (deficiency in cytochrome oxidase, absence of incompatibility of the mitochondria with gene $cl_1$) while others conform to the $cl_1^+/cl_1^+$ genotype (thermoresistance, growth rate).

be, the just-described phenomena still belong to the "unclear" class and may turn out to be nucleic acid-based.

## B. Cellular Inheritance of Functional States

Functional differences between cell lines of identical genotype are of at least two types: (1) differences which involve the mutually exclusive expression of isofunctional gene products, as is the case for the genes coding for the $\beta$, $\gamma$, and $\delta$ chains of hemoglobin, each one being expressed at a different developmental stage; (2) more complex differences associated with the specialization of cells, as observed in the various differentiated tissues of higher organisms. An example of cellular inheritance of the first category of functional differences, the inheritance of surface antigenic type, will be again borrowed from *Paramecium's* anthology; the inheritance of hepatic differentiation in hepatoma cells will illustrate the second category of functional differences.

### 1. Surface Antigens of Paramecium

The inheritance of surface antigens (s. ags.) in this organism was, in fact, one of the earliest discovered unclear phenomena which challenged the imagination of biologists back in the late 1940's and for which M. Delbrück proposed his "steady state" model (1949) (Section III,b,2).

**a. General.** The four basic features of the system (for a detailed account, see Beale, 1957; Sommerville, 1970; Finger, 1974) can be summarized as follows: (1) The surface of *Paramecium* is coated with a high molecular weight, ca. 250,000 daltons (Peer, 1959b; Bishop, 1961), protein. First thought to be composed of several subunits (Steers, 1965; Jones, 1965), this protein has been more recently shown to consist of a single polypeptide chain (Reisner *et al.*, 1969; Hansma, 1975). This protein is highly immunogenic, and injection of whole cell breis into rabbits induces the formation of antibodies that react with the surface antigen *in vivo* (causing immobilization of cells coated with the same antigen as the cells of the injected brei) and *in vitro* (with the formation of precipitate). (2) Each strain of *Paramecium* is genetically capable of synthesizing a number (about 10) of such different, immunologically distinct, s. ags. In a few cases, the immunological differences have been shown to be correlated with differences in the peptide fingerprints of the antigens (Steers, 1961, 1962; Jones and Beale, 1963). In many other cases, the differences are apparent in immunodiffusion tests (Peer, 1959c; Bishop, 1963; Finger and Heller, 1962). In most cases, specific antibodies prepared against an antigen will not immobilize animals coated with another one. (3) Despite

the presence of genetic information for a number of immunologically distinct proteins, each animal generally exhibits on its surface only one s.ag.
species and, if maintained under stable culture conditions, all descendants
of a *Paramecium* clone are alike. This is why specific antibodies against
one particular antigen can be prepared. However, *Paramecia* will generally react to a range of environmental changes by switching to the expression of another s.ag., which may in turn be inherited through cell
divisions as long as environmental conditions are kept constant. This is
how an inventory of the genetic capabilities of a strain can be established.
Moreover, different strains of *Paramecium* are characterized by different
arrays of s.ags. (serotypes). Therefore, by appropriate crosses between
different strains, a limited but significant inventory could be made which
supports the conclusion that the specificity of each surface antigen is
controlled by a single nuclear gene. All of the loci identified thus far are
not linked to each other. At some loci, two or more different alleles are
known. (4) Aside from its clonal inheritance, the serotype is in various
cases cytoplasmically inherited (as in Fig. 3B) in crosses between cells
expressing different serotypes. Despite their identical nuclear genotype,
the clones derived from the two exconjugants will be phenotypically different and continue to express the serotype previously expressed in the
cytoplasmic parent, as for instance in the crosses 51A × 51B studied by
Sonneborn and Le Suer (1948).

This sketchy outline shows that, as often has been stressed by others
(see, for instance, Nanney, 1968a; Sonneborn, 1970), the s.ag. system
constitutes an example of *differentiation*, i.e., of controlled selection of a
particular pattern of gene expression from an invariant genome. As
pointed out by Capdeville (1971) and Finger (1974), this situation of
clonally inherited phenotype exclusion among different loci coding for
isofunctional proteins is similar to what is observed for the genes coding
for immunoglobulins (Mage, 1975).

**b. The Differentiation of the Macronucleus.** To appreciate the
significance of this system as a model of differentiation, it is necessary to
describe the development and functioning of *Paramecium* nuclei. The
ciliates posses two types of nuclei: one or more diploid micronuclei, and a
highly "polyploid" macronucleus. In *Paramecium tetraurelia*, the two
micronuclei each contain about 40–60 very small chromosomes (Dippell,
1954) and divide mitotically, developing very long spindles without dis-
ruption of the nuclear membrane. The macronucleus which arises from a
micronucleus contains about 800 times the DNA of the micronucleus
(Behme and Berger, 1970; Allen and Gibson, 1972). There is no evidence
that extensive amplification or elimination of DNA sequences accompanies this quantitative increase (Allen and Gibson, 1972) and therefore

the macronucleus of *Paramecium* presumably contains 800 times the number of all of the genes of the micronucleus. The organization of the genetic material in the macronucleus of *Paramecium* and *Tetrahymena*, genetically the most studied species of ciliates, remains unknown except for the ribosomal cistrons which are amplified (Yao *et al.*, 1974) and free (Gall, 1974) in *Tetrahymena* and in *Paramecium* and display a palindromic structure. Furthermore, in *Tetrahymena*, macro- and micronuclei differ by their histone content (Gorovsky and Keevert, 1975) and only the macronucleus contains methylated bases (Gorovsky, 1973). The division of macronuclei, which used to be called "amitotic," involves a diffuse bundle of microtubules and efficiently transmits all the genetic information over at least a hundred divisions.

Practically all transcription is carried out on the macronuclear templates and almost none (Pasternak, 1967) in the micronuclei. Although the dispensability of the micronuclei in the vegetative life remains questionable for *Paramecium*, it has been shown that the phenotype is exclusively controlled by the macronuclear genes (Sonneborn, 1946, 1954a; Pasternak, 1967). Therefore, an understanding of the history of a macronucleus is essential for an appreciation of its pattern of expression. This history, depicted in Fig. 5 and its legend, starts anew at each sexual step (conjugation or autogamy) from a single diploid zygotic nucleus derived from the meiotic products of the micronuclei. This diploid zygotic nucleus undergoes two successive mitoses, yielding four diploid nuclei. Two of these nuclei will retain their diploid and transcriptionally inactive condition and divide mitotically at each subsequent division. The other two nuclei will differentiate into macronuclei, i.e., undergo a rapid "polyploidization" (Berger, 1973), amplifly their ribosomal cistrons (Yao *et al.*, 1974), and acquire the capacity for active transcription. What determines these alternative and thereafter inherited functional properties of genetically identical nuclei is for the most part unknown, except that these different fates seem to be controlled in part by the localization of the zygotic nucleus and of its first division products. This local effect involves some unidentified interaction of nuclei with a particular area of the plasma membrane or cortex (Sonneborn, 1954b) probably at the first postzygotic division (J. Beisson, unpublished). Displacement of the zygotic nucleus induced by centrifugation (Sonneborn, 1954b) or caused by mutation (Ruiz *et al.*, 1976; Beisson *et al.*, 1976b) may result in abnormal numbers of micro- and macronuclei.

The development of the macronucleus is accompanied by its functional differentiation; for those functions, e.g., mating type, for which the genome contains alternative information, a choice is made that establishes which one will be expressed. As shown in Fig. 5, in each exconju-

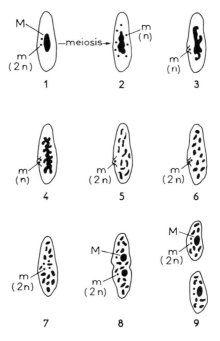

**Fig. 5.** Nuclear reorganization during sexual processes (conjugation or autogamy) in *P. aurelia*. (1–2) The two diploid micronuclei (m) undergo meiosis, yielding eight haploid products, seven of which degenerate. (3–4) The surviving haploid micronucleus, located in the "paroral cone" undergoes one mitotic division while the macronucleus (M) begins to fragment. (5) Formation of the zygotic diploid nucleus. This step marks the only difference between conjugation and autogamy. In conjugation, the two paired cells of complementary mating type have exchanged one of the haploid nuclei and the zygotic nucleus results from the fusion of one migratory "♂" pronucleus and one stationary "♀" pronucleus, while at autogamy the zygotic nucleus results of the fusion of two genetically identical haploid nuclei. (6–7) The zygotic nucleus undergoes two successive mitoses, while the fragmentation of the macronucleus is completed. (8) Two out of the four diploid nuclei begin to differentiate into macronuclei (M). (9) The first division of the exconjugant or autogamous cell is achieved after mitosis of the two micronuclei, while each daughter cell (karyonide) receives one of the two newly developed macronuclei. The DNA content of these new macronuclei has not yet reached its final value, but their differentiation (with respect to mating type or surface antigenic type) is generally established and may be different in the two. The fragments of the old macronucleus will be progressively degraded within a few divisions.

gant or autogamous cell, two new macronuclei develop. At the first division they will be distributed to the two daughter cells, called *karyonides*. The two new macronuclei may differentiate similarly, and in particular cases, this differentiation is regularly identical to that previously expressed in the cytoplasmic parent. In such cases, the phenotype is cytoplasmically inherited. In other cases, the new macronuclei may display different differentiations, implying that their differentiation is not determined by the preexisting cytoplasmic state but by unknown more discrete factors. Accordingly, the phenotypes of sister karyonidal clones may be different. Various examples of karyonidal inheritance have been described, for instance, mating type inheritance in several *Paramecium* species (reviewed by Preer, 1969), the inheritance of resistance to calcium (Genermont, 1966), and the inheritance of several properties in *Tetrahymena* where macronuclear development displays similar features.

In summary, local cytoplasmic influences which can be compared to the effect of polar cytoplasm on the nuclei of polar cells in *Drosophila* (Illmensee and Mahowald, 1974) are involved in the functional differentiation of a diploid micronucleus into a macronucleus. Generally established at an early stage of its development, the differentiation of the macronucleus will be clonally inherited in each karyonidal clone. But as will be shown below, macronuclear differentiation itself (i.e., some stable change in macronuclear DNA) does not suffice to explain the inheritance of the antigenic type.

**c. Expression of the Surface Antigens Controlled by the $g$ Locus in *Paramecium primaurelia*.** In *P. primaurelia* (previously called *P. aurelia,* syngen 1) three main loci, $s$, $g$, and $d$, controlling s. ags. are known the expression of which depends on the temperature. At low temperature, e.g., below 15°–18°C, $s$ is expressed, but as the temperature is raised, $s$ expression is turned off and $g$ becomes expressed, and the cells eventually switch to the expression of $d$ above 30°–32°C. A series of alleles is known at locus $g$, designated according to the names of the wild-type strains of different geographic origin in which they were first identified: 156, 168, 90, 60, 33, etc. Most of the corresponding s. ags. (156G, 168G) can be distinguished immunologically (Capdeville, 1971), and they differ also by the range of temperature within which their expression is stable; for instance, 156G is stable up to 32°C, while 168G is replaced by a D antigen above 28°C. When homozygotes *156g/156g* and *168g/168g* are grown at 30°C, the latter will express s. ag. D while the former will retain the phenotype 156G: if these two homozygotes are crossed and maintained at 30°C, the two $F_1$ heterozygous clones will generally be phenotypically different, one expressing D, the other expressing G, each retaining the phenotype of its cytoplasmic parent, despite their identical nuclear genotype, as previously observed in other systems (Beale, 1952).

Because of the immunological differences between allelic G antigens, the study of other aspects of s. ags. expression has been possible and has shown that not only intergenic but also interallelic exclusion can take place. In crosses between cells homozygous for different $g$ alleles and expressing antigen G, both exconjugant ($F_1$) clones will continue to express this antigen if maintained at the permissive temperature. However depending upon the alleles brought together in the heterozygotes, a variety of situations is regularly observed (Table I). The two extreme cases are (1) coexpression of the two alleles and (2) phenotypic exclusion of one allele by the other (Capdeville, 1971). Either kind of phenotypic expression, once established, will be maintained through vegetative multiplication of heterozygous clones. In crosses between homozygotes *156g/156g* and *168g/168g*, both exconjugants are genetically alike, *156g/168g*, and express s. ag. 156G only in 99% of the cases. No 168G antigen can be detected either on the cell surface or in cell extracts. However in 1% of the cases, heterozygotes exhibit both 156G and 168G s. ags., and in this case, only one karyonidal clone from a pair of exconjugants is phenotypi-

**TABLE I**

**Patterns of Allelic Exclusion at the _g_ Locus in
_Paramecium primaurelia_ [a]**

| Cross | Phenotype of $F_1$ karyonidal clones | |
|---|---|---|
| $\dfrac{156g}{156g} \times \dfrac{33g}{33g}$ | 156G | |
| $\dfrac{156g}{156g} \times \dfrac{513g}{513g}$ | 156G | |
| $\dfrac{156g}{156g} \times \dfrac{168g}{168g}$ | 156G (90%)<br>156G (>99%)[b] | 156G–168G (10%)<br>156G–168G(<1%)[b] |
| $\dfrac{156g}{156g} \times \dfrac{60g}{60g}$ | 156G (90%) | 156G–60G (10%) |
| $\dfrac{168g}{168g} \times \dfrac{60g}{60g}$ | 60G (70%) | 168G–60G (30%) |
| $\dfrac{168g}{168g} \times \dfrac{513g}{513g}$ | 168G–513G (75%) | 513G (25%) |

[a] After conjugation, for each pair of heterozygotes, the two exconjugants and then the two products of their first division are separated, yielding four karyonidal clones. In spite of its identical heterozygous genotype and regardless of its cytoplasmic origin, each karyodinal clone may differentiate with a predictable frequency (depending upon the two alleles brought together) into a stable phenotype that is either "exclusive" or hybrid. The relative frequency of the two alternative phenotypes can be influenced by the culture medium.

[b] Corresponds to cells grown in baked lettuce infusion, while in all the other cases, the cells were grown in "Scotch grass" infusion.

cally hybrid, the three others being 156G. Sister karyonidal clones of identical genotype, expressing either 156G or 156G + 168G and kept under the same culture conditions, will maintain their phenotypic differences, just as heterozygotes *156g/168g*, kept at 30°C, maintain a G or D phenotype depending upon the phenotype of their cytoplasmic parent. Furthermore, in contrast to the generally reversible expression of s. ags. genes, the two phenotypes 156G and 156G + 168G are irreversibly *fixed* in the two types of clones. If these two phenotypically different cell lines are transferred to 32°C, after a few divisions, their progeny will shift to D antigen, but when brought back to 24°C, they will reexpress their distinctive phenotype 156G or 156G + 168G according to their previous phenotype. The particular interest of this situation is twofold. Since sister karyonidal clones grown under the same condition can display irreversibly different phenotypes, it appears that (1) the temperature is not the only factor regulating the expression of s. ags. loci, and (2) the coexpression or phenotypic exclusion reflects a macronuclear heritable differentiation.

Macronuclear differentiation alone however does not explain the following facts. (1) The threshold temperature for transformation from G to D or D to G is different for the two types of cell lines. Heterozygotes *156g/168g* of phenotype 156G shift to D at 32°C, like homozygotes *156g/156g*, while heterozygotes of phenotype 156G + 168G switch to D above 28°C as do homozygotes *168g/168g*. (2) When transferred back to 24°C, both types of clones return rapidly to G, within a few fissions, as do homozygotes *156g/156g*, while homozygotes *168g/168g* need more than 15 fissions to do so. (3) In either transformation of the cells lines determined for coexpression of 156G and 168G, both s. ags. are lost or acquired *simultaneously* (Capdeville, 1977).

All these data on the expression of *156g* and *168g* alleles certainly do not clarify the problem of s. ags., whose cytoplasmic inheritance has been puzzling for nearly 30 years. The fundamental question to be resolved remains at which level regulation operates. But whatever mechanism underlies these phenomena of intergenic and interallelic phenotypic exclusion, two salient properties of the system must be taken into account: (1) The capacity for switching from one phenotype to another is channeled by a cellular memory which is apparent either in the cytoplasmic inheritance or in the reexpression of a previous determination and therefore seems to depend both on nuclear and on cytoplasmic factors. (2) Since in cell lines determined for coexpression of *156g* and *168g*, the two alleles are turned on or off *simultaneously*, under conditions that are different from those operating when one allele only is expressed, the s. ags. molecules themselves seem to play a role in the cellular response to temperature changes, and it is tempting to assume that the s. ags. molecules are instrumental in the phenotype switch.

## 2. Differentiated States in Mammalian Cells

**a. Determination and Differentiation.** In higher organisms, differentiated states represent the final stage of the stepwise evolution of initially undifferentiated and totipotent cells. Although, during development, each step comes into gear with the next one in an apparently continuous fashion, each step may correspond to a switch from one state to another one among a series of potentially stable and mutually exclusive modes of functioning. Three steps along the developmental history of cells have been shown to be stable through vegetative multiplication. (1) Clonal lines of embryonic carcinoma cells, similar in many respects to early embryo cells, can be grown *in vitro* without losing their pluripotentiality, as demonstrated by their capacity to differentiate later into a variety of tissues (see Martin, 1975, for review). (2) Determined states (in which commitment to a further final differentiation is established) are heritable. Thus, myogenic cell lines have been kept in exponential growth over months without losing their capability to differentiate eventually into muscle cells (Yaffee, 1968), and imaginal discs of *Drosophila* have been cultivated *in vivo* in adult flies over several years without changing their capacity to differentiate normally when finally placed in a metamorphosing insect (see Gehring, 1968). (3) A variety of differentiated cells, permanent cell lines derived from tumors such as hepatomas, melanomas, etc., or established cell lines issued from normal tissues, cultivated *in vitro* and more or less easily cloned, maintain their characteristic morphology and continue expressing a number of tissue-specific genes (for reviews, see Green and Todaro, 1967; Wigley, 1975).

For all three types of systems, the stability (under appropriate conditions) of the cellular functional state does not seem to involve any irreversible change in the cell's capacity to switch to another equally stable functional state, but the range of possible alternative stable states becomes progressively restricted. Embryonic carcinoma cells can differentiate into a number of cell types, but determined cells can generally switch only to the differentiated state corresponding to their determination (although transdeterminations are occasionally observed, at least in *Drosophila*). In general, differentiated cells have only the potentiality to maintain or to arrest the expression of their specific differentiated or "luxury functions." The loss of differentiated functions is sometimes observed under experimental conditions, although the cells may reexpress their original differentiation upon reestablishment of adequate conditions. Therefore the stability of differentiated states presents two facets. The first one, common to any cell type, and possibly trivial, corresponds to the maintenance of a particular pattern of gene expression. The second facet corresponds, as pointed out by Ephrussi (1972), to the heritable determination underlying each type of differentiation. The central problem is to understand on what the establishment and stability of determination is based.

Although the problem remains unresolved, one can gain some insight through the study of somatic cell hybrids. The opportunities offered by somatic hybrids (reviewed by Ephrussi, 1972; Davis and Adelberg, 1973; Davidson, 1974) are described in Chapter 5. A priori, somatic hybrids might seem the worst possible material to look for non-nucleic acid inheritance, since it has been previously pointed out that the favorable systems are provided by persistent differences observed between cell lines of identical genotype. In somatic cell hybrids, all the observed changes in phenotype are correlated with qualitative and/or quantitative genetic changes (addition or loss, rearrangements of chromosomes). However somatic hybrids provide a range of experimental situations to challenge the potentialities of differentiated genomes, and to find out to what extent differentiation depends upon heritable changes in nucleic acids.

**b. A Model System: Expression of Differentiated Functions in Hybridized Rat Hepatoma Cells.** The rat hepatoma cells used in one series of hybridization studies are heteroploid cells of a permanent line. They maintain in culture a number of liver-specific characteristics, in particular, a distinctive morphology and the production of several liver proteins, including serum albumin, tyrosine aminotransferase (TAT), alanine aminotransferase (AAT), and particular isozymes of alcohol dehydrogenase (liver ADH) and fructose 1,6 diphosphate aldolase (aldolase B). TAT and AAT are produced at a relatively low level but higher levels of the enzymes can be induced *in vitro* (Thomson *et al.*, 1966) as *in vivo* by corticosteroid hormones. Extensive studies of hybrids between rat hepatoma cells and a number of other cell types displaying none of these liver-specific features have defined under what conditions the liver-specific functions are expressed or not. (1) Generally, the differentiated features (liver-specific enzymes and cell morphology) are *extinguished* in the hybrids; that is what happens in hybrids between rat hepatoma cells and diploid rat epithelial cells (Weiss and Chaplain, 1971) or between rat hepatoma cells and pseudodiploid Chinese hamster fibroblasts (Weiss *et al.*, 1975). (2) Extinction depends upon gene dosage. When 1s and 2s* rat hepatoma cells are hybridized, respectively, with subdiploid mouse lym-

---

* As all these cell lines are aneuploid and their chromosomes complement fluctuates, the karyotype is designated by the modal number of chromosomes in the stem-line, symbolized as 1s. 2s corresponds to a doubled modal number of chromosomes.

phoid cells, the liver functions which are extinguished in the first case (1s hybrids) are not extinguished in the second one (2s hybrids) (Brown and Weiss, 1975). (3) From hybrid lines in which the liver-specific functions are extinguished, clones can be isolated that *reexpress* the extinguished functions as a result of the loss of chromosomes from the other parent. The various functions can be reexpressed independently of one another (Sparkes and Weiss, 1973; Bertolotti and Weiss, 1974). (4) Genes which were inactive in the nonhepatoma parental cells can be *activated* in hybrids, especially when the hepatoma cells are 2s (Peterson and Weiss, 1972; Malawista and Weiss, 1974; Brown and Weiss, 1975).

The phenotype of particular hybrids between 1s or 2s rat hepatoma cells and mouse lymphoid cells (LC), in which the rat or mouse origin of several enzymes can be identified by their distinctive physicochemical properties, is described in Table II, in which are reproduced a limited sample of the data from Brown and Weiss (1975). Three features only are considered: hepatocyte cellular morphology, serum albumin production, and TAT (basal and inducible) production. The two salient facts are (1) in the first three hybrids (LCF$_1$, LCF$_5$, LCF$_\alpha$) to which the hepatoma

**TABLE II**

**Expression of Liver Functions in Rat Hepatoma–Mouse Lymphoid Cell Hybrids**[a]

| | | Phenotype | | | | | | Karyotype | |
|---|---|---|---|---|---|---|---|---|---|
| | | | | | TAT | | | | |
| Cell line | Hepato-cytelike morphol-ogy | Albumin | | Total | | Mouse | | % Expected chromosome number | Ratio rat/mouse chromosomes |
| | | Rat | Mouse | B | I | B | I | | |
| Parent 1s | + | ++ | | 30 | 350 | | | 100 | — |
| Parent 2s | + | ++ | | 39 | 478 | | | 100 | — |
| Parent Lc | − | − | − | 1 | 1 | − | − | 100 | — |
| LCF$_1$ | − | + | + | 2 | 18 | − | − | 88 | 1.0 |
| LCF$_5$ | − | + | + | 1 | 5 | − | − | 88 | 1.1 |
| LCF$_\alpha$ | − | + | + | 4 | 20 | − | + | 86 | 1.5 |
| LC2F$_3$ | + | ++ | ++ | 137 | 300 | ++ | ++ | 94 | 2.8 |
| LC2F$_{19}$ | + | ++ | ++ | 55 | 230 | ++ | ++ | 91 | 3.1 |
| LC2F$_{25}$ | + | ++ | ++ | 16 | 190 | − | ++ | 68 | 4.2 |

[a] Expression of liver functions in hybrids between 1s or 2s rat hepatoma and near diploid mouse lymphoid cells. "+" for cell morphology indicates a hepatomalike morphology. The figures for TAT (B, Basal; I, Induced) represent specific activity in extracts of uninduced or induced hybrids in which mouse enzyme detected by electrophoresis and heat inactivation is absent or uncertain (−) or present (+ or ++).

parent contributed a 1s chromosome complement, all rat liver characteristics are totally or largely extinguished, while production of some mouse serum albumin is activated, corresponding to residual level of rat albumin production; (2) in the last three hybrids (LC2F$_3$, LC2F$_{19}$, LC2F$_{25}$) to which the hepatoma parent contributed a 2s chromosome complement, not only are all rat liver characteristics fully expressed but production of high levels of albumin and TAT by the *mouse* genome is activated.

The two groups of three hybrid clones described in Table II correspond to clones for which the ratio of rat/mouse chromosomes remained similar to the input ratio. In other clones from the same crosses, strong deviations were observed; for instance, one clone of the 2s series lost predominantly rat chromosomes and was karyologically and phenotypically identical to a clone from the 1s series. Therefore, it is clear that the pattern of gene expression in hybrids does not depend upon some preexisting characteristics of the parents but "on the continuous production of diffusible regulatory factors specified by one or both parental genomes," as pointed out by Brown and Weiss (1975).

Further examples of the observed modulations of gene expression are provided by the study of "dedifferentiated" clones from a rat hepatoma line. Seven independent dedifferentiated clones obtained from a cloned line of rat hepatoma cells were studied by Deschatrette and Weiss (1974). These clones were first recognized among a very large number of normally differentiated clones by their distinctive morphology: loose colonies, pale cells lacking fat droplets, and the presence of several small nucleoli instead of one or two large ones. Out of the seven, two still produced albumin and reduced but significant amounts of one or more of the four liver-specific enzymes analyzed, while five failed to produce any detectable activity of these enzymes. No albumin production was detected in the three clones tested for that trait. These dedifferentiated clones did not show significant changes in chromosome number. Three of these clones display an unexpected common property: they synthesize a high level of glucose-6-phosphate dehydrogenase (G6PD); no such activity is ever present in differentiated hepatoma cells of this particular line, although it is present in liver cells. This "dedifferentiated" state is highly stable, although variants reexpressing all of the liver functions were eventually obtained from one of these clones.

*In conclusion*: What do these facts demonstrate? Despite the real or theoretical drawbacks of the system (uncontrolled changes in the genome associated with the changes in phenotype, "abnormal" properties of the cells, interspecific nature of the hybrids, etc), one striking fact stands out. Whether unexpressed after extinction in hybridized rat hepatoma cells, or in "dedifferentiated" clones, or unexpressed in cell lines of different histogenetic origin not determined for liver differentiation, the structural genes coding for the various liver-specific functions remain responsive to diffusible regulatory products and can be reexpressed or activated. This fact excludes the possibility that the stability of differentiated phenotypes is based on more or less irreversible genic changes corresponding to the simple situation depicted in Fig. 1B, at least as far as the structural genes are concerned. Another significant fact is the correlation between cell morphology and the expression or nonexpression of the few functions studied, showing that each cellular state actually involves the differential activity of many genes or gene products. It seems likely therefore that the stability of the various alternative phenotypes is based upon either very complex regulatory mechanisms acting simultaneously on many genes or upon an integrated network of structural and metabolic constraints; in the latter case, a non-nucleic acid basis of this stability is possible, although it remains to be demonstrated.

## III. MECHANISMS

According to our present state of knowledge, all cell components seem to be made of molecules which are either nucleic acids or specified by nucleic acids. Therefore, the mechanisms controlling the expression of the genome could be thought of as totally programmed in the genome leaving no room for non-nucleic acid inheritance. However, the existence of non-nucleic acid inheritance is suggested by some ''unclear'' aspects of cell heredity. In the phenomena presented above, in which cellular phenotypes are inherited under alternative stable states, the expression of the genome can be modulated by structural or functional constraints that seem to escape direct gene control and are somehow self-perpetuated. Since the molecular basis and mode of action of these constraints remain unknown, a discussion of the possible mechanisms underlying non-nucleic acid inheritance is purely speculative. Its major aim is to explore to what extent molecules or molecular complexes can control the maintenance of their own synthesis and/or activity and establish closed circuits whose function can be durably modified independently of the overall metabolic and structural network.

### A. Possible Mechanisms Involved in Inheritance of Structural Organization

### 1. General Significance of Structural Inheritance

One is led to ask first why there are so few examples known, if structural inheritance has a general importance. There seems to be one principal explanation. The range of possible variations in structural arrangement of a particular structure is probably very limited, as shown by the following example. In *Paramecium*, several mutations induce errors in the positioning of kinetosomes such that a number of kinetosomes are located between the rows and their polarity (shown by the position of the attached kinetodesmal fiber) is random. Through divisions, this disorder increases and eventually leads to very abnormal animals that stop dividing. But in nearly all mutant cells, complete rows of inverted kinetosomes similar to those produced in the grafting experiments (Section II,A) are found, and their origin is easily traced back to single or to a few kinetosomes which happened to be rotated by 180°. Because of the mode of growth of the cortex, which proceeds mostly through longitudinal elongation of the cortical membranes, only normal kinetosomes or those which are rotated by 180° can have a stable ''progeny'' integrated in the overall pattern. Erratic kinetosomes with various other polarities may tend to prime new cortical units of similar polarity, but the structural constraints exerted by the adjacent rows do not permit development of transverse or oblique rows, and the only possible results are either an increased disorder or integration into normal or reversed rows. Similarly, if a structure such as a mitochondrial membrane or a cell wall can display in particular cases two different patterns of molecular organization, it is most likely an exceptional phenomenon. The genetically determined tridimensional configuration and physicochemical properties of molecules can be expected to obey, in general, stringent constraints in their interaction and association. This may be the main reason why functional and stable variations of structural patterns are so rare. However, the basic significance of such variations should not be underestimated because of their rarity.

A second question one is led to ask is, why such variations are found mostly in ciliates? An obvious explanation lies in the fact that ciliates in general and *Paramecium* in particular have a highly complex and easily observable structural organization, whereas in other organisms with less ordered structures, modifications equivalent to inverted rows of kinetosomes might be much less conspicuous. Detection of patterns at the molecular level requires sophisticated techniques such as those used by Boyse *et al.* (1968) or Stackpole *et al.* (1971) for working out the geography of surface antigens. Variations of patterns, if they occur, would most probably either be undetected or would appear as functional changes and be very difficult to trace back to structural changes. In the case of inverted rows of kinetosomes in *Paramecium,* the structural change has also a functional counterpart; all cilia formed on inverted kinetosomes beat in a direction opposite to that of the other cilia (Tamm *et al.*, 1974) so that the swimming style of the animal is abnormal. If the cilia were the size of bacterial flagella, the structural basis of the abnormal behavior would certainly have escaped notice.

For these reasons, it is reasonable to assume that structural inheritance is not an atypical phenomenon limited to ciliates but has indeed a more general significance. The examples described show that for some cellular structures the same component molecules, molecular complexes, or organelles can be arranged in two or more different configurations which, once established, are equally well maintained in the presence of an unchanged genome. Two different mechanisms may underly this self-maintenance: (1) assembly of a new organelle or structural complex is primed by a preexisting identical structure; (2) the presence of a preexisting structure is not a prerequisite, and the site of assembly and development of the new structure is determined by specific interactions between the molecules composing the developing structure and a particular framework of different molecular nature acting more or less like a template. In the latter case, the "self-maintenance" of the pattern merely reflects the properties of the surrounding molecular milieu. These two types of mechanisms will be discussed successively.

## 2. The Role of Primers in Structural Inheritance

The idea that primers are necessary to the development of complex structures mainly stems from some examples that are briefly reviewed below.

(a) In most animal cells, centrioles (see Pitelka, 1969) are endowed with a sort of structural continuity as are kinetosomes in most protozoa. Each new centriole develops in close vicinity to a preexisting one at an extremely precise place with respect to it. Although no material link can be demonstrated by electron microscopy between the old and new structures, "something" emanating from or correlated with the old structures triggers and guides the development of the new one. As discussed earlier, this is particularly evident for the multiplication of kinetosomes in ciliates. However, it is now equally clear that many (and possibly all) cells are capable of producing such organelles *de novo*. For instance in *Naegleria*, development of cilia is induced by environmental changes in amoeboid cells devoid of preexisting kinetosomes (Dingle and Fulton, 1966). In *Oxytricha*, Grimes (1973) has shown that all kinetosomes disappear during encystment and reappear at the same sites upon excystment. In another ciliate, *Urostyla*, new cirri (tight groups of kinetosomes and their cilia) are formed at their normal location even when the old cirri close to which they normally develop have been removed (Jerka-Dziadosz and Frankel, 1969; Jerka-Dziadosz, 1972).

(b) The case of flagellar growth in *Salmonella* has also been cited frequently as an example of primed macromolecular assembly. These flagella are composed of a single protein species (flagellin) and the conditions for *in vitro* assembly of flagellin have been studied. Nothing happens unless fragments of flagella (seeds) are added to the monomers in solution; then the monomers can assemble into flagella-like structures, and the growth from the seed is polarized as it is *in vivo*. However, it was later shown that "self-assembly" of monomers was possible under different and particular physicochemical conditions (see Iino, 1974, for review).

(c) A similar story can be told about the bacterial cell wall. In *Bacillus subtilis*, when the cell wall is completely removed by enzymatic treatments, protoplasts are formed, some of which can multiply as stable L forms incapable of spontaneous regeneration of the wall (Landman and Halle, 1963). In this case also, environmental conditions (namely, physical contact of the protoplasts with solid gelatin or membrane filters) were eventually found to enable L forms to reconstruct normal cell walls (Clive and Landman, 1970).

Two conclusions can be drawn from these facts. (1) Provided that the necessary building blocks are present and that particular physicochemical and/or physiological conditions are met, any complex organelle can probably be assembled in the absence of a preexisting similar organelle. (2) Preexisting structures, although dispensable in principle, obviously do *contribute* to the establishment of conditions permitting assembly of similar structures in a rapid, reliable, and predictable way.

If structural inheritance is not based upon priming processes, it remains to be understood what are, at the molecular level, the conditions that trigger and guide the assembly of new structures. This is, in fact, a basic biological problem raised by any structural pattern. Discussing the determinism of pattern formation in general is beyond the scope of this chapter [for stimulating speculations on this subject, see, for instance, Crick (1970), Wolpert (1971), Frankel (1974), and Meinhardt and Gierer (1974)]. But with respect to the non-nucleic acid basis of structural inheritance, there is one point that needs discussion, what tells the new structures where they are to be assembled? Whether favored by the vicinity of preexisting structures or induced by other particular physicochemical and physiological conditions, the development of a new structure implies that some relevant information or signal becomes available at a precise place. Consider the case of microtubules. Their importance in the determination of cell shape (see, for instance, Brown and Bouck, 1973; Tucker, 1970), in the structural changes accompanying cell division, and in the determination of surface patterns is now amply documented (reviewed by Olmsted and Borisy, 1973), although the mechanisms responsible for their assembly, oriented growth, disassembly, etc. are only beginning to be revealed. The theoretical necessity for some basic signals is obvious, whether called "foci" (Porter, 1966), "orienting centers" (Inoué and Sato, 1967), "nucleating centers" (Tilney, 1968), or microtubule organizing centers (MTOC) (Pickett-Heaps, 1969). A more precise suggestion by Hartman (1975) is that specific RNA molecules serve as primers for microtubules, kinetosomes, and centrioles; some cytochemical evidence supports this view (Hartman *et al.*, 1974; Dippell, 1976). But whatever the biochemical nature of the postulated nucleating centers, how they are replicated and most importantly how they come to settle in the right place, remains to be determined. This leads back to the necessary preexistence of some kind of framework in which specific building blocks can find their place and interact with other molecules under conditions that leave them practically no degree of freedom. Because of their structural continuity, cell membranes represent good candidates for this necessary framework.

### 3. The Role of Membrane Continuity in Structural Inheritance

While DNA is reproduced in a semiconservative fashion and most organelles (such as ribosomes or kinetosomes) are multiplied in a conservative way, cell membrane reproduction seems to be of the dispersive type; a cell membrane can be roughly visualized as a patchwork of old and new areas. The kinds of experimental support for this view are not very numerous, mostly because the demonstration requires stable landmarks well anchored over the entire cell surface. Fortunately, bacterial flagella and cilia in ciliates provide such landmarks.

In *Bacillus subtilis*, about 8 flagella are anchored in the plasma membrane and regularly spaced over the bacterium. In a strain that can divide but not form new flagella at high temperature, the transmission of preexisting flagella to daughter cells was followed by Ryter (1971) who showed that flagella were segregated by *groups* on either side of one or two main zones of membrane growth (most likely responsible for the separation of replicated chromosomes). The same conclusion is supported by data on the segregation of molecular markers, Lac-permease and Mel-permease (Autissier and Kepes, 1971).

In ciliates, the probable site of membrane growth can be deduced from the localization of new kinetosomes. Each new kinetosome is inserted next to an old one, and then is moved further away, the separation being achieved by membrane elongation between them. As newly inserted kinetosomes are devoid of cilia, the topography of ciliated versus non-ciliated kinetosomes could be established in *Tetrahymena*: new kinetotomes are more or less regularly inserted everywhere over the cell surface throughout the cell cycle (Perlman, 1973; Nanney, 1975). In *Paramecium*, many more new kinetosomes are inserted in the equatorial zone than at the poles of the cell, but everywhere new ciliary territories alternate regularly with old ones.

It is not clearly established how this integration of new membrane components is achieved at the ultrastructural and molecular levels and there is no evidence that the new components are integrated at specific sites in the membrane. Different mechanisms may operate in different cell types and in different cell membrane systems. An instructive example may be the salt gland of the duck in which the epithelial secretory cells respond to salt stress by a considerable increase of the plasma membrane surface. The process has been studied by electron microscopy (Levine *et al.*, 1972) and the growth of the membrane has been found to result from integration of numerous small "vesicles" similar in structure to the plasma membrane. These vesicles are integrated in the plasma membrane by fusion as in exocytosis [for the mechanisms of membrane fusion, see Lucy (1970), Poste and Allison (1973), and Satir (1975)]. A similar process has been described in rat cells for the integration of cell coat glycoproteins (Leblond and Bennett, 1974). In other systems, the operation of this mechanism might not be as conspicuous as it is in the case of duck salt gland where the increase in cell surface is particularly rapid and massive. Other types of studies show that major proteinic components as well as phospholipids are gradually integrated in the plasma membrane throughout interphase (Graham *et al.*, 1973).

Membrane growth, therefore, seems to result either from localized insertion of preformed membrane vesicles or from the addition of individual molecules. Both types of mechanisms are equally well supported by the facts (reviewed by Bretscher and Raff, 1975) and by the present theory of membrane structure—the fluid mosaic model—formally presented by Singer and Nicholson (1972). Two conclusions seem reasonable: (1) during their growth, the membranes remain continuous and their overall framework is preserved; (2) newly formed daughter cells inherit the plasma membrane of the mother cell, especially if membrane growth mostly takes place before division and cytokinesis is predominantly a physical process (stretching and constriction).

The main problem here is whether the continuity of the membrane framework provides the molecular organization that ensures the precise localization of specific membrane sites. *A priori*, the "fluidity" of the membrane does not seem to support this possibility. The rapid intermixing of surface antigens observed by Frye and Edidin (1970) after the formation of mouse + human heterokaryons would suggest that the distribution of the protein components is rather random. Similarly, freeze-fracture faces of cell membranes rarely show any pattern in the distribution of the protein "particles" within the lipid bilayer. However, organized arrays of particles, demonstrating that relatively stable local differentiations do exist, have been described in various cell types, for instance, the arrays involved in cell-to-cell contacts in metazoa (Staehelin, 1974).

A most impressive case of local membrane differentiation occurs in the plasma membranes of *Tetrahymena* and *Paramecium* which display two types of repetitive particle arrays (ciliary "necklaces" and trichocyst or mycocyst insertion sites), regularly alternating all over the cell surface. Trichocyst sites have been shown to differentiate as an intrinsic property of the plasma membrane (Beisson *et al.*, 1976b). Other evidence that the cell surface, even if it is a fluid mosaic, is nevertheless ordered, has been provided by the demonstration that the different cell surface antigens on lymphoid cells show a characteristic topography (Boyse *et al.*, 1968), which can be different in different cell types (Stackpole *et al.*, 1971). Finally, although it is established that the specificity of cell surface components is not fixed once and for all but, on the contrary, is changing throughout the cell cycle, it has also been shown that far from being random, these changes are sequential (Fox *et al.*, 1971) and predictable. Much work is presently focused on the correlation between membrane changes and the various intracellular events of the cell cycle. A very attractive hypothesis that not only reconciles the fluid mosaic model of membrane structure with the probably precise topography of surface components, but that also may provide an explanation for the correlation between surface changes and intracellular events, has been proposed by Berlin *et al.* (1974). The model postulates that various surface proteins interact with underlying microtubules (and possibly microfilaments). A similar model has been more recently extensively documented by Nicolson (1976).

*In conclusion*, there are two ways of considering the basic problem of the mechanisms underlying structural inheritance: a formal and static view based on the concept that stable structures can generate identical structures, and a dynamic view based on the idea that each structure or its vicinity is the seat of more or less continuous stepwise changes, each step providing the specific conditions required for triggering and controlling particular processes of structural assemblies. The foregoing discussion certainly supports the second point of view, it seems therefore very tempting to postulate that many features of cell structure are based on the properties of the cell membrane, on its molecular organization and its interactions with intracellular organelles. However, experimental confirmation remains difficult to obtain.

### B. Possible Mechanisms Involved in Cellular Inheritance of Functional States

The phenomena of functional inheritance presented above pertain to the general problem of cell differentiation for which there is at present no explanation.

A variety of models have been proposed to account for differentiation, in general, on the basis of more or less stable modifications of the genome: changes in DNA superstructure (Cook, 1974) or specific methylation processes (Holliday and Pugh, 1975), a complex network of regulatory sequences (Britten and Davidson, 1969), interactions of DNA with chromatin proteins (Tsanev and Sendov, 1971; Paul, 1972), or DNA modifications and restrictions similar to those existing in bacterial systems (Sager and Kitchin, 1975). Although these theories are based on reasonable assumptions and may approximate mechanisms that contribute to cellular differentiation, data acquired in recent years indicate that the phenotype is not necessarily controlled at the level of DNA or chromatin. Examples of these data are the following. In avian erythrocytes, more of the genome is transcribed than is actually translated on polysomes (Scherrer and Marcaud, 1968), and in sea urchin embryos, 5–10 times more unique DNA sequences are represented in nuclear RNA than in the mRNA of polysomes (Smith et al., 1974; Galau et al., 1974). The existence and stability of untranslated messengers, already known in unfertilized eggs (cf. Gross, 1967), also have been demonstrated for myosin messenger in myoblasts (Buckingham et al, 1974). In addition, the quantity of RNA synthesized in the nucleus in growing and resting fibroblasts is the same (Johnson et al, 1975). Posttranscriptional control of gene expression, at the level of processing and transport to the cytoplasm or at the level of translation, thus appears as a possible mechanism of cell differentiation. However the generality of such phenomena, as well as the mechanisms responsible for posttranscriptional controls remain unknown.

Therefore, in attempting to discuss the possible contribution of non-nucleic acid inheritance to cell differentiation, it seems currently more constructive to assume that the activity of most genes is controlled at the chromatin level. Accordingly, the questions to answer are the following: (1) Does the *stability* of active versus inactive genic states lie only in the chromatin itself or elsewhere, more precisely, in the DNA itself or elsewhere, since the pattern of association of histones and DNA is probably reset at each chromosome duplication (Jackson et al., 1975)? (2) Do metabolic states present some intrinsic stability or inertia based on the self-maintained synthesis or activity of particular gene products?

## 1. The Stability of Genic States

As shown by the examples of functional inheritance described above, the expression of differentiation can be modulated by environmental or intracellular factors, whereas the determination underlying each type of differentiation is maintained even when conditions prevent the expression of the corresponding differentiation. Therefore, the main problem is whether the stability of determination is based on heritable changes in the DNA. Irreversible DNA modifications can reasonably be ruled out as demonstrated in various systems: activation in somatic hybrids with hepatoma cells of genes which were inactive in the nonhepatoma parental cells; reactivation of chick erythrocyte nuclei; transdeterminations in *Drosophila*; capacity of differentiated nuclei to direct development of *Xenopus* eggs; redifferentiation of iris cells into lens cells, etc. As for reversible DNA changes, their occurrence, nature, and mechanisms of maintenance remain speculative, and determination might be explained without assuming any change in the DNA itself in the following way.

It is conceivable that regardless of its individual mode of regulation (whether constitutively expressed, positively or negatively controlled), each gene retains in *all* cell types, all of its functional properties, i.e., all of its capacities to respond to specific regulatory molecules. Accordingly, (a) genes would be active or inactive depending only upon the signals provided by their intracellular environment or by the extracellular milieu, and filtered and relayed by surface and cytoplasmic components; or (b) the stability of patterns of gene expression would mostly be ensured by the physicochemical constraints resulting from the network of interactions between all the gene products present at a given time. The stability of determined and differentiated states would not be achieved by a few successive major switches in the patterns of gene expression but by progressive changes in the nature and relative amount of many gene products yielding stepwise qualitative and quantitative changes not easily reversible. This dialectic view of cell differentiation is basically similar to the model of "control circuits" proposed by Kaufman (1973) to account for the characteristics of transdetermination patterns in *Drosophila* (reviewed by Gehring, 1968) and extended by Wolpert and Lewis (1975) to the whole process of development.

To what extent are these two sets of assumptions supported by the facts?

**a. Genes Passively Obey Regulatory Signals.** A rather convincing piece of evidence is provided by experiments on hybridized rat hepatoma cells for which extinction, reexpression, and activation of several structural genes seem to be controlled by the presence or absence of particular diffusible regulatory molecules. In other systems where no such demonstration exists, this explanation may be the simplest one. For instance, the genes that are activated by a particular hormone in a target cell need not be in a heritable functional state different from that of the same genes in a nontarget cell but their activity may only be controlled by hormone receptors whose presence or absence results from the previous history of the cell.

Further support for this hypothesis may be found in a number of situations for which the patterns of gene expression seem to depend on primary changes at the cell surface: some surface component first triggered by the external signal then triggers some "second messenger" (Sutherland, 1972) acting on gene expression. As stressed by Pardee (1975) "difficulty arises when we ask what is cause and what is effect; what is primary and what are secondary events." However, in a particularly favorable system where many surface proteins can be identified, the differentiation of prothymocytes into thymocytes, the immediate response to induction of the differentiation is the appearance of several new surface antigens, eventually leading through several cycles of divisions to functionally competent immunocytes with still another pattern of surface antigens (Boyse and Abott, 1975). In some cases, the action of BUdR, either triggering differentiation of neuroblastoma cells into neurons (Schubert and Jacob, 1970), or inhibiting myogenesis in myoblasts (Rogers et al., 1975), has been shown to act not directly on the nucleus as reflected by incorporation of the analog into DNA, but at the cell surface, most probably by modification of the specificity of a cell surface component via interaction with glycosyltransferases. Egg activation (Mazia et al., 1975; Johnson and Epel, 1975), the action of hormones, cell interactions in the differentiation of tissues or, for unicellular organisms, cell interactions at the onset of sexual processes, and cell division (Pardee, 1971) all provide examples of primary effects of inducing agents on the cell surface resulting in changes in gene activity, metabolic activity, and cell structure, possibly, as discussed by Johnson and Epel (1975), by inducing cytoplasmic rearrangements making substrates available to enzymes, etc. Changes in the nuclear program or activity are also known to be determined by local cytoplasmic properties as, for instance, in the case of nuclear differentiation in *Paramecium*, in the determination of germ cells by polar cytoplasm in *Drosophila* (Illmensee and Mahowald, 1974), or in the control of DNA synthesis by cortical properties in *Stentor* (De Terra, 1975).

The probable general significance of these facts is that, in many cases, the cell surface and the cytoplasm constitute relays that mediate and channel the interactions between the milieu and the genes which may remain unchanged and passively obey the signals released by the relays.

**b. Physicochemical Constraints within the Cell Stabilize Gene Expression.** The existence of stringent physicochemical constraints creating obligate correlations between many cellular functions is strongly suggested by the fact that each differentiated state is characterized not only by the production of some specific proteins, corresponding to the "luxury function," but also by a particular cellular organization and particular surface properties. In hybridized rat hepatoma cells, extinction of differentiated functions is correlated with modifications of cell shape and surface morphology, and conversely, reexpression of the differentiated functions is accompanied by reestablishment of all the previous features. Furthermore, the conspicuous differentiated features seem correlated to less conspicuous changes in the basic metabolism. For instance, the control of glucose transport and the rate of its utilization seem to be extensively modulated in different cell types (Elbrink and Bihler, 1975). It is also well known that various "household" enzymes exhibit different isozyme patterns in different cell types. Therefore, determination and differentiation involve correlated changes in the expression and activity of *many* genes and functions.

In bacterial systems, the correlated expression of several genes is generally limited to the few genes of an operon or to a few genes responsive to the same regulatory molecules. Since the existence of operons has not been demonstrated in eukaryotes and since different determined and differentiated states are mutually exclusive, two hypotheses can be considered. Either each complex set of genes responsible for mutually exclusive phenotypes is genetically programmed to respond in a coordinated way to specific conditions, or all the genes remain individually responsive to regulatory molecules and the observed correlations depend on interactions and constraints beyond the DNA level.

While there is as yet no evidence supporting the former hypothesis, two kinds of observations fit better with the latter hypothesis. First, after extinction in hybridized rat hepatoma cells, the various luxury functions can be reexpressed independently of one another (Sparkes and Weiss, 1973; Bertolotti and Weiss, 1974). This suggests that the normally coordinated expression of all the hepatic functions is not programmed at the DNA level. Second, if the primary control mechanism of the observed functional correlations were in the genes themselves, one would expect that phenotypic switches would be triggered only by rather specific factors. On the contrary, if the primary control of gene activity lies in the constraints existing within the metabolic and structural network, one would expect that many unrelated factors could trigger a particular phenotypic switch by modification of any one of the numerous interacting molecules or structures. It is indeed striking that many unspecific and chemically unrelated products can trigger or inhibit the same differentiation.

Finally, in contrast to bacterial systems in which induction or repression can result in an abrupt (100-fold or 1000-fold) increase or decrease in enzyme activity, in eukaryotic cells, the variations in gene expression seem to be modulated within a much narrower range. For instance, induction of TAT by adrenal steroid hormones in hepatoma cells results in a 5- to 15-fold increase in enzyme activity. Furthermore, within a particular cell line, the amount of various proteins may vary progressively throughout developmental processes, as described by Hauschka (1968) for thirteen proteins during the development of chick skeletal muscle. These smooth modulations of gene expression suggest a fine control of the *rates* of transcription and/or translation rather than the occurrence of stable changes in the DNA.

In conclusion, although the existence of heritable changes in gene activity, based on self-reproducing but reversible modifications in the structure of parts of the genome cannot be ruled out, it seems more likely that the differential expression of many genes in different cell types depends in part upon the properties of the overall metabolic network.

## 2. The Stability of Metabolic States

The fine adjustment of numerous coupled or interacting enzymatic and structural systems certainly plays an important role in cell homeostasis. Nevertheless, this sort of self-stabilizing property of the metabolic and structural network does not seem sufficient to account for the stability of differentiated states and particularly for the stability of determination under changing conditions. However, a stronger stabilization of cellular functional states might be ensured in the following conditions. If a structural or enzymatic molecule or complex, once present, can contribute to the maintenance of its own synthesis or activity, it establishes a closed circuit whose functioning will escape direct control from the rest of the metabolic and structural circuitry. The phenomenon of cortical inheritance in *Paramecium* reflects such a mechanism: each cortical unit represents a closed circuit for kinetosome duplication within the overall cortical pattern. The following examples show that closed circuits can exist also at the metabolic level.

Surprisingly, the first situation of this type was described long ago in *Escherichia coli*. As is well known, addition of a galactoside to the culture medium induces the activity of the *lac* operon. However, the response of the bacteria (synthesis of $\beta$-galactosidase and galactoside permease) depends on the concentration of the inducer. In the presence of *low* concentrations of inducer, an all-or-none response is observed; some bacteria become fully induced, while others are not induced, and the two states are maintained side by side through divisions (Novick and Weiner, 1957; Cohn and Horibata, 1959b). A similar situation was studied by Cohn and Horibata (1959a). Addition of glucose normally prevents $\beta$-galactosidase induction. However, glucose has no effect on bacteria which have been induced *before* glucose addition; in the same medium containing glucose *and* low concentration of inducer, both states (induced and noninduced) can be perpetuated over many cell generations. In both cases, the explanation lies in the properties of the galactoside permease which permits intracellular accumulation of galactoside and which is induced by the same galactoside. External factors such as glucose or low concentration of inducer can prevent the induction of the permease, but once it has started, its synthesis and activity is maintained because the presence of one or a few permease molecules permits intracellular concentration of galactosides which, in turn, induce the gene coding for the permease. The permease system, therefore, displays autocatalytic properties.

In *Paramecium* which can be considered as a "simple higher organism," the stability of the surface antigenic type (expressed by its maintenance through vegetative growth and cytoplasmic inheritance) could be explained by a mechanism similar to the one just described in *E. coli*, although for *Paramecium* its molecular basis remains unknown. The similarities of the situations are particularly evident in the following case: synthesis of antigen D is controlled in part by temperature and induced above 30°–32°C, while antigen G, also controlled in part by temperature, is rapidly induced below 28°C. Cell lines of identical genotype, respectively induced by growth at 24° and 32°C to express either G or D, will continue to maintain their preexisting phenotype if grown at 30°C. The phenomenon can be explained by assuming that once present, each particular s. ag., which, like galactoside permease, is a membrane protein, creates or maintains intracellular conditions favoring its own synthesis.

It is worth recalling that to explain this and other features of the inheritance of s. ags. in *Paramecium*, Delbrück (1949) proposed the model illustrated in Fig. 6. The model assumes that two alternative phenotypes, c and c', depend on the functioning of two mutually exclusive pathways. The transitory influence of a stimulus (temperature, for instance) can switch the system from one state to another one. But once put into gear,

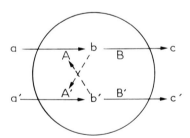

**Fig. 6.** Delbruck's model of alternative steady states. Two alternative steady states are supposed to occur within a cell, respectively, involving enzymes A, B . . . and A', B'. . . . If an intermediary metabolite, b', of one chain inhibits enzyme A which catalyzes a reaction of the other chain, and if conversely b inhibits A', either one of the two chains that functions at a given time will tend to maintain its activity. Temporary removal or decreased concentration of b' (or b) would permit the functioning of the other chain.

either state will be maintained by its own activity and/or by the inactivation of the other one. The possible mechanisms underlying this very general model involve self-maintenance of either the *synthesis* of c/c', or of its *activity*. The just-cited example of the galactoside permease of *E. coli* provided the first example of the synthesis of an enzyme depending on the presence of the enzyme itself. The following example suggests that the activity of an enzyme can be maintained by its own functioning.

The first step in the differentiation of *Dictyostelium discoideum* consists in the aggregation of amoebae, which is a chemotactic response to the presence of cAMP in the environment. The first aggregated cells excrete cAMP which is detected by other cells. An analysis (Rossomundo and Sussman, 1973) of the properties of the two enzymes involved in cAMP production (adenyl cyclase and cAMP-dependent ATPase) suggests the existence of a metabolic loop. Adenyl cyclase is activated by 5'-AMP and produces cAMP, while the ATPase is activated by cAMP and yields 5'-AMP. The activity of the loop is first triggered by exogenous cAMP which activates the ATPase, while the resulting 5'-AMP activates the adenyl cyclase yielding endogenous cAMP that relays the effect of the exogenous one and maintains this closed circuit in activity.

These examples indicate that both the synthesis and the activity of enzymes can display autocatalytic properties. To what extent such mechanisms are actually involved and play a significant role in cellular homeostasis remains questionable, but their existence should be kept in mind. In particular, in view of the importance of surface proteins, it is tempting to assume that the synthesis of some of them, including hormone receptors, might behave like galactoside permease or *Paramecium's* s. ags. A particularly attractive example could be provided by the T antigen which is present on spermatozoa of mouse and man *and* on the developing morula (Artzt *et al.*, 1973), and therefore could be "transmitted" to the surface of the egg by the spermatozoan at fertilization, its integration in the egg membrane triggering its further synthesis.

However, a single step switch mechanism does not seem sufficient to explain stable differences in cellular phenotypes, persisting long after the primary inducing conditions have ceased to exist; but it must not be overlooked that cell differentiation represents the result of *several* successive steps or switches and that each new step, made possible by the preceding one, may, in turn, modify the cellular state in such a way that returning to the initial state is no longer possible by reestablishment of previous conditions. "Cell memory" may simply mean inertia and metabolic constraints.

*In conclusion*, it is fair to say that we are still ignorant of the mechanisms which ensure the remarkable stability of differentiated phenotypes.

If the just-discussed mechanisms exist, they pertain to non-nucleic acid inheritance in that their stability escapes direct gene control, but the proof of their existence awaits a full and convincing demonstration in molecular terms. It is, however, tempting to speculate that if such systems exist, they might be preferentially located in cell membranes (at the surface or inside the cell) where addition of new proteins, loss of some proteins (Burger, 1969; Johnson and Epel, 1975), or modifications of the environment of one protein are most susceptible to trigger changes in other proteins and in the intracellular milieu.

## IV. CONCLUSION

There is still no clear answer to the question "Do the nongenic parts of the cell provide some kind of transmissible information superimposed on the basic genetic information encoded in nucleic acids?" The central difficulty of the problem lies in the circular relationship existing between the nucleic acids and the non-nucleic acid components of the cell. The latter regulate or modulate the expression of the genome at various levels, while the genome controls the specificity of all cellular components. The capacity of these cellular components to interact with the expression of the genome is, therefore, also programmed in the nucleic acid sequences themselves. It is thus conceptually and experimentally difficult to delineate the respective roles of nucleic acids and non-nucleic acid components in cell heredity.

However, if this particular version of the problem of the hen and the egg has no solution, a provisional answer can be proposed. Undoubtedly, each cell as a whole, whether a *Paramecium* or a hepatoma cell, is endowed with a strong inertia expressed by the stability of its phenotype and by its ability to undergo clear-cut switches from one stable state to another. Does the inertia lie in the DNA, in the cytoplasm, or in both? The tentative answer is that a part of this inertia lies in the cytoplasm. The basis of the inertia contributed by the genes is understood, as far as the stability of nucleic acid sequences is concerned. As for the inertia contributed by the non-nucleic acid components, it does not seem to involve any specific supplementary *information*, but only the stabilization of alternative structural or functional states, whenever they happen to be equally compatible with the gene controlled specificity of the concerned molecules. Non-nucleic acid inheritance, therefore, would reflect the rare degrees of freedom left to molecular interactions within the structural and metabolic network.

Although the distinction between structural and functional inheritance is probably artificial, they have been presented separately for one main reason. Evidence for structural inheritance exists, while the role of non-nucleic acid inheritance in functional stability remains a question.

It is clear that gene products can be arranged in different structural patterns and that such alternate arrangements may be, in some cases, equally stable and somehow self-maintained, independently of direct gene control. Such self-maintenance implies that superimposed on genetic information are influences of local conditions in the cytoplasm.

Functional inheritance is manifested as a commitment to a particular state, whether determined or differentiated. A memory of this commitment is maintained; it has strongly restricted possibilities of responses to environmental changes. However, both the memory and the restrictions appear much less compelling when whole nuclei or particular genes (as in somatic hybrids, or isolated nuclei injected into an enucleated egg) are confronted with a new cytoplasmic environment than they are when the whole intact cell is challenged by external factors. Therefore, it seems extremely likely that cell memory and functional restrictions depend more on the intrinsic stability of the established networks of interconnected metabolic functions than on heritable changes in the DNA or on genetically programmed alternative patterns of gene expression.

Finally, as both structural and functional inheritance lead one to postulate the prominent importance of the cell membrane, which is also more and more recognized in developmental processes (Bennet *et al.*, 1972), it may be assumed that a great deal of the inertia of biological systems has its seat in the membranes which (as discussed by Hershey, 1970) may be the only parts of the cell, aside from the DNA, that maintain a physical continuity.

## ACKNOWLEDGMENTS

I am particularly indebted to Boris Ephrussi for his constant encouragement and many stimulating discussions during the preparation of this manuscript. I am also very grateful to Piotr P. Slonimski for his comments and helpful suggestions and to Mary C. Weiss for a thorough critical reading of the manuscript.

## REFERENCES

Aigle, M., and Lacroute, F. (1975). Genetical aspects of [URE 3] a nonmitochondrial, cytoplasmically inherited mutation in yeast. *Mol. Gen. Genet.* **136**, 327–335.

Allen, S. L., and Gibson, I. (1972). Genome amplification and gene expression in the ciliate macronucleus. *Biochem. Genet.* **6**, 293–313.

Artzt, K., Dubois, P., Bennett, D., Condamine, H., Babinet, C., and Jacob, F. (1973). Surface antigens common to cleavage embryos and primitive teratocarcinoma cells in culture. *Proc. Natl. Acad. Sci. U.S.A.* **70**, 2988–2992.

Autissier, F., and Kepes, A. (1971). Segregation of membrane markers during cell division in *E. coli*. II. Segregation of Lac-permease and Mel-permease studied with a penicillin technique. *Biochim. Biophys. Acta* **249**, 611–615.

Avner, P. R., and Griffith, D. E. (1973). Studies on energy-linked reactions. Genetic analysis of oligomycin-resistant mutants in *Saccharomyces cerevisiae*. *Eur. J Biochem.* **32**, 312–321.

Baker, W. K. (1968). Position effect variegation. *Adv. Genet.* **14**, 133–169.

Beale, G. H. (1952). Antigen variation in *Paramecium aurelia*, variety 1. *Genetics* **37**, 62–74.

Beale, G. H. (1957). The antigen system of *Paramecium aurelia*. *Int. Rev. Cytol.* **6**, 1–23.

Behme, R. J., and Berger, J. (1970). The DNA content of *Paramecium aurelia* stock 51. *J. Protozool.* **17**, Suppl., 20.

Beisson, J., and Sonneborn, T. M. (1965). Cytoplasmic inheritance of the organization of the cell cortex in *Paramecium aurelia*. *Proc. Natl. Acad. Sci. U.S.A.* **53**, 275–282.

Beisson, J., Sainsard, A., Adoutte, A., Beale, G. H., Knowles, J., and Tait, A. (1974). Genetic control of mitochondria in *Paramecium*. *Genetics* **78**, 403–413.

Beisson, J., Lefort-Tran, M., Pouphille, M., Rossignol, M., and Satir, B. (1976a). Genetic analysis of membrane differentiation in *Paramecium*. Freeze-fracture study of the trichocyst cycle in wild-type and mutant strains. *J. Cell Biol.* **69**, 126–143.

Beisson, J., Rossignol, M., Ruiz, F., and Adoutte, A. (1976b). Genetic analysis of morphogenetic processes in *Paramecium tetraurelia*. II. A mutation affecting cortical pattern and nuclear reorganization. *J. Protozool.* **23**, 3A.

Beisson-Schecroun, J. (1962). Incompatibilité cellulaire et interactions nucléo-cytoplasmiques dans les phénomènes de "barrage" chez le *Podospora anserina*. *Ann. Genet.* **4**, 3–50.

Bennet, D., Boyse, E. A., and Old, L. J. (1972). Cell surface immunogenetics in the study of morphogenesis. *In* "Cell Interactions" (L. G. Silvestri, ed.), pp. 247–263. North-Holland Publ., Amsterdam.

Berger, J. D. (1973). Selective inhibition of DNA synthesis in macronuclear fragments in *P. aurelia* exconjugants and its reversal during macronuclear regeneration. *Chromosoma* **44**, 33–48.

Berlin, R. D., Oliver, J. M., Ukena, J. E., and Yin, H. H. (1974). Control of cell surface topography. *Nature (London)* **247**, 45–46.

Bertolotti, R., and Weiss, M. (1974). Expression of differentiated functions in hepatoma cell hybrids. V. Reexpression of aldolase B *in vitro* and *in vivo*. *Differentiation* **2**, 5–17.

Bishop, J. O. (1961). Purification of an immobilization antigen of *Paramecium aurelia*, variety 1. *Biochim. Biophys. Acta* **50**, 471–476.

Bishop, J. O. (1963). Immunological assay of some immobilizing antigens of *Paramecium aurelia*, variety 1. *J. Gen. Microbiol.* **30**, 271–280.

Boyse, E. A., and Abbott, J. (1975). Surface reorganization as an initial inductive event in the differentiation of prothymocytes to thymocytes. *Fed. Proc., Fed. Am. Soc. Exp. Biol.* **34**, 24–28.

Boyse, E. A., Old, L. J., and Stockert, E. (1968). An approach to the mapping of antigens on the cell surface. *Proc. Natl. Acad. Sci. U.S.A.* **60**, 886–893.

Bretscher, M. S., and Raff, M. C. (1975). Mammalian plasma membranes. *Nature (London)* **258**, 43–49.

Brink, R. A. (1965). Genetic repression of *R* action in maize. *In* "The Role of Chromosomes in Development" (M. Locke, ed.), pp. 183–230. Academic Press, New York.

Britten, R. J., and Davidson, E. H. (1969). Gene regulation for higher cells: A theory. *Science* **165,** 349–357.

Brown, D. L., and Bouck, G. B. (1973). Microtubule biogenesis and cell shape in *Ochromonas*. II. The role of nucleating sites in shape development. *J. Cell Biol.* **56,** 360–378.

Brown, J. E., and Weiss, M. C. (1975). Activation of production of mouse liver enzymes in rat hepatoma-mouse lymphoid cell hybrids. *Cell* **6,** 481–494.

Buckingham, M. E., Caput, D., Cohen, A., Whalen, R. G., and Gros, F. (1974). The synthesis and stability of cytoplasmic messenger RNA during myoblast differentiation in culture. *Proc. Natl. Acad. Sci. U.S.A.* **71,** 1466–1470.

Burger, M. M. (1969). A difference in the architecture of the surface membrane of normal and virally transformed cells. *Proc. Natl. Acad. Sci. U.S.A.* **62,** 994–1001.

Capdeville, Y. (1971). Allelic modulation in *Paramecium aurelia* heterozygotes: Study of G serotypes in syngen 1. *Mol. Gen. Genet.* **112,** 306–316.

Capdeville, Y. (1977). Expression of s. antigens genes in heterozygotes of *P. primaurelia*: Effect of temperature. (In preparation.)

Clive, D., and Landman, D. E. (1970). Reversion of *B. subtilis* protoplasts to the bacillary form induced by exogenous cell wall, bacteria and by growth on membrane filters. *J. Gen. Microbiol.* **61,** 233–243.

Cohn, M., and Horibata, K. (1959a). Inhibition by glucose of the induced synthesis of the β-galactoside enzyme system of *E. coli*. Analysis of maintenance. *J. Bacteriol.* **78,** 601–612.

Cohn, M., and Horibata, K. (1959b). Analysis of the differentiation and of the heterogeneity within a population of *E. coli* undergoing induced β-galactosidase synthesis. *J. Bacteriol.* **78,** 613–623.

Cook, P. R. (1970). Species specificity of an enzyme determined by an erythrocyte nucleus in a interspecific hybrid cell. *J. Cell Sci.* **7,** 1–3.

Cook, P. R. (1974). On the inheritance of differentiated traits. *Biol. Rev. Cambridge Philos. Soc.* **49,** 51–84.

Crick, F. (1970). Diffusion in embryogenesis. *Nature (London)* **225,** 420–422.

Davidson, R. L. (1974). Genetic expression in somatic cell hybrids. *Annu. Rev. Genet.* **8,** 195–218.

Davies, D. R. (1972). Cell wall organization in *Chlamydomonas reinhardi*: The role of extranuclear systems. *Mol. Gen. Genet.* **115,** 334–348.

Davies, D. R., and Lyall, V. (1973). The assembly of a highly ordered component of the cell wall: The role of heritable factors and of physical structure. *Mol. Gen. Genet.* **124,** 21–34.

Davies, D. R., and Plaskitt, A. (1971). Genetical and structural analysis of cell wall formation in *Chlamydomonas reinhardi*. *Genet. Res.* **17,** 33–43.

Davis, F. M., and Adelberg, E. A. (1973). Use of somatic cell hybrids for analysis of the differentiated state. *Bacteriol. Rev.* **37,** 197–214.

Delbrück, M. (1949). Comment. *In* "Unités biologiques douées de continuité génétique," pp. 33–34. CNRS, Paris.

Deschatrette, J., and Weiss, M. C. (1974). Characterization of differentiated clones from rat hepatoma. *Biochimie* **56,** 1603–1611.

De Terra, N. (1975). Evidence for cell surface control of macronuclear DNA synthesis in *Stentor*. *Nature (London)* **258,** 300–303.

Dingle, A. D., and Fulton, C. (1966). Development of the flagellar apparatus of *Naegleria*. *J. Cell Biol.* **31,** 43–54.

Dippell, R. V. (1954). A preliminary report on the chromosomal constitution of certain variety 4 races of *P. aurelia*. *Caryologia, Suppl.* pp. 1109–1111.

Dippell, R. V. (1968). The development of basal bodies in *Paramecium*. *Proc. Natl. Acad. Sci. U.S.A.* **61**, 461–468.

Dippell, R. V. (1976). Effects of nuclease and protease digestion on the ultrastructure of *Paramecium* basal bodies. *J. Cell Biol.* **69**, 622–637.

Elbrink, J., and Bihler, I. (1975). Membrane transport: Its relation to cellular metabolic rates. *Science* **188**, 1177–1184.

Ephrussi, B. (1972). "Hybridization of Somatic Cells." Princeton Univ. Press, Princeton, New Jersey.

Fincham, J. R. S., and Sastry, G. R. K. (1974). Controlling elements in maize. *Annu. Rev. Genet.* **8**, 16–50.

Finger, I. (1974). Surface antigen of *Paramecium aurelia*. *In* "Paramecium: A Current Survey" (W. J. Van Wagtendonk, ed.), pp. 131–164. Elsevier, Amsterdam.

Finger, I., and Heller, C. (1962). Immunogenetic analysis of proteins of *Paramecium*. I. Comparison of specificities controlled by alleles and by different loci. *Genetics* **47**, 223–239.

Fox, J. O., Sheppard, J. R., and Burger, M. M. (1971). Cyclic membrane changes in animal cells: Transformed cells permanently display a surface architecture detected in normal cells only during mitosis. *Proc. Natl. Acad. Sci. U.S.A.* **68**, 244–247.

Frankel, J. (1974). Positional information in unicellular organisms. *J. Theor. Biol.* **47**, 439–481.

Frankel, J. (1975). Pattern formation in ciliary organelle systems of ciliated Protozoa. *Cell Patterning, Ciba Found. Symp.* No. 29, pp. 25–49.

Frye, L. D., and Edidin, M. (1970). The rapid intermixing of cell surface antigens after formation of mouse-human heterokaryons, *J. Cell Sci.* **7**, 319–335.

Galau, G. A., Britten, R. J., and Davidson, E. H. (1974). A measurement of the sequence complexity of polysomal messenger RNA in sea urchin embryos. *Cell* **2**, 9–20.

Gall, J. F. (1974). Free ribosomal genes in the macronucleus of *Tetrahymena*. *Proc. Natl. Acad. Sci. U.S.A.* **71**, 3078–3081.

Gehring, W. (1968). The stability of the determined state in cultures of imaginal disks in *Drosophila*. *In* "The Stability of the Differentiated State" (W. Beerman, J. Reinert, and H. Ursprung, eds.), pp. 136–154. Springer-Verlag, Berlin and New York.

Genermont, J. (1966). Le déterminisme génétique de la vitesse de multiplication chez *Paramecium aurelia* (syng. 1). *Protistologica* **2**, 45–51.

Gorovsky, M. A. (1973). Macro- and micronuclei of *Tetrahymena pyriformis*: A model system for studying and structure and function of eukaryotic nuclei. *J. Protozool.* **20**, 19–25.

Gorovsky, M. A., and Keevert, J. B. (1975). Absence of histone $F_1$ in a mitotically dividing genetically inactive nucleus. *Proc. Natl. Acad. Sci. U.S.A.* **72**, 2672–2676.

Graham, J. M., Sumner, M. C. B., Curtis, D. H., and Pasternak, C. A. (1973). Sequence of events in plasma membrane assembly during the cell cycle. *Nature (London)* **246**, 291–295.

Green, H., and Todaro, G. J. (1967). The mammalian cell as differentiated microorganism. *Annu. Rev. Microbiol.* **21**, 573–600.

Grimes, G. W. (1973). Morphological discontinuity of kinetosomes in *Oxytricha fallax*. *J. Cell Biol.* **57**, 229–232.

Gross, P. R. (1967). The control of protein synthesis in embryonic development and differentiation. *Curr. Top. Dev. Biol.* **2**, 1–46.

Guérineau, M., Slonimski, P. P., and Avner, P. R. (1974). Yeast episome: Oligomycine resistance associated with a small covalently closed nonmitochondrial circular DNA. *Biochem. Biophys. Res. Commun.* **61**, 462–469.

Gurdon, J. B. (1970). Nuclear transplantation and the control of nuclear activity in animal development. *Proc. R. Soc. London, Ser.* **176,** 303–314.

Gurdon, J. B., Laskey, R. A., and Reeves, O. R. (1975). The developmental capacity of nuclei transplanted from keratinized skin cells of adult frog. *J. Embryol. Exp. Morphol.* **34,** 93–112.

Hansma, H. G. (1975). The immobilization antigen of *Paramecium aurelia* is a single polypeptide chain. *J. Protozool.* **22,** 257–259.

Harris, H., Sidebottom, E. M., Grace, D. M., and Bramwell, M. E. (1969). Expression of genetic information: A study with hybrid animal cells. *J. Cell Sci.* **4,** 499–525.

Hartman, H. (1975). The centriole and the cell. *J. Theor. Biol.* **51,** 501–509.

Hartman, H., Puma, J. P., and Gurney, T. (1974). Evidence for the association of RNA with the ciliary basal bodies of *Tetrahymena*. *J. Cell Sci.* **16,** 241–259.

Hauschka, S. D. (1968). Clonal aspects of muscle development and the stability of the differentiated state. *In* "The Stability of the Differentiated State" (H. Ursprung, ed.), pp. 37–57. Springer-Verlag, Berlin and New York.

Heckman, K., and Frankel, J. (1968). Genic control of cortical pattern in *Euplotes*. *J. Exp. Zool.* **168,** 11–38.

Hershey, A. D. (1970). Genes and hereditary characteristics. *Nature (London)* **226,** 697–701.

Hills, G. J., Gurney-Smith, M. G., and Roberts, K. (1973). Structure, composition and morphogenesis of the cell wall of *Chlamydomonas reinhardi*, II. Electron microscopy and optical diffraction analysis. *J. Ultrastruct. Res.* **43,** 179–192.

Holliday, R., and Pugh, J. E. (1975). DNA modifications mechanisms and gene activity during development. *Science* **187,** 226–232.

Iino, T. (1969). Genetics and chemistry of bacterial flagella. *Bacteriol. Rev.* **33,** 454–475.

Iino, T. (1974). Assembly of *Salmonella* flagellin *in vitro* and *in vivo*. *J. Supramol. Struct.* **2,** 372–384.

Illmensee, K., and Mahowald, A. P. (1974). Transplantation of posterior polar plasm in *Drosophila*. Induction of germ cells at the anterior pole of the egg. *Proc. Natl. Acad. Sci. U.S.A.* **71,** 1016–1020.

Inoé, S., and Sato, H. (1967). Cell motility by labile association of molecules. The nature of mitotic spindle fibers and their role in chromosome movements. *J. Gen. Physiol.* **50,** 259–288.

Jackson, V., Granner, D. K., and Chalkley, R. (1975). Deposition of histones into replicating chromosomes. *Proc. Natl. Acad. Sci. U.S.A.* **72,** 4440–4444.

Jerka-Dziadosz, M. (1972). An analysis of the formation of ciliary promordia in the hypotrich ciliate *Urostyla weissei*. II. Results from ultraviolet microbeam irradiation. *J. Exp. Zool.* **179,** 81–88.

Jerka Dziadosz, M., and Frankel, J. (1969). An analysis of the formation of ciliary primordia in the hypotrich ciliate *Urostyla weissei*. *J. Protozool.* **16,** 612–637.

Johnson, J. D., and Epel, D. (1975). Relationship between release of surface proteins and metabolic activation of sea urchin egg at fertilization. *Proc. Natl. Acad. Sci. U.S.A.* **72,** 4474–4478.

Johnson, L. F., Williams, J. G., Abelson, H. T., Green, H., and Penman, S. (1975). Changes in RNA in relation to growth of the fibroblast. III. Posttranscriptional regulation of m RNA formation in resting and growing cells. *Cell* **4,** 69–75.

Jones, I. G. (1965). Studies on the characterization and structure of the immobilization antigens of *Paramecium aurelia*. *Biochem. J.* **96,** 17–23.

Jones, I. G., and Beale, G. H. (1963). Chemical and immunological comparisons of allelic immobilization antigens in *Paramecium aurelia*. *Nature (London)* **197,** 205–206.

Kaufman, S. (1973). Control circuits for determination and transdetermination. *Science* **181**, 310–317.

Landman, O. D., and Halle, S. (1963). Enzymically and physically-induced inheritance changes in *Bacillus subtilis*. *J. Mol. Biol.* **7**, 721.

Leblond, C. P., and Bennett, G. (1974). Elaboration and turnover of cell coat glycoproteins. *In* "The Cell Surface in Development" (A. A. Moscona ed.), pp. 29–49. Wiley, New York.

Levine, A. M., Higgins, J. A., and Barrnett, R. J. (1972). Biogenesis of plasma membranes in salt glands of salt-stressed domestic ducklings: Localization of acetyl transferase activity. *J. Cell Sci.* **11**, 855–873.

Linnane, A. W., Haslam, J. M., Lukins, H. B., and Nagley, P. (1972). The biogenesis of mitochondria in microorganisms. *Annu. Rev. Microbiol.* **26**, 163–198.

Lucy, J. A. (1970). The fusion of biological membranes. *Nature (London)* **227**, 815–817.

Lynn, D. H., and Tucker, J. B. (1976). Cell size and proportional distance assessment during determination of organelle position in the cortex of the ciliate *Tetrahymena*. *J. Cell Sci.* **21**, 35–46.

Lyon, M. F. (1972). Chromosome inactivation and developmental patterns in mammals. *Biol. Rev. Cambridge Philos. Soc.* **47**, 1–35.

Mage, R. G. (1975). Normal and altered phenotypic expression of immunoglobulin genes. *Fed. Proc., Fed. Am. Soc. Exp. Biol.* **34**, 40–46.

Malawista, S. E., and Weiss, M. C. (1974). Expression of differentiated functions in hepatoma cell hybrids: high frequency of induction of mouse albumin production in rat hepatoma-mouse lymphoid hybrids. *Proc. Natl. Acad. Sci. U.S.A.* **71**, 927.

Martin, G. R. (1975). Teratocarcinomas as a model system for the study of embryogenesis and neoplasia. *Cell* **5**, 229–243.

Mazia, D., Schatten, G., and Steinhardt, R. (1975). Turning on of activities in unfertilized sea urchin eggs: Correlation with changes of the surface. *Proc. Natl. Acad. Sci. U.S.A.* **72**, 4469–4473.

Meinhardt, H., and Gierer, A. (1974). Applications of a theory of biological pattern formation based on lateral inhibition. *J. Cell Sci.* **15**, 321–346.

Nanney, D. L. (1958). Epigenetic control systems. *Proc. Natl. Acad. Sci. U.S.A.* **44**, 712–717.

Nanney, D. L. (1966). Corticotypes in *Tetrahymena pyriformis*. *Am. Nat.* **100**, 303–318.

Nanney, D. L. (1968a). Ciliate genetics: Patterns and programs of gene action. *Annu. Rev. Genet.* **2**, 121–140.

Nanney, D. L. (1968b). Cortical patterns in cellular morphogenesis. *Science* **160**, 496–502.

Nanney, D. L. (1975). Patterns of basal bodies addition in ciliary rows in *Tetrahymena*. *J. Cell Biol.* **65**, 503–512.

Nicolson, G. L. (1976). Transmembrane control of the receptors on normal and tumor cells. I. Cytoplasmic influence over cell surface components. *Biochim. Biophys. Acta* **457**, 57–108.

Novick, A., and Weiner, M. (1957). Enzyme induction, an all-or-none phenomenon. *Proc. Natl. Acad. Sci. U.S.A.* **43**, 553–566.

Olmsted, J. B., and Borisy, G. G. (1973). Microtubules. *Annu. Rev. Biochem.* **42**, 507–540.

Pardee, A. B. (1971). The surface membrane as a regulator of animal cell division. *In Vitro* **7**, 95–104.

Pardee, A. B. (1975). The cell surface and fibroblast proliferation. Some current research trends. *Biochim. Biophys. Acta* **417**, 153–172.

Pasternak, J. (1967). Differential genic activity in *Paramecium aurelia*. *J. Exp. Zool.* **165**, 395–417.

Paul, J. (1972). General theory of chromosome structure and gene activation in eukaryotes. *Nature (London)* **238**, 444–446.

Perlman, B. S. (1973). Basal bodies addition in ciliary rows of *Tetrahymena pyriformis*. *J. Exp. Zool.* **184**, 365–368.

Peterson, J. A., and Weiss, M. C. (1972). Expression of differentiated functions in hepatoma cell hybrids: Induction of mouse albumin production in rat hepatoma-mouse fibroblast hybrids. *Proc. Natl. Acad. Sci. U.S.A.* **69**, 571–572.

Pickett-Heaps, J. D. (1969). The evolution of the mitotic apparatus: An attempt at comparative ultrastructural cytology in dividing plant cells. *Cytobios* **3**, 257–280.

Pitelka, D. R. (1969). Centriole replication. *In* "Handbook of Molecular Cytology" (A. Lima-de-Faria, ed.), pp. 1199–1218. North-Holland Publ., Amsterdam.

Porter, K. R. (1966). Cytoplasmic microtubules and their functions. *Princ. Biomol. Organ., Ciba Found. Symp., 1965* pp. 308–345.

Poste, G., and Allison, A. C. (1973). Membrane fusion. *Biochim. Biophys. Acta* **300**, 421–465.

Preer, J. R. (1959a). Nuclear and cytoplasmic differentiations in the Protozoa. *In* "Developmental Cytology" (D. Rudnick ed.), pp. 3–20. Ronald Press, New-York.

Preer, J. R. (1959b). Studies on the immobilization antigens of *Paramecium*. III. Properties. *J. Immunol.* **83**, 378–385.

Preer, J. R. (1959c). Studies on the immobilization antigens of *Paramecium*. IV. Properties of the different antigens. *Genetics* **44**, 803–814.

Preer, J. R. (1969). Genetics of the Protozoa. *Res. Protozool.* **3**, 130–278.

Reisner, A. H., Rowe, J., and Macindoe, H. M. (1969). The largest known monomeric globular protein. *Biochim. Biophys. Acta* **188**, 196–206.

Rizet, G. (1952). Les phénomènes de barrage chez *Podospora anserina*. I. Analyse génétique des barrages entre souches S et s. *Rev. Cytol. Biol. Veg.* **13**, 51.

Rogers, J., NG, S. K. C., Coulter, M. B., and Sanwal, B. D. (1975). Inhibition of myogenesis in a rat myoblast line by 5-bromodeoxyuridine. *Nature (London)* **256**, 438–439.

Rossomondo. E. F., and Sussman, M. (1973). A 5'-adenosine monophosphate-dependent adenylate cyclase and an adenosine 3',5'-cyclic monophosphate-dependent adenosine triphosphate pyrophosphorylase in *Dictyostelium discoideum*. *Proc. Natl. Acad. Sci. U.S.A.* **70**, 1254–1257.

Ruiz, F., Adoutte, A., Rossignol, M., and Beisson, J. (1976). Genetic analysis of morphogenetic processes in *Paramecium tetraurelia*. I. A mutation affecting trichocyst formation and nuclear divisions. *Genet. Res.* **27**, 109–122.

Ryter, A. (1971). Flagella distribution and a study on the growth of the cytoplasmic membrane in *B. subtilis*. *Ann. Inst. Pasteur, Paris* **121**, 271–288.

Sager, R., and Kitchin, R. (1975). Selective silencing of eukaryotic DNA. *Science* **189**, 426–433.

Sainsard, A., Claisse, M., and Balmefrezol, M. (1974). A nuclear mutation affecting structure and function of mitochondria in *Paramecium*. *Mol. Gen. Genet.* **130**, 113–125.

Satir, B. (1975). The final step in secretion. *Sci. Am.* **233**, 29–38.

Scherrer, K., and Marcaud, L. (1968). Messenger RNA in avian erythroblasts at the transcriptional and translational levels and the problem of regulation in animal cells. *J. Cell. Physiol.* **72**, Suppl. 1, 181.

Schubert, D., and Jacob, F. (1970). 5-Bromodeoxyuridine-induced differentiation of a neuroblastoma. *Proc. Natl. Acad. Sci. U.S.A.* **67**, 247–254.

Singer, S. J., and Nicholson, G. L. (1972). The fluid mosaic model of the structure of cell membranes. *Science* **175**, 720–731.

Smith, M. J., Hough, B. R., Chamberlin, M. E., and Davidson, E. H. (1974). Repetitive and nonrepetitive sequence in sea urchin heterogeneous nuclear RNA. *J. Mol. Biol.* **85,** 103–126.

Sommerville, J. (1970). Serotype expression in *Paramecium. Adv. Microbiol. Physiol.* **4,** 131–178.

Sonneborn, T. M. (1946). Inert nuclei: Inactivity of micronuclear genes in variety 4 of *Paramecium aurelia. Genetics* **31,** 231.

Sonneborn, T. M. (1954a). Is gene K active in the micronucleus of *Paramecium aurelia*? *Microb. Genet. Bull.* **11,** 25–26.

Sonneborn, T. M. (1954b). Patterns of nucleocytoplasmic integration in *Paramecium. Caryolgia, Suppl.* pp. 307–325.

Sonneborn, T. M. (1963). Does preformed cell structure play an essential role in cell heredity? *In* ''The Nature of Biological Diversity'' (J. M. Allen, ed.), pp. 165–221. McGraw-Hill, New York.

Sonneborn, T. M. (1970). Gene action in development. *Proc. R. Soc. London, Ser. B* **176,** 347–366.

Sonneborn, T. M. (1975). The *Paramecium aurelia* complex of 14 sibling species. *Trans. Am. Microsc. Soc.* **94,** 155–178.

Sonneborn, T. M., and Dippell, R. V. (1960). The genetic basis of the difference between single and double *Paramecium aurelia. J. Protozool.* **7,** Suppl., 26.

Sonneborn, T. M., and Dippell, R. V. (1961). Self-reproducing differences in the cortical organization of *Paramecium aurelia,* syngen 4. *Genetics* **46,** 900.

Sonneborn, T. M., and Le Suer, A. (1948). Antigenic characters in P. *aurelia* (variety 4): Determination, inheritance and induced mutations. *Am. Nat,* **82,** 69–78.

Sparkes, R. S., and Weiss, M. C. (1973). Expression of differentiated functions in hepatoma cell hybrids: Alanine aminotransferase. *Proc. Natl. Acad. Sci. U.S.A.* **70,** 377–381.

Stackpole, C. W., Aoki, T., Boyse, A., Old, L. J., Lumley-Franck, J., and de Harven, E. (1971). Cell surface antigens: serial sectioning of single cells as an approach to topographical analysis. *Science* **172,** 472.

Staehelin, L. A. (1974). Structure and function of intracellular junctions. *Int. Rev. Cytol.* **39,** 191.

Steers, E., Jr. (1961). Electrophoretic analysis of immobilization antigens in *Paramecium aurelia. Science* **133,** 2010–2011.

Steers, E., Jr. (1962). A comparison of the tryptic peptides obtained from immobilization antigens of *Paramecium aurelia. Proc. Natl. Acad. Sci. U.S.A.* **48,** 867–874.

Steers, E., Jr. (1965). Amino acid composition and quaternary structure of an immobilization antigen from *Paramecium aurelia. Biochemistry* **4,** 1896–1901.

Sutherland, E. W. (1972). Studies on the mechanism of hormone action. *Science* **177,** 401–408.

Tamm, S. L., Sonneborn, T. M., and Dippell, R. V. (1974). The role of cortical orientation in the control of the direction of ciliary beat in *Paramecium. J. Cell Biol.* **64,** 98–112.

Tartar, V. (1956). Pattern and substance in *Stentor. In* ''Cellular Mechanisms in Differentiation and Growth'' (D. Rudnick, ed.), pp. 73–100. Princeton Univ. Press, Princeton, New Jersey.

Tartar, V. (1961). ''The Biology of *Stentor.*'' Pergamon, Oxford.

Thomson, E. B., Tomkins, G. M., and Curran, J. F. (1966). Induction of tyrosine α-ketoglutarate transaminase by steroid hormones in a newly established tissue culture cell line. *Proc. Natl. Acad. Sci. U.S.A.* **56,** 296–303.

Tilney, L. G. (1968). The assembly of microtubules and their role in the development of cell form. *Symp. Soc. Exp. Biol.* **27,** Suppl. 2, 63–102.

Tsanev, R., and Sendov, B. L. (1971). Possible molecular mechanism for cell differentiation in multicellular organisms. *J. Theor. Biol.* **30,** 337–393.

Tucker, J. B. (1970). Initiation and differentiation of microtubule patterns in the ciliate *Nassula. J. Cell Sci.* **7,** 793–821.

Weiss, M. C., and Chaplain, M. (1971). Expression of differentiated functions in hepatoma cell hybrids: reappearance of tyrosine aminotransferase inducibility after the loss of chromosomes. *Proc. Natl. Acad. Sci. U.S.A.* **68,** 3026–3029.

Weiss, M. C., Sparkes, R. S., and Bertolotti, R. (1975). Expression of differentiated functions in hepatoma cell hybrids. IX. Extinction and reexpression of liver-specific enzymes in rat hepatoma-Chinese hamster fibroblast hybrids. *Somatic Cell Genet.* **1,** 27–40.

Whittle, J. R. S., and Chen-Shan, L. (1972). Cortical morphogenesis in *Paramecium aurelia*: Mutants affecting cell shape. *Genet. Res.* **19,** 271–279.

Wigley, C. B. (1975). Differentiated cells *in vitro. Differentiation* **4,** 25–55.

Wolpert, L. (1971). Positional information and pattern formation. *Curr. Top. Dev. Biol.* **6,** 183.

Wolpert, L., and Lewis, J. H. (1975). Towards a theory of development. *Fed. Proc., Fed. Am.* **34,** 14–20.

Yaffe, D. (1968). Retention of differentiation potentialities during prolonged cultivation of myogenic cells. *Proc. Natl. Acad. Sci. U.S.A.* **61,** 477.

Yamada, T., and McDevitt, D. S. (1974). Direct evidence for transformation of differentiated iris epithelial cells into lens cells. *Dev. Biol.* **38,** 104–119.

Yao, M. C., Kimmel, A. R., and Gorovsky, M. A. (1974). A small number of cistrons for ribosomal RNA in the germinal nucleus of a eukaryote *Tetrahymena pyriformis. Proc. Natl. Acad. Sci. U.S.A.* **71,** 3082–3086.

# Index